D1277402

The
Other
Space
Race

Transforming War

Paul J. Springer, editor

To ensure success, the conduct of war requires rapid and effective adaptation to changing circumstances. While every conflict involves a degree of flexibility and innovation, there are certain changes that have occurred throughout history that stand out because they fundamentally altered the conduct of warfare. The most prominent of these changes have been labeled "Revolutions in Military Affairs" (RMAs). These so-called revolutions include technological innovations as well as entirely new approaches to strategy. Revolutionary ideas in military theory, doctrine, and operations have also permanently changed the methods, means, and objectives of warfare.

This series examines fundamental transformations that have occurred in warfare. It places particular emphasis upon RMAs to examine how the development of a new idea or device can alter not only the conduct of wars, but their effect upon participants, supporters, and uninvolved parties. The unifying concept of the series is not geographical or temporal; rather, it is the notion of change in conflict and its subsequent impact. This has allowed the incorporation of a wide variety of scholars, approaches, disciplines, and conclusions to be brought under the umbrella of the series. The works include biographies, examinations of transformative events, and analyses of key technological innovations that provide a greater understanding of how and why modern conflict is carried out, and how it may change the battlefields of the future.

The
Other
Space
Race

Eisenhower and the
Quest for Aerospace Security

Nicholas Michael Sambaluk

Naval Institute Press ǀ Annapolis, Maryland

Naval Institute Press
291 Wood Road
Annapolis, MD 21402

Library of Congress Cataloging-in-Publication Data

Sambaluk, Nicholas Michael, author.
  The other space race : Eisenhower and the quest for aerospace security / Nicholas Michael Sambaluk.
      pages cm— (Transforming War)
  Includes bibliographical references and index.
  ISBN 978-1-61251-886-2 (alk. paper) — ISBN 978-1-61251-887-9 (ebook)
  1. Space race—History—20th century. 2. Astronautics and state—United States—History—20th century. 3. Aeronautics and state—United States—History—20th century. 4. National security—United States—History—20th century. 5. Astronautics, Military—History—20th century. 6. Eisenhower, Dwight D. (Dwight David), 1890–1969. I. Title.
  TL789.8.U5S26 2015
   358′.8—dc23
                              2015025439

♾ Print editions meet the requirements of ANSI/NISO z39.48-1992 (Permanence of Paper).
Printed in the United States of America.

23 22 21 20 19 18 17 16 15   9 8 7 6 5 4 3 2 1
First printing

For Mom and Dad

In preparation for modern war it is essential that the possibility of new weapons be considered in strategy.
    —IRVIN STEWART, EXECUTIVE SECRETARY, OFFICE OF SCIENTIFIC RESEARCH
    AND DEVELOPMENT[1]

What we can afford least is to define the problem of future war as we would like it to be, and by doing so introduce into our defense vulnerabilities based on self-delusion.
    —MAJ. GEN. H. R. MCMASTER, USA, COMMANDER, U.S. ARMY
    MANEUVER CENTER OF EXCELLENCE[2]

CONTENTS

ILLUSTRATIONS

ACKNOWLEDGMENTS

O n September 18, 2014, I finally came face to face with the space weapon that never was. I was at Purdue University on a business trip, walking with colleagues through the second floor of the school's engineering building. I had signed the contract to produce this book with Naval Institute Press three weeks earlier, and with the manuscript drafted, the polishing phase was under way. I had not expected to see a mockup of the Air Force's Dyna-Soar hanging on display, but it was there for a good reason. Boeing sponsored Purdue's new engineering building, named it for one of the school's more famous alumni (Neil Armstrong), and displayed a full-scale mockup of the North American Apollo 11 that Armstrong commanded in 1969, as well as the half-size display of the Boeing Dyna-Soar he had been selected in 1962 to fly for the Air Force. Seeing the craft brought a moment of wonder and reflection.

In addition to dedication, a little inspiration, and a lot of toil, books require the encouragement that comes from a vast array of others' help, good wishes, support, and advice.

Serious history starts at the archive, and thanks go there first. Tom Hargenrader at the National Aeronautics and Space Administration (NASA) Headquarters History Office in Washington, DC, unearthed a folder of documents on hypersonic research that proved both fascinating and critical to this study. Joe Caver provided particular help navigating the Air Force Historical Research Agency's extensive holdings; the Library of Congress and the Eisenhower Presidential Library holdings are superb and the staffs very capable.

Even beloved projects can sometimes feel heavy, and good friends do much to lighten the load. Dr./Maj. Bill Nance, USA, Maj. Joe Scott, USA, and Dr. Gian Gentile (Col., USA, Ret.), were the first people I met at West Point and are supportive and encouraging friends. The friendship, encouragement, and advice of Jean Bennett of Texas are cherished and mean much to me. Kimla Shelley and Martha Killough have been supportive voices for most of my life, and my sixth-grade teacher Cindy Dussia made the generous mistake of letting me teach my classmates about World War II for "five minutes" . . . and not reminding me of the time until half an hour had passed.

Direction is an important element as well. This book started as a dissertation, and my committee, co-chaired by Adrian Lewis and Ted Wilson of the University of Kansas, deserves thanks. So does Roger Spiller, in whose class I first noticed the Dyna-Soar program. My thesis adviser, the late Al Hurley, showed a talent, dedication, and innate goodness that impressed me as a graduate student and that still remain an inspiration. The history department chair Dr./Col. Ty Seidule, deputy chair Dr./Col. Gail Yoshitani, and military division chief Dr./Lt. Col. Jason Musteen all earned my esteem and thanks while I was teaching at West Point in the days and polishing the manuscript in the evenings. I enjoyed meeting Dr. Alex Roland during his year as distinguished visiting professor at West Point; a skilled former NASA historian, he had met many of the people about whom I had written. I am also indebted to Dr. Alexander Ibe of Weatherford College.

Colleagues and comrades are a blessing, and I am very fortunate in the ones I have had. Friends know me as cheerful, upbeat, and willing to help; at the University of Kansas, I learned that Frank Cai is not only the world's best officemate, he is also the most positive person I know. From my very first day of graduate school, Dr. Greg Ball (now Air Force Cyber Command's historian) has been a close friend and remains the one person I know who drafted his dissertation in less time than I did. Marion Mealey, Gates Brown, and Kevin Lee stand tall among the friends I met in graduate school. West Point chooses its faculty well, and I want to especially thank Lt. Col. Dave Siry, Dr. Dan Franke, and Dr. Charlie Thomas (now at the Air Force Command and Staff College) for their true friendship.

I want to thank Dr. Adrienne Harrison (Maj., USA, Ret.), for her camaraderie, encouragement, and friendship, as well as for her professional insights when reading the manuscript. If this book explains as much about Eisenhower and policy as hers does about Washington and learning, I've succeeded.

My parents, Nick and Rose, and my brother, Eric, gave me an upbringing and an environment that fostered wonder, encouraged learning, and valued history, making mine a happy childhood. Good books were always on the lower shelves within easy reach, and I discovered my career passion even before my first day of first grade. Preparing for that day meant buying a lunchbox, and I still remember what struck me as an uninspiring assortment of cartoon superheroes emblazoned on the boxes at the store. None interested me. I remember the patience that my parents showed during this extended process. And I remember the happy solution they devised: finding a camping box and marking it with the decals of the first Apollo missions that my godmom Carolyn had given me. I was delighted.

It was a perfect fit for a young historian—and good reason to give thanks.

TERMS AND ACRONYMS

AEC   Atomic Energy Commission. Although familiar with
            advanced science and technology, the organization lacked
            familiarity with flight research and was ultimately rejected
            as either a core or a model for the establishment of a new
            civilian space agency.

ARDC   Air Research and Development Command. Established in
            1950 to streamline Air Force science and technology devel-
            opment.

ARPA   Advanced Research Projects Agency. Established in
            February 1958, it reviewed military research projects and
            served as a temporary holding place for most military
            space projects prior to their being returned to either the
            armed services or the National Aeronautics and Space
            Administration after it became operational in October 1958.

ARS   American Rocket Society. During the 1950s, the organiza-
            tion represented nonpolicymakers interested in rockets
            and space travel. It urged civilian oversight of U.S. space
            projects and called for the establishment of a civilian space
            agency coming from a reorganized National Advisory
            Committee on Aeronautics. This coincided with the steps
            that the Eisenhower administration would take.

BoMi   Bomber Missile. Born in an Air Force decision in June 1956,
            it would be redesignated Rocket Bomber (RoBo) and ulti-
            mately would become the Step III aspect of Dyna-Soar.

Brass Bell   Air Force space glider reconnaissance project that originated in May 1955. It would form the basis for Step II of the Dyna-Soar program in 1957.

CIA   Central Intelligence Agency. Established in 1947 to pursue intelligence and security priorities overseas. Because it was a civilian, rather than a military, organization, President Dwight D. Eisenhower determined that it and not the Air Force would operate the U-2 reconnaissance plane.

DDR&E   Director for Defense Research and Engineering. Established in 1958 during Eisenhower's reorganization of the Department of Defense. As DDR&E, Herb York impeded development of Dyna-Soar.

DOD   Department of Defense. Established in 1947 during a major reorganization of the U.S. defense establishment, its creators intended to provide civilian control over a more closely coordinated direction of military, air, and naval forces.

Dyna-Soar   a flat-bottomed glider that the Air Force intended to pioneer the space portion of the aerospace realm. It was to reach space atop booster rockets, fly in space, and glide through the atmosphere to accomplish piloted but unpowered landings. In space, it would carry out a number of missions, although the most important envisioned mission would be to carry and "drop" nuclear weapons in the event of general war. The program was consolidated from the HYWARDS, Brass Bell, and RoBo projects in 1957 and slowly gathered momentum during the remainder of the Eisenhower administration. In 1961, project spending dramatically grew, and President John F. Kennedy initially lent his support. Late in 1961, the administration accelerated the vehicle development while adversely (from the Air Force perspective) reorienting the project's objectives. The Dyna-Soar project was cancelled in the first days of Lyndon Johnson's presidency.

HYWARDS   Hypersonic Weapons Research and Development Supporting System. This was the initial boost-glide project undertaken by Bell Aircraft in 1952 and embraced by the Air Force in 1954. It would form the heart of the Dyna-Soar Step I phase of the Dyna-Soar project when the U.S. boost-glide initiatives were consolidated in 1957.

NACA	National Advisory Committee on Aeronautics. Established in 1915 and transformed in 1958 into the National Aeronautics and Space Administration. Much of its effort went toward assisting the Air Force in mastering a range of technical aeronautics challenges.
NASA	National Aeronautics and Space Administration. Its creation from NACA in 1958 was partly a result of Eisenhower's willingness to mollify advocates of greater activity in space. In large measure, NASA stood as a rival, rather than as an assisting entity, to the Air Force in its space efforts.
NRO	National Reconnaissance Office. The entity, organizationally within the Air Force but not controlled by it, operated the Corona reconnaissance satellite.
NSC	National Security Council. In the Eisenhower era, the president and his principal advisers who constituted the NSC met on a weekly basis.
OCB	Operations Coordinating Board. It supported the NSC by providing oversight of agencies to ensure compliance with NSC directives.
OSD	Office of the Secretary of Defense, who directs DOD.
PSAC	President's Scientific Advisory Committee. Organized in the first days after Sputnik I, the panel provided advice that helped Eisenhower resist the work of arms advocates.
R&D	research and development.
RoBo	Rocket Bomber, a successor name of the BoMi and the component of Dyna-Soar that became Step III when the programs were consolidated in the second half of 1957.
S&T	science and technology. "Science" refers generally to the theoretical and experimental study relating to hypotheses, theories, and laws of science. "Technologies" are tools emanating from the application of scientific premises, which are then applied by engineers seeking to address particular problems or requirements.
Sputnik	name for the first Soviet satellites, translated from the Russian for "fellow traveler." Sputnik I, a 184-pound satellite, was launched in October 1957 and decayed in January 1958. Sputnik II weighed half a ton and carried the first live creature into orbit in November 1957; it decayed in April 1958.

USAF	United States Air Force. Institutionally independent as of 1947, its airpower doctrine carried forward in the late 1950s and into the 1960s as aerospace doctrine.
USIA	United States Information Agency. Essentially a Cold War propaganda agency, it urged greater activity in space as a means of answering the Soviet propaganda gains that came with early space triumphs.
Weaponized/ weaponiza- tion	referring to the arming or armed condition of a technology. Whereas "military" craft are used by the armed forces and may or may not be armed, "weaponized" vehicles are by definition armed systems.
WS-117L	the USAF's first major satellite program. The Corona project was separated from it and became the start of the primary technological reconnaissance platform used during the Eisenhower, Kennedy, Johnson, and Nixon administrations.

CHRONOLOGY OF EVENTS

1945–1946		Physicist Richard Leghorn, the U.S. Navy, the RAND Corporation, and others hypothesize about the long-term potential for scientific, communications, reconnaissance, or weaponized applications of satellites
1952	April	Bell Aircraft proposes a manned bomber-missile (BoMi) to the USAF
1954	April	USAF and Bell Aircraft arrange a contract to study an advanced bomber-reconnaissance system
	November	Eisenhower receives the first reports of the Killian Committee (including calls for "immediate" work to develop a reconnaissance satellite), and he approves James Killian's and Ed Land's proposal for a specialized, stop-gap reconnaissance plane (the U-2) to provide intelligence
1955	January	ARDC initiates studies for a reconnaissance system capable of 100,000-foot altitudes
	May	USAF Headquarters (HQ) announces a requirement for a piloted high-altitude reconnaissance system to be available by 1959; National Security Council document 5520 lays down national space policy, noting that "a small scientific satellite will provide a test of the principle of 'freedom of space'"

	July	Union of Soviet Socialist Republics (USSR) rejects Eisenhower's Open Skies proposal; U.S., and then Soviet, representatives commit their countries to launching a scientific satellite during the coming International Geophysical Year
	December	USAF requests study of the potential for a piloted, hypersonic vehicle capable of bombardment and reconnaissance
1956	March	ARDC Research and Target Systems Division develops an abbreviated plan for developing the hypersonic test bed HYWARDS; USAF and Bell Aircraft negotiate a study for the reconnaissance system Brass Bell
	June	ARDC outlines the requirement for RoBo
	July	First U-2 mission occurs over the USSR
	November	Eisenhower is reelected; the Suez Crisis and the Hungarian Revolution culminate
1957	February	NACA officials concur with ARDC commander Lt. Gen. Thomas S. Power, who supports work toward hypersonic weapons systems
	April	USAF HQ directs ARDC to establish a development plan that encompasses all hypersonic projects
	August	USSR demonstrates the first intercontinental ballistic missile; ARDC Research and Target Systems Division submits a plan for consolidation of HYWARDS, Brass Bell, and BoMi/RoBo into Dynamic Soarer (Dyna-Soar)
	October	Sputnik I (first satellite) is launched
	November	Sputnik II (first living creature in orbit) is launched; Eisenhower announces creation of the PSAC; USAF Chief of Staff Gen. Thomas D. White approves consolidation of the Dyna-Soar program
	December	First Vanguard launch attempt fails; USAF attempt to preempt ARPA by creating a Directorate of Astronautics fails
1958	February	Explorer I (first U.S. satellite) is launched; ARPA is established

	May	Sputnik III is launched
	June	NSC 5814/1 declares being a "recognized leader" in space a U.S. interest
	October	NASA becomes operational
	December	Gordo the monkey is launched into space but perishes in an unsuccessful reentry; the Project SCORE (Signal Communications by Orbiting Relay Equipment) mission broadcasts the first relayed human voice
1959	February	Discoverer I test fails
	April	DDR&E Herb York approves $14.5 million for Dyna-Soar but directs that the primary mission is suborbital hypersonic flight
	May	ARDC directs a study of Dyna-Soar's potential as a weapons system with global capabilities
	November	Assistant Secretary of the Air Force Victor Charyk initiates the Phase Alpha technical study
	December	Boeing signs a contract for Dyna-Soar, with the work statement covering Step I and limited to Phase Alpha; Charyk agrees to release additional funds to allow Alpha to proceed
1960	February	ARPA Order 84 is cancelled by Order 84–60, Amendment #1, indicating that "ARPA has no further responsibility for the Dyna Soar project," which was then undergoing Phase Alpha
	March	USAF affirms a medium lift-to-drag ratio craft such as Dyna-Soar as appropriate for manned reentry
	April	Charyk and York review Phase Alpha and approve the Dyna-Soar program; USAF negotiates a contract for Step I with Boeing
	June	USAF gives Martin Company responsibility for Dyna-Soar's booster airframe
	August	Discoverer XIII is recovered post-mission and Discoverer XIV is recovered in a mid-air grab during reentry, opening the way to satellite reconnaissance

	October	USAF HQ approves work on Step II and Step III studies; Soviet strategic rocket forces commander Mitrofan Nedelin and one hundred others are killed in a launchpad disaster that the Soviet news agency, TASS, censors
	November	John F. Kennedy wins the 1960 presidential election
	December	Dyna-Soar office completes a stand-by plan ready to accelerate development by using the same booster design for both suborbital and orbital flights
1961	January	USAF HQ announces Titan II to be the suborbital Step I booster; Eisenhower makes his farewell address
	February	Defense Secretary Robert McNamara inadvertently discloses to the press that the missile gap (a key issue in Kennedy's campaign three months earlier) does not exist
	March	Kennedy promotes Dyna-Soar in a message to Congress
	April	Cosmonaut Yuri Gagarin's orbital mission
	May	Boeing offers a "streamline" approach for accelerating Dyna-Soar by eliminating the suborbital Step I; in a Mercury capsule, Alan Shepard makes the first suborbital U.S. human spaceflight; Kennedy proposes a human lunar landing by decade's end
	December	USAF HQ approves a development plan drawn up the previous month
1962	January	Air Force Space Command halts further consideration of Step III
	February	In a Mercury capsule, John Glenn makes the first U.S. orbital spaceflight
	June	DOD officially designates Dyna-Soar the "X-20;" the experimental designation undercuts the Air Force concept of Dyna-Soar Step I as a "conceptual test vehicle" toward an armed system
	November	Kennedy confides to aides, "I'm not that interested in space"

1963   March   McNamara orders a comparison of the military potentials of Dyna-Soar and Gemini

        December   McNamara announces Dyna-Soar's cancellation and exploration of a manned orbital laboratory (cancelled in 1969) to use Gemini as a USAF space crew vehicle

The
Other
Space
Race

1

"What's a Heaven For?"
Setting Security Policy in Space

Gently in manner, strongly in deed.
—Motto in Dwight Eisenhower's Oval Office

Ah, but a man's reach should exceed his grasp, or what's a heaven for?
—Robert Browning, quoted in a 1960 Air Force history of its space program

The Space Policy Contest

The world typically remembers the space race as the Cold War competition between the Union of Soviet Socialist Republics (USSR) and the United States beginning with the Soviet launch of the first satellite (Sputnik I) in 1957 and culminating in the U.S. landing of astronauts on the moon in 1969. But even prior to the Sputnik launch, the United States had already taken important steps outside of the public eye. The Dwight D. Eisenhower administration quietly worked to establish a precedent for peaceful satellite flight in space over international borders and to use space-based surveillance systems to derive information that would minimize the chance of a surprise Soviet attack. The Air Force, convinced that space would ultimately become a realm of combat, made initial studies about hypersonic armed vehicles that could skim the upper atmosphere or even orbit the planet. For the supporters of armed space exploitation, as for many Americans in general, deterrence dictated preeminence. High-profile Soviet space accomplishments suddenly made these issues seem more urgent and transformed the degree to which the debate was public.

The two perspectives on the utilization of space were drifting on a collision course in the mid-1950s. The president's objectives required

1

secrecy, while Air Force leaders and other advocates of armed space exploitation openly went forward with their flagship craft concept: the Dynamic Soarer space glider bomber, known as Dyna-Soar.

The Air Force intended Dyna-Soar to reach orbital space atop a booster rocket, conduct reconnaissance and, potentially, bombardment missions, and return to Earth by gliding through the atmosphere to make an unpowered landing. Its envisioned flight pattern resembled that later used by the space shuttle (1981–2011), but its weaponized missions were significantly different. *Niagara Aerospace Museum*

No Farewell to the Dyna-Soar Program?

On the afternoon of January 3, 1961, Dwight Eisenhower had barely two weeks remaining in his presidency to persuade the American public of the continued merits of his long-haul approach to military and economic security. His final State of the Union Address on January 12 and his last budget message five days later were two opportunities to use the presidential bully pulpit. But ultimately, his most famous speech would be the farewell address he and his writers had been planning since the summer of 1959. In this address, the president would again urge Americans to undertake an alert and engaged form of citizenship—and also to be wary of the power of the military-industrial complex and the influence of a scientific-technological elite pushing for more ponderous technologies with little appreciation of true defense needs.

Gordon Gray entered the president's office that afternoon with several pages from the newest draft of the upcoming budget speech in hand. The aide apologized that the speech required yet another interruption of the president's other work. A proposed phrase in the draft speech declared that "Pending further testing and experimentation, there appears to be no valid scientific *or security* [emphasis added] reason at this time for extending manned space flight beyond the Mercury program."[1] Gray noted that the inclusion of the words "or security" signaled the administration's rejection of the Air Force's Dynamic Soarer space glider.

An Air Force priority since the time of the Sputnik launch, Dyna-Soar was intended to complement and ultimately succeed the B-52 Stratofortress and the Mach 3 B-70 Valkyrie as a nuclear weapon delivery system. It would diversify the U.S. nuclear deterrent and thereby, in Air Force eyes, promote national security by providing an armed presence in space. Dyna-Soar was also to conduct strategic targeting reconnaissance from space, and its planners believed that this small (thirty-five feet) vehicle covered in special heat-resistant alloys would open the era of space flight, and eventually, interplanetary travel.

The president considered Gray's observation. Four days earlier, he had told Gray that he "questioned whether we should proceed with the DINOSAUR [*sic*]."[2] However, Eisenhower and his aide had agreed that a critical remark in the outgoing president's budget address was not the proper means of ending the program. Advisers, in step with Eisenhower's perspective, asserted in late 1960 that Dyna-Soar was of "no military value." The president himself had insisted as much in a tense and extended exchange with the Air Force Chief of Staff during a National Security Council (NSC) meeting in the period between John F. Kennedy's election victory and his upcoming inauguration, now just days away.[3] Gray indicated that the Defense Department, the State Department, the director of the National Aeronautics and Space Administration (NASA), the new civilian space administration, and the chairman of the president's own scientific advisory board unanimously advised against language in the budget speech that would "raise issues rather than put them to rest."[4]

Perhaps with a sigh, the old general, a policymaker for the last eight years, consented to wording that would not challenge the Air Force's dream vehicle. As he had confided to advisers earlier in his tenure, Eisenhower did not want to completely flout the public will on space issues because "psychology on this matter has proven to be tremendously important— even if it is not too well informed."[5] The farewell address would be

Eisenhower's last opportunity to persuade his countrymen to adopt a long-haul approach to the Cold War. Throughout his first term in office, the country had supported him, and he had employed a hidden-hand model of leadership that relied on loyal and like-minded subordinates to take action—and to receive credit or blame, thus shielding the president as he played the avuncular non-politician. The style fit Eisenhower's own character while addressing the difficulty he had with the mental image of himself as a general working as a politician. A major trouble during his second term had been that critics increasingly condemned his security policy. Eisenhower's speeches to the nation in November 1957, just after the first Soviet satellite launches, sought both to reassure Americans of their country's sufficient defense and to call upon them to act as individuals to revitalize technological curiosity and know-how for the long haul. But defense critics continued to decry his seeming inactivity.

Despite his historic experience in the Second World War, and because Cold War necessities and his executive approach restricted his ability to respond to detractors, Eisenhower faced a convergence and growth in his critics' volume and number and their influence on security and defense. Kennedy's election had been their triumph and symbolized the rejection of Eisenhower's policies, and the outgoing president understood it. Now, by January 1961, much had changed. The Soviet interception of a U-2 the previous May had revealed the existence of Eisenhower's secret spy plane. In a few days, President-elect Kennedy would receive a wealth of intelligence data from the brand-new and top-secret Corona reconnaissance satellite. The photographs would show that the "missile gap," which had been a major feature of Kennedy's criticism and campaign, was in fact completely nonexistent.

Eisenhower had laid the foundation for a wise space security policy in the middle of the 1950s. Once space entered the public spotlight after the Soviets lifted Sputnik into orbit in 1957, however, the shaping of space policy had been challenged and contorted, principally by the competing visions of Air Force officers and Democrats in Congress with the aid of some shrill voices in the media. Eisenhower's leadership style was well attuned with setting strong, long-term space policy, but it was less well suited to the public debate occurring during his last years as president. That poor fit eroded public perceptions of him and reduced his ability to maneuver. Dyna-Soar spending was accelerating, and the project was poised to become the largest single item in the entire national space budget, dwarfing all else on both the military and civilian sides.[6] With Eisenhower leaving office, his successor an unknown quantity, and myriad national problems including deficit spending, strife over civil rights, and the prospect of rising tension in the Cold War through renewed nuclear tests and revolutionary warfare, the president felt compelled to hone his parting statement.

Throughout his presidency, Eisenhower refused to divulge the top-secret intelligence information that would have quieted his increasingly vocal critics and benefited him politically in the short term. He had his eyes on a long-endurance struggle against the USSR, and he would not tip his hand for the sake of short-term advantage. The upshot of Eisenhower's approach was that he ceded dominance of space security policymaking. In the process, it revealed the political vulnerability of the hidden-hand presidency.

The genesis and trajectory of U.S. space policy were the result of a struggle between policymakers and military thinkers to answer the nation's security needs. The development of national space policy was a contest between these groups to define the parameters—and, to an extent, even the very nature—of the Cold War.

This book focuses primarily on the conflicting perspectives and initiatives of the Eisenhower administration with the outlooks and actions of the U.S. Air Force between 1954 and 1961. Pivotal trends and programs shaped the country's early space policy during this seven-year period. While early policy had been laid beyond the glaring and sometimes distorted spotlight of national discourse, the Soviet launch of Sputnik satellites threw these issues and the image of a space "race" onto the public stage. This affected the interactions of civilian and military elites who sought to articulate and enact the space policies that they expected would promote national security.

The Policymaking Context

Though obscure today, the Air Force's Dynamic Soarer space bomber project was once the flagship of the nation's expected aerospace future, and it made a significant (if subsequently unrecognized) mark on U.S. security policy-making in space.[7] Airpower enthusiasts, within and outside the Air Force, sought to preserve national security through the capacity to dominate the sky and project power to strike devastating blows against an enemy. This philosophy was not new in the 1950s, and its endurance owed something to an ability to adapt.

Despite being an early and vocal figure in airpower theory, Italian general Giulio Douhet did not speak for all advocates of airpower. Some contemporaries, such as Brig. Gen. Billy Mitchell, USA, agreed that gaining control of contested skies was a prerequisite for victory; however, Mitchell (particularly in his earlier works) propounded the need for a mix of fighter and bomber planes as well as reconnaissance aircraft. Where Douhet imagined a single decisive stroke, others were prepared to expect a longer and bitterer struggle.[8] Initially organized within the Signal Corps (a support branch rather than a combat branch of the U.S. Army), the Army Air Corps was established to give the air arm greater stature in 1926.

U.S. targeting differed from Douhet's vision. Rather than wanton attacks on cities to terrorize populations, U.S. theorists wanted to conserve their planes, bombs, and aircrew by instead striking only at the vital nodes of an enemy's warmaking potential. Within this context, however, U.S. airpower thought grew increasingly hard line. By the 1930s, the Air Corps Tactical School was teaching what amounted to a shadow doctrine that emphasized the war-winning potential of bombing, and instructors such as Capt. Claire Chennault, who advocated a balanced mix of aircraft and tasks, were brushed to the wayside.[9] Better bombers, it was believed, would penetrate enemy air space and attack precise targets, dealing crushing blows.

A serious shortcoming of airpower thought in the interwar era related to the advance of technology. Aircraft performance improved radically during the 1930s, particularly because of increases in engine power, reductions in drag, and sturdier airframes using metal construction. However, the technology still did not keep pace with the bold expectations of theorists who saw decisive potential and the need for service independence. Theory lacked evidence for several reasons. One was the general absence of major wars between industrial powers between 1918 and 1939. Another was related to neutrality legislation; airpower doctrine called for precision bombing of vital targets, but legislation prevented the U.S. military from collecting the information it needed so that it could select and destroy enemy targets. A further, avoidable problem was that theorists neglected to pay enough attention during the interwar years to the empirical evidence gained through their own practice of precision bombing techniques.[10]

Several factors explain the warm U.S. reception of the airpower philosophy. The combination of wide oceans and long-range bombers seemed to insulate the United States, especially when bombers were expected to destroy hostile fleets at sea. This in turn promised to avert the human and financial costs of raising large ground forces in the event of a future war. Furthermore, targeting enemy factories also sounded morally preferable to deliberate killing of civilians in urban centers. The exploits of heroic aviators gave airpower, whether military or civilian, added panache, and the Army Air Corps (the Army Air Force [AAF] as of 1941) recognized this and encouraged high-profile feats. Articles in popular readership journals such as *Reader's Digest* maintained public awareness of the promise of airpower. For those without the time or inclination to read articles by airpower advocate and aircraft designer Alexander de Seversky, Walt Disney released a film version of his book, *Victory Through Air Power*, just nineteen months after U.S. entry into the Second World War.[11]

Doctrine is not supposed to be rigid.[12] The rapid development of Air War Plans Division-1 and its successor air war plans owed something to the commitment of airmen to their doctrine.[13] However, doctrinal rigidity meant that AAF bomber crews suffered horrific losses while long-range escort fighters (rejected as doctrinally irrelevant) were hurriedly and belatedly developed. Bombing had dealt damage to Nazi industry and had helped attrit the Luftwaffe and divert artillery production to antiaircraft guns, but these were not the outcomes promised by interwar doctrine. The appearance and impressive performance of German jet fighters late in the war also haunted AAF commanders. The gruesome incendiary raids over Japan in the conflict's final months reflected a focus on results that forced a change in method. Elaborate studies indicating the effectiveness of firebombing, coupled with the invalidation of high-altitude attempts at precision because of the jet stream over Japan, led precision bombing to give way to repeated incendiary attacks.[14]

In 1947, the Air Force gained independence as a separate branch of the U.S. military, and it had already established a command for research and development (R&D), which Maj. Gen. Curtis LeMay would lead. Air commanders had bowed to realities when, in the last phases of the war, they accepted that contemporary technology could not allow unescorted entry of bombers or precision delivery of munitions. The Air Force reacted to the postwar environment by trying to anticipate and prepare for the techno-logical promise and threat of the distant future. Air Force Chief of Staff Gen. Henry "Hap" Arnold's work with scientist Theodore von Kármán reflected an interest in preparing for the distant future. Others, such as R&D chief LeMay, remained convinced that a functioning and modern operational force was more useful than a futuristic prototype.[15]

Philosophically, the two ideas were complementary rather than in competition. Fiscally, dollars could be spent only once. Technologies and doctrines needed to coordinate with the national policy and strategy that the military exists to uphold. Perilously, this vital point can be forgotten if the implications of a technology or doctrine are unclear, if a policy or strategy is not yet determined or enunciated, or if the creators of a particular technology or doctrine presume that national policy and strategy will fit the contours of their own design. All of these forces affected the early formation of U.S. space security policy during the crucial years of the Dyna-Soar program. The exact capabilities and technological challenges in making Dyna-Soar work remained unclear, while national policy on space developed without detailed awareness of Air Force leadership. Air Force planners expected that the relentless advance of technology would force the United States to pursue peace through readiness, which would mean increasingly sophisticated flying weapons systems.

The president's responsibilities include formulating and enacting national policy in a variety of areas, including national security. The art of policymaking means determining a national course of action, communicating that intent to subordinate elements of the governmental structure, and confirming that actions are in keeping with the policy. It means anticipating that unforeseen developments will arise and coping with new political environments by knowing how, when, and why to adjust policy. Cabinet members, staff, and advisers have always been necessary to enable the chief policymaker to fulfill his Constitutional role, but this supporting apparatus did not change the fact that by the twentieth century, the president was the chief policymaker. Harry S. Truman's motto, "The Buck Stops Here," spoke truth. The job of the president's staff is in part to be an extension of his mind and hands, supervising delegated tasks in a manner in keeping with overall policy. Scholars contemporary to Eisenhower observed that the initiative of subordinates was useful only when it "reflect[ed] the loyal identification with, and understanding of, the desires of the superior."[16] When problems prove intractable and cannot be solved at the level of assistants, the questions are passed up the chain until they are resolved by other assistants or, failing that, they reach the president for decision.[17]

Eisenhower was a man who "'liked to think out loud' with people he trusted," because a vetted and circumspect idea was more likely to lead to an effective decision and a more coherent policy than an unquestioned, preconceived notion could do.[18] Eisenhower preferred, at least in the public eye, to be in charge but not to be monopolistically "hands on." This redoubled the requirement that his subordinates truly understand his vision and perspective. But he could be misunderstood even by those who praised and supported him, and he was certainly misinterpreted by detractors.

The New Look, a commitment to power projection, nuclear weapons possession, and strategic alliances, was Eisenhower's vehicle for preserving national security and solvency. The constant awareness of both the military and economic sides of the security dilemma was a hallmark of Eisenhower's New Look, and he attempted to explain these views to confidants.[19] Adm. Arthur Radford, Eisenhower's first Chairman of the Joint Chiefs of Staff, explained to the National Press Club in the winter of 1953 that "our 'new look' prepares for the long pull. . . . It is aimed at providing a sturdy military posture which can be maintained over an extended period of uneasy peace." It was a workable and pragmatic defense posture fitted to its specific time.[20]

However, to the casual ear, these statements sounded like stodgy fiscal conservatism in the midst of a booming economy. It was also very tempting to interpret the "more bang for the buck" feature of Eisenhower-era massive retaliation as a wholesale endorsement of Air Force attitudes. Air Force planners wanted to dominate the national defense structure and to pursue successive generations of increasingly spectacular weapons systems. Although this was not Eisenhower's purpose in favoring the Air Force portion of the overall Defense budget, airpower advocates easily drew the tempting conclusion, and they felt encouraged to pursue their agenda.

Eisenhower's presidential style was a double-edged sword. Commonly misinterpreted and misunderstood as avuncular detachment, the popular image was a palpable liability in time of crisis, because it fed the notion that no one was at the helm when the country was at sea. This impression of Eisenhower lay generally unchallenged for decades, although a few observers, such as journalist and biographer Murray Kempton, suspected that Eisenhower had been a more complicated figure than he appeared to be. Eisenhower came to office as a career soldier. He was ill at ease with transitioning to open involvement in politics, yet he was well versed in politics and policymaking. He wanted to avoid being seen as a partisan figure, both because his popularity depended on it and because part of him genuinely wanted to avoid becoming enmeshed with politics.[21]

Eisenhower was not as apolitical a president as he deliberately aimed to appear. He was determined to wage a "total cold war" that leveraged instruments for psychological advantage in an ongoing global competition with the Soviet Union, while domestically he aimed to build a more mainstream and competitive Republican party, rejecting the moniker of "Eisenhower Republican" because he meant to leave his stamp on the *entire* party.[22] He drew his political strength from his popularity, rooted in the wartime experience that transformed him into a household name and a man perceived as admirable, approachable, and apolitical. For Eisenhower, successfully participating in politics meant appearing not to do so at all. He negotiated this obstacle with enough success that contemporary political scientists deemed him a "friendly, folksy, [and] easygoing soldier who reflects the ideals of a democratic and industrial civilization."[23] Both in his 1952 election and his reelection in 1956, pollsters confirmed that Eisenhower was a widely admired man but that the traits that had earned him such acclaim had more to do with perceptions of his personality than with reflections on his leadership or his strategic experience.

Eisenhower felt that his professional reputation commanded a stature that needed little further aggrandizement.[24] He had overseen the triumph of Allied forces in North Africa and Western Europe, juggling strong Allied personalities in the process, and he been responsible for the Normandy landings. But people *liked* Ike without necessarily comprehending the extent of his accomplishments or capacities, despite the war being in recent memory. This situation permitted critics of his security policy to gain traction in the wake of the Sputnik launch, and it threw space security policy into limbo for the remainder of his presidency.

For his part, John F. Kennedy exuded energy. He drew inspiration from his conviction in America's place as a world leader, his belief in the inherent activity of great powers, and his own personal efforts to stave off nascent rumors of his poor health. Speechwriter and confidant Ted Sorensen noted that, regarding space exploration, Kennedy "was more convinced than any of his advisers that a second-rate, second-place effort was inconsistent with this country's security." Kennedy strove to "epitomize strength and good health" as well as project the image of a thinker (through his books *Why England Slept* and *Profiles in Courage*) and of a man of action, youthful war hero, and family man.[25]

The presidency symbolized a seductive and tantalizing power that appealed to Kennedy. Perceptive biographers recognized his tendency toward bilateral personal relationships, which were a facet of his complex and compartmentalized personality. On the campaign trail, he confided "that the life of a legislator was much less satisfying than that of a chief executive," since a congressman's years of toil could be upended "by a president in one day and one stroke of the pen." Biographer Robert Dallek explained that "Jack [Kennedy] believed that effective leadership came largely from the top" because "being president provided opportunities to make a difference no senator could ever hope to achieve."[26]

True as this may be, however, Kennedy struggled, in space and in other areas, to provide that clear leadership. Despite having relatively high favorability ratings as president, his legislative record was the poorest of any twentieth-century chief executive. On civil rights, although convinced of the irrationality and evil of prejudice, Kennedy failed through the first two years of his interrupted tenure to truly follow up on his convictions. In office, he was startled to discover the intransigence and complexity of national and global problems that he had identified as a candidate, and numerous early crises slowed his pace on the space issue, where he had promised concerted action.[27] Kennedy's frustrations and challenges in office speak to the complexity (and the centrality) of converting ideals into action and nesting action within pragmatic policy.

Even large defense budgets are finite, and decisions about which weapons to buy and how to use them are bound to be complex. The security challenge in the 1950s was serious and multifaceted. There was a credible chance of a Soviet nuclear attack on the United States or, more likely, on U.S. allies more easily within Soviet reach. There was also the possibility that Soviet forces would attack members of the North Atlantic Treaty Organization (NATO) alliance. Soviet or Chinese forces might invade the decolonizing lands of the "third world." Additionally, Communists might conceivably seek to subvert the character and political friendship of any country on the globe. Wisconsin Senator Joe McCarthy rose briefly but meteorically to political stardom by capitalizing on this unease. The U.S. military had to be prepared for any or all of these possibilities, and R&D and acquisition decisions had to be made in this environment.

Militaries invest in technologies that answer perceived needs.[28] The expression about wanting more "bang for the buck," coined in 1954 by Defense Secretary Charles "Engine Charlie" Wilson, tends to be associated with the nuclear retaliatory theme in Eisenhower's New Look, but historian Alex Roland has observed this trend "as far back as we can see in human history." It is about more than the literal bang, however: the effective use of military technologies leverages psychological effects in addition to purely physical ones.[29] The physical instruments of war are of little usefulness without the doctrines—the formal expressions that outline how armed forces approach operations—that employ them to best effect. The weapons, and the ideas behind their use, must align with the visions of the national policymakers.[30] Meanwhile, policy itself can be subject to change, and military planners must be prepared accordingly.

Science and technology (S&T) had been advancing quickly and continued to do so at a breathtaking pace. Rapid advances had already changed warfare and transformed life in industrialized countries. After the nuclear bombs helped end the Second World War, some people were prepared to discard the study and purported lessons of all earlier military history. A technological edge appeared an obvious necessity in keeping the nation safe, and because emerging S&T brought new possibilities in all realms of life (including warfare), many people interpreted progress through a deterministic lens. Promoters of the hydrogen bomb and the long-range thermonuclear-armed ballistic missile argued that what was technologically possible was therefore necessary. Soviet technological advances reinforced the confidence and assertiveness of these advocates, who promoted notions of a "bomber gap" in the mid-1950s and of a "missile gap" at the turn of the decade. The U.S. Air Force's Strategic Air Command (SAC) original motto was, "War is our profession, peace is our product."[31]

The Soviet air force had a distinct geostrategic requirement and therefore a different structure and focus than those of the United States. With long borders and a history of hostile neighbors, the USSR could not rely as disproportionately on airpower at the expense of ground power. The atomic era, however, dictated that the Soviets could not afford to let their long-range bomber force become antiquated, as it had by the start of the Second World War. Efforts to acquire U.S. long-range bombers gave way to the expedient of reverse-engineering the B-29 Superfortress and then supplanting it with a later generation of indigenous bomber designs. The USSR did build large numbers of nimble, short-range jet fighters from the late 1940s onward, and Nikita Khrushchev spent the second half of the 1950s insisting that Soviet ballistic missiles were similarly ubiquitous. Soviet feats that aimed to imply an unassailable superiority—including the launch of Sputnik I in 1957 and hearsay the following year of a nuclear-powered airplane and a glide-bomb with a 3,500-mile range—instead prodded many in the United States to reinvigorate research, development, and acquisition of even more superlative systems.[32] The Soviets' paranoia and hostility played a part in creating the world that made them fearful and paranoid.

The apparent solution, from the U.S. technologists' perspective, lay in anticipating the future and reaching it before the Soviets did. As Air Force Chief of Staff, Gen. Thomas White urged planning estimates within the branch to look "into the future as far as possible . . . to maintain a qualitatively superior Air Force." The Air Force used new organizational models as well as new technologies to accelerate the process of innovation during the Cold War. The service not only conducted a series of futures studies but also presumed that policymaking would be the cat's paw of technological development, on the premise that constantly improved technology was a prerequisite to that security.[33] If policymakers did at first fail to understand the crucial importance of one or another military technology, then the solution lay in informing them, or the broader public, of the imminent but underestimated dangers of complacence. Sudden and shocking events assist advocates in making their cases to policymakers and the public.[34] For example, preparedness advocates in 1940 saw the swift victories of Nazi armies in Europe as both a threat to the United States and as a way to demonstrate the urgent need for stronger national security efforts. The middle of the twentieth century saw airpower (and then aerospace) advocates eager to drum up support for the programs they deemed vital to national security, and these aerospace advocates enthusiastically utilized news of events and trends useful toward that end.

Interpreting a chaotic sea of ideas to define the will of "the people" is a harrowing task, because it is a complex amalgam of myriad peoples, often without definite or consistent ideas about a given issue. Although public figures may declare that they hear and understand the public's supposedly unified voice, it is not one group, and it does not speak with one voice. Yet keeping the desires of "the people" in mind is simultaneously pragmatic, obligatory, and elusive.

A prevailing national public culture is an aggregate of many ideas and emotions that are shared and taught by a political community. Several factors—a need for government secrets, editorial decisions by outlets that convey news, and the pressing concerns of people's daily lives—mean that constituents tend to lack a fully informed picture of issues that nonetheless impact their security. Cultural attitudes evolve; they are partially malleable moving targets. Jackson Lears accurately explained that "a dominant culture is a continuous process, not a static 'superstructure,'" marked by "tensions and idiosyncrasies."[1] Cultures influence, rather than determine, peoples' ideas, and in turn people each force the larger culture to adapt in small ways as well. Whereas opinions may be transient, cultural tenets exist and adapt with a culture, and policymakers cannot safely ignore deeply ingrained values any more than leaders can refer blindly to popular sentiment. The "bully pulpit" provides presidents with a unique but finite platform from which to mold the contours of prevailing cultural tenets so that enough people deem new policies to be in alignment with social norms.

Note
1. Thomas J. Jackson Lears, *No Place of Grace: Antimodernism and the Transformation of American Culture, 1880–1920* (Chicago: University of Chicago Press, 1994), xvii–xviii.

As long as the public trusted Eisenhower on defense issues, he could quietly shape space policy without serious challenge. The administration laid the groundwork for its basic space policy (its overall principles of action) and its space activity (the space vehicle projects themselves) prior to Sputnik. With the advice of the National Security Council, these first steps emphasized strategic, but not weaponized, use of space for reconnaissance. As secret decisions, they took place out of the public eye but nonetheless dovetailed with the New Look posture that Eisenhower maintained and that the American public generally supported and implicitly ratified in 1956 with Eisenhower's reelection.

The Air Force's commanders were not privy to the extent of American reconnaissance satellite development occurring through the Corona program during the Eisenhower era, which had in fact been hidden from their view because of concerns that the leak-prone Air Force would disclose the existence of the top-secret program. (As a relatively new independent branch, the Air Force defended its status through extensive collaboration with the media; during the Cold War, the service was notorious for leaking information to the press, to the consternation of Eisenhower and then later also to John F. Kennedy's defense secretary, Robert McNamara.) Furthermore, commanders dismissed hopes of preventing an arms race and eventual combat in space as dangerously shortsighted and extremely improbable. "Sputnik has finally killed the 'Ike knows all about defense' myth," wrote defense hawk and commentator Stewart Alsop to Democratic senate majority leader Lyndon Johnson.[35] Eisenhower's critics had been aching for a chance to neutralize his public credibility on national security, and for the last three years of his tenure, Eisenhower fought on the defensive in that area. Many journalists, defense hawks, and Democratic hopefuls filled the air with noise that convinced the president that the nation's public demanded more high-profile space activity than he favored.

During this early juncture, vocal supporters of Dyna-Soar were motivated more by ideas of national security and personal ambition than by a quest for constituencies. Existing programs mean defense jobs,[36] but Dyna-Soar's contractor (Boeing) was not selected until mid-1959, and Eisenhower officials delayed the kind of metal-bending work that would translate to a congressional constituency whose loyalty could be ascribed to jobs in a representative's district. Congressional and media support for Dyna-Soar and aerospace power derived from Cold War security concerns.

During his remaining time in office, Eisenhower felt constrained by the impression of a national consensus for space. This propelled projects such as Signal Communications by Orbiting Relay Equipment (SCORE)—the first human broadcasts from orbit—and the Mercury program, and it led Eisenhower to feel unable to cancel the Air Force's drive to build the Dyna-Soar space glider. From late 1957 to the close of his presidency, Eisenhower's subordinates worked to prevent the Dyna-Soar program from endangering the Space for Peace policy, which aimed to provide a diplomatic shield for reconnaissance satellites. But the Air Force retained control of its Dyna-Soar program and kept it sputtering forward so that Kennedy's incoming administration could lend the support that he had vowed to give to American space initiatives.

In the waning weeks of his tenure, Eisenhower told confidants that he wanted to cancel Dyna-Soar. National discourse had pressured him to pursue space projects and had precluded his opting to cancel the program earlier. Now, in his last month, there was no purpose in canceling a program that seemed likely to be favored by President-elect Kennedy, whose successful campaign had included spurious criticism of Eisenhower's allowing a "missile-space" gap. Dyna-Soar was one of the projects troubling the president's mind as he prepared and delivered his farewell address, with its dual warnings about the military-industrial complex and the scientific-technological elite, and its plea to the American public to embrace a civic engagement that could counteract these forces. This appeal, and the mixed reception to it, reflected a partial shortcoming in Eisenhower's approach to policymaking.

Kennedy, for his part, spent his pivotal first year in office dealing with a series of challenges, including the Bay of Pigs fiasco, the orbit of a Soviet cosmonaut, a less than stellar summit at Vienna, the construction of the

Berlin Wall, and the resumption of nuclear weapons testing, which threatened to reduce the contemporary U.S. nuclear arms advantage. Kennedy's position on space appeared much more strident than that of his predecessor, but the relationship of military and civilian space efforts was not yet clear. In March 1961, prior to deciding to start a lunar venture, Kennedy called for boosting Dyna-Soar program funding. His May address announcing a lunar quest did not explicitly declare its civilian character, and the remainder of his presidency was marked by alternating interests.

Consistently, Kennedy expected the Apollo program to provide national prestige at home and on the world stage, but the context was flexible: sometimes he sought to use the lunar venture as a realm of potential U.S.-Soviet cooperation, while at other times he framed the effort as a superpower competition. In November 1961, the administration dramatically revised Dyna-Soar's objectives and potentially undercut its reason for existence in Air Force eyes, yet the following month the administration announced approval of an acceleration in Dyna-Soar's development schedule. The supporters of aerospace projects, confident of the inevitability of their vision and eager to ensure that U.S. technology continued to outpace its Soviet counterparts, interpreted such developments selectively. Not surprisingly, the Air Force continued to pursue the vehicle that they anticipated would one day be called upon in its originally envisioned armed role of extending the realm of airpower past the atmospheric sky into the heavens beyond.

In his study of the railroad and the atomic bomb, historian Geoffrey Herrera aptly noted that "technology is both a product of the international system and domestic institutions as well as a driver of international system change."[37] President Eisenhower and President Kennedy and the Air Force's leaders were acutely aware of this. "Aerospace" advocates in the Air Force (and in the media and Congress) emphasized that the technological pace of the twentieth century and the international dynamic of the Cold War made space exploitation imperative and inevitable. These blocs therefore built up a domestic momentum, which Eisenhower (privy to intelligence data and skilled in strategic practice) worked to counteract. Both presidents recognized that the development and introduction of Dyna-Soar as a weapons system would alter the international landscape. Kennedy, who incorporated security issues into his larger campaign thesis about vigor, investment, and action on domestic and international fronts in the Cold

War, found that he had to answer as president to the power groups he had courted as a candidate. Vowing to give "equal attention to both" armed and peaceful initiatives, Kennedy discovered that in space, both paths were expensive; furthermore, they were difficult to reconcile with each other. Between 1954 and 1961, in the shadow of the Cold War, the United States developed its approach and its first major projects for space.

2

"Symbol of . . . Longing and Hope"
Dwight Eisenhower and America, 1952–1956

You like Ike, I like Ike, everybody likes Ike.
—EISENHOWER CAMPAIGN SLOGAN, 1952

Our military superiority may never be so great again.
—KILLIAN COMMITTEE REPORT

The Transformative Power of World War

By 1956, President Dwight David Eisenhower—the man, the persona, and his policies—had met the needs and facilitated the dreams of much of the American public for four years. Eisenhower's personality had been honed through a disciplined upbringing followed by decades of service in the U.S. Army. That experience had culminated with his coordination of a wartime coalition that liberated Western Europe and his oversight of the Western alliance arrayed to deter Soviet aggression in the early postwar years.

The Second World War had transformed Eisenhower's career and stature. When fighting first erupted in Europe in 1939, Eisenhower was a lieutenant colonel whose military career appeared to be nearing its end. By the close of 1942, Eisenhower was a newly minted major general, having earned three promotions in the space of thirteen months by performing well within a fast-expanding army. By the end of 1952, Eisenhower was a household name, a retired five-star general famous for his beaming grin and the success of the Normandy landings on D-day. Furthermore, Dwight Eisenhower was the President-elect of the United States.

The war had a similarly profound impact on the country itself. The First World War established the nation as the leading global creditor, and many Americans explored consumer credit and mass-produced housing in the 1920s before the stock market crash and start of the Great Depression hit the United States and the world.[1] A second world war defeated both the Depression and the Axis enemy, but it left many questions unanswered. Much of this uncertainty dealt with economics. It was American participation in the Second World War, not the programs of what President Franklin D. Roosevelt called "Doctor New Deal," that had relieved the agony of the Depression. As inflation rose in the early postwar years, it was not clear whether the era would see prosperity or a return of Depression-style crisis.

Technological issues posed their own challenges and opportunities and altered the government's relationship with science and academia and also changed interactions between the military and the rest of the government as well as industry. War had prompted enormous government investment in scientific and technological research, and Vannevar Bush, Roosevelt's scientific adviser and the head of the Office of Scientific Research and Development (OSRD), declared in 1946 that the wartime situation had "demanded a closer linkage among military men, scientists, and industrialists than had ever before been required." As Army chief of staff, Eisenhower himself had endorsed ongoing cooperation between civilian experts and military planners; this path, embraced during the Second World War, had been a break from the past.[2] OSRD's official history asserted that "the chief reason for success was the extraordinary degree of cooperation evinced by the services, the civilian scientists, and industry."[3] This military-industrial complex, which Eisenhower would later famously identify as a source of potential danger to American society, appeared to have been vindicated by the victory over Nazi Germany and Imperial Japan. In the eyes of many, the new technologies meant that the past had no lessons to impart about the future. One high-ranking but anonymous airman suggested in December 1950 that "the weapons and the methods of all [past wars] are out of date."[4]

The two most expensive S&T projects in the world up to that time were the Boeing B-29 Superfortress aircraft and the Manhattan Project, which yielded the atomic bomb. The employment of the two combined systems against Hiroshima and Nagasaki in August 1945 coincided with (and arguably caused) the war's end. Understandably, Americans were quick to conclude that nuclear power had averted the horrific prospect of assaulting tenaciously defended Japanese beaches. In the early postwar period, four in every five Americans approved of the nuclear attacks on Japan.[5] For many Americans, a shortened war did not just mean a return to normalcy; it likely saved many of them personally from a death in battle. Inventions such as penicillin, the Higgins boat, and DDT had existed prior to the war, but their mass production and industrial-scale use aimed to save lives and help make victory possible.[6]

Global crisis and human frailty raised uncomfortable and complex problems. Some Americans were also uncomfortable with the implications of nuclear warfare, and most were disturbed by the ugly barbarity uncovered in places such as Auschwitz. Despite the wartime consensus that the Axis enemies needed to be crushed, many Americans, particularly in the Republican Party, disapproved of a more activist or even interventionist role in the world. But prominent opinion leaders, including such Republicans as Senator Henry Cabot Lodge, had abandoned isolationism with news of the Pearl Harbor attack in 1941, and the wife of the Republican editor of the *New York Herald Tribune* declared in October 1945 that "science has killed isolationism."[7] As President Harry Truman began his on-the-job-learning about diplomacy, he and others quickly found the alliance with the Soviet Union's "people's democracy" unraveling as Red Army troops brutalized civilians, looted, and facilitated the establishment of puppet regimes. In the fresh wake of a war that had begun for both the United States and the Soviet Union with a surprise attack and during which fifth-column traitors and armed partisans had been rampant threats, it was easy to view the present world in the context of the recent past. Adversaries could seem to be new Hitlers, and suspicions were mutual. The revolutionary vitriol of Marxist theory did not help matters; nor did Soviet reluctance to depart from Iran, Soviet efforts to blockade West Berlin, or Marxist violence in Greece.

America in the Cold War

People generally react to new events by contextualizing them within the ideas, perceptions, and practices with which they are comfortable, and Americans in the late 1940s and early 1950s were no exception. "Traditions, values, attitudes, patterns of behavior, habits, [and] symbols" all matter when coping with new environments, including those relating to security issues.[8] The 1950s have been frequently but inaccurately remembered as a vacuous and naively idyllic time, an age oblivious to the pressing issues that would blossom in the 1960s. Civil rights, women's rights, environmentalism, and antiproliferation each came to the fore of the national consciousness in subsequent decades, but their seeds were sown in the 1950s.

Anticommunism was another product of the times. World war, fifth columns, extermination camps, and atomic bombs taught Americans the cost of appeasement and the dangers of totalitarian systems. Wisdom appeared to stand with zealous application of the lesson. Joseph McCarthy's maniacal hunt for communists in the State Department and the Army, like Congressman Richard Nixon's eager search for them in Hollywood, thrived in this environment largely because the national culture was prepared for it to. The entertainment industry briefly cooperated.[9] Consensus about confronting communism, and the presumption of a monolithic antagonist, propelled such initiatives as the Truman Doctrine and the Marshall Plan, which sought to bolster beleaguered governments and to assist the economic recovery (and political rehabilitation) of non-communist societies in Europe. Fear that war—*real* war—in the future would mean nuclear attacks prompted involvement by the United Nations (UN) in a bloody "police action" to stymie communist aggression in Korea.

Depression and war had deferred much gratification and enjoyment, and America's profound wealth, both in actual terms and relative to the rest of the planet at the end of the Second World War, facilitated the consumerism that was another reaction to the era. The "Rosie the Riveter" image indicated the dramatic turn from the Depression years, in which there were too many men to do too few jobs, to the war years, when there was more paid work available than men to do it. With some commodities rationed and others unavailable, wartime earnings accrued unspent. Many, eager to banish the paucity of the previous 15 years, also bought today by using tomorrow's earnings. By 1955, nearly a third of automobile purchasers had not yet paid off the amount they had borrowed on their previous car, and installment credit expanded about tenfold in the second half of the decade.[1] Possession felt like ownership. The introduction of the annual model change tempted Americans to consume—to possess the *new* version of major consumer goods—and this was even heralded as a source of American strategic strength. High employment, good wages, and low inflation prompted the head of the newly unified American Federation of Laborers–Congress of Industrial Organizations to declare that "American labor has never had it so good."[2]

The American suburb proliferated, as 15 million new homes were built during the 1950s, and the urban or suburban proportion of the population, which had been just under six percent at the start of the decade, nearly doubled by 1960. Daniel Boorstin decried the "community of consumers,"[3] but, obviously, there was something in the suburbs—in their reality, their image, or both—that Americans did genuinely enjoy and desire. People clamored to purchase new homes and the freedom they represented. Betty Freidan's critiques notwithstanding, many Americans were enthusiastic about the nuclear family suburban model, precisely because it represented a step toward greater freedom, independence, and prosperity. Domestic policy accelerated consumerism when tax laws rewarded debt (through tax deductions on mortgage interest) and penalized thrift.[4]

An empowered adult generation and a burgeoning generation of children shifted consumer culture in favor of youth.[5] Toys such as Play-Doh, the hula hoop, Barbie dolls, and GI Joe figures sold well and were in tune with the era.[6] They facilitated creative play while reinforcing popular postwar notions about what femininity and masculinity should look like. The scale of the consumerist wave was unprecedented, and by 1956 the largely disposable income of the country's 13 million teens had risen 25 percent in just three years.[7]

An awareness of prosperity and a yearning for a dream that seemed tantalizingly close but still just out of reach defined a context not only in consumerism but in politics, from civil rights to national defense. Technologies that implicitly (or explicitly) endowed the owner with the freedom, independence, and prosperity of the American dream drew more attention and interest from potential customers. Companies responded to this tendency and reinforced it, advertising "the next big boom."[8]

Technology does not exist in a vacuum, and Americans were prepared to see the technologies of the Second World War adapted to civilian uses through technology transfer. Food was "dehydrated, vacuum packed, and bombarded with electrons" to be longer lasting, more convenient, and less expensive. Chocolate that didn't melt in war-torn deserts or jungles translated to candy bars that also stayed pristine on a postwar picnic. The television, high-fidelity record player, and tape recorder expanded from strategic applications to enhance postwar entertainment.[9]

Notes

1. Jezer, *The Dark Ages*, 125–27; Lizabeth Cohen, *A Consumer's Republic: The Politics of Mass Consumption in Postwar America* (New York: Vintage Books, 2003), 124.
2. Stephen E. Ambrose, *Eisenhower: Soldier and President* (New York: Simon and Schuster, 1990), 386; David Halberstam, *The Fifties* (New York: Villard Books, 1993), 117–18, 120; C. F. Kettering, "The Future of Science," *Science* 104, no. 2713 (December 27, 1946), 611.

3. Daniel Boorstin, *The Democratic Experience* (New York: Random House, 1973), 426.

4. Cohen, *A Consumer's Republic*, 7–8, 144.

5. W. T. Lhamon, *Deliberate Speed: The Origins of a Cultural Style in the American 1950s* (Washington, DC: Smithsonian Institution Press, 1990), 8.

6. Samuel, *Future: A Recent History*, 79.

7. Ibid.; Lhamon, *Deliberate Speed*, 8; "Historical U.S. Population Growth," *NPG Facts & Figures*, accessed September 7, 2013, http://www.npg.org/facts/us_ historical_pops.htm.

8. Halberstam, *The Fifties*, 135, 165; Rome, *The Bulldozer in the Countryside*, 65, 67.

9. Lhamon, *Deliberate Speed*, 12.

Contemporary America celebrated consumerism, speed, and convenience, but these trends coexisted with angst, due in part to what journalist Norman Cousins described in mid-August 1945 as the "new age" of atomic power.[10] The psychological need for a "silver lining" of nonweaponized applications of nuclear technology created a cognitive space for initiatives aiming to demonstrate peaceful uses of nuclear technology.[11] However, as historian Paul Boyer aptly noted, "The atomic threat was simply less immediate than one's job, one's family, the cost of living, even the reviving rhythms of domestic politics and political affairs." Cousins noticed "a standardization of catastrophe,"[12] and some people reacted to atomic fears by urging the establishment of interception capability, retaliatory power, or shelters.[13]

Anxieties could be expressed or repressed in other ways, such as in science fiction works that offered ways to temporarily elude the proximity of real troubles while commenting obliquely about contemporary events. Sometimes, as with the 1951 film *The Day the Earth Stood Still*, the commentary was quite explicit, as aliens and robots came to warn against nuclear testing. Hostile forces of nature, angered or empowered by nuclear tests, factored into 1950s movies like *Them!*, *The Beast from 20,000 Fathoms*, and *Attack of the Crab Monsters*.[14]

Space had already been popular terrain for science fiction. H. G. Wells' *The War of the Worlds* first appeared in the late nineteenth century, and Isaac Asimov's *Foundation* was serialized by the 1940s. The early 1950s saw a throng of space-themed movies and television programs, and the show *Space Patrol* appealed to broad audiences.[15] People were prepared to see characters dealing within the context of an exotic space setting, albeit addressing recognizable issues and challenges. Futuristic science advocacy appeared alongside science fiction. *Collier's* magazine proved a substantial platform for such writing. In March 1952, it publicized ex-Nazi scientist Wernher von Braun's notions about a space station of the future. Von Braun's work with *Collier's* was the byproduct of an unpublishable science fiction novel set in 1980, following a lethal attack in which a nuclear-armed space station named Lunetta destroyed the Soviet Union.[16] *Collier's* editor, Cornelius Ryan, delayed an issue focused on Mars until April 1954 so that articles about potential space stations and astronauts could build up reader interest. By 1955, Disney had begun airing a television series that described possible space projects ranging from enormous rockets to winged spacecraft to crewed satellites and lunar exploration.[17]

Actual space research had proceeded in obscurity in the United States and in secrecy in the Soviet Union. Under Harry Truman, ballistic missile development had essentially halted, but between 1948 and 1952, the United States had launched biological experiments aboard captured V-2 missiles and Aerobee sounding rockets, exposing mice and monkeys to an average of $2\frac{1}{2}$ minutes of weightlessness. Across the Iron Curtain, the Soviets began biological studies on mice, rats, and dogs in 1951. The Soviets never intended to recover live specimens, and the Americans, who did try to recover them, succeeded in only two of their eight attempts.[18] American tests lapsed for the next six years, until 1958. Serious thought about space possibilities, such as the concept of geosynchronous communications satellites espoused in an article in the October 1945 issue of *Wireless World*, was rare.[19]

This era, in short, was one of many competing threats, anxieties, and ambitions. As president, Dwight Eisenhower's task was to produce calm during an era of dynamic economic and scientific change, social pressure, geopolitical instability, and cold war. Fortunately, Eisenhower was uniquely equipped for this task. He saw calm confidence, both in the demeanor of the nation's chief executive and in the posture of its national government, as key. He had suffered a frustrating knee injury while on the West Point football team; it reinforced his proactive verve and determination and his belief in teamwork, and it also caused him to adopt his famous grin, which helped him overcome challenges and prevent opponents from knowing what was on his mind.[20] Composure characterized Eisenhower's approach to his own career in national security. "Anger cannot win," he noted early in the Second World War. "It cannot even think clearly."[21]

Eisenhower, while serving as Army chief of staff and as commander of NATO forces in Europe, believed that U.S. policymakers, including President Harry Truman, displayed little of the confidence he felt was crucial in the high-stakes Cold War environment. Despite both men rising through hard work from humble backgrounds, Eisenhower and Truman never developed a close friendship, and they would grow to become bitter opponents. Eisenhower regarded Truman as "a fine man who, in the middle of a stormy lake, knows nothing of swimming."[22] Both men distrusted the Soviet Union by the end of the 1940s, and both men were wary of rampant budgets. However, Truman's determination to shrink defense spending to barely $13 billion on the eve of the Korean War seemed totally at odds with the growing geopolitical menace of communism and the administration's interest in containing it. The outbreak of war in Korea in 1950, on the other hand, prompted an immediate four-fold increase in defense spending. Truman appeared to be zigzagging recklessly from one extreme to another, ultimately embracing out-of-control spending after inadvertently inviting communist aggression.

The outlook for future policy seemed grim. As the 1952 elections approached, the Democratic Party had held the presidency for twenty uninterrupted years while controlling the Senate and House of Representatives most of that time as well. An incoming generation of voters essentially had not seen a competitive two-party system in operation during their lifetime, and the most likely Republican nominee for 1952 would be the defeatable isolationist old guard leader Robert Taft.[23]

Convincing Eisenhower to run also meant convincing him that the country demanded his candidacy. In the first postwar years, Gallup and Roper polls showed Eisenhower, whose political affiliations were unknown, leading Truman among Democrats and leading Thomas Dewey and Taft among Republicans. By early 1952, Eisenhower saw "suddenly that he himself may be the symbol of that longing and hope" felt by the nation, and he joined the race. About 19 million televisions were in U.S. homes, with barely one hundred television stations broadcasting. Still, the Republic campaign in 1952 pioneered national television advertising in an election, ramping up particularly in the final weeks prior to election day.[24] Eisenhower had campaigned against communism, government corruption, and the agonizing stalemate in Korea, but his persona was a more important factor in his victory than the issues. He was simply popular with an extreme range and vast number of Americans, and traditional Democrats willingly crossed party lines to vote for Ike. It was a fact captured in animated television ads that showed a parade of Americans, led by Uncle Sam, representing what was meant to depict men and women of various backgrounds and professions.[25] Eisenhower coasted to an enviable victory with 55 percent of the popular vote and a staggering 442 votes in the electoral college. Journalist Arthur Krock explained during the inauguration what Americans saw in their new president: "His manner is genial, his ways and reflexes are kindly; his bearing is soldierly, yet his well-tailored civilian clothes never seem out of character. His smile is attractively pensive, his frequent grin is infectious. . . . He fairly radiates 'goodness,' simple faith and [his] honest background."[26]

Foreign policy and national security topped the new president's immediate agenda. Truman's adoption of containment-on-the-cheap followed by dramatic defense spending with the outbreak of the Korean War set little positive precedent and left many loose threads. Eisenhower noted that "nobody in war or anywhere else ever made a good decision if he was frightened to death. You have to look facts in the face, but you have to have the stamina to do it without just going hysterical."[27] However, Truman did bequeath to his successor an impressive array of nuclear munitions and development programs for aircraft that could drop them on the USSR and return home. He had substantially increased the country's nuclear arsenal, from three hundred nuclear weapons in mid-1950 to about one thousand when he left office. November 1952 also saw the test of an experimental thermonuclear device.[28]

What Truman had not done, however, was set a coherent policy for confronting communism in the long term. He had addressed a series of emergencies—in Greece, Turkey, Berlin, Korea—but continuity had been lacking. National Security Council Paper 68 (NSC-68) embodied the intellectual foundation of containment policy in calling for facing communist aggressions wherever they might occur.[29] Such a wildly ambitious and assertive policy would require substantial armed forces. But until the outbreak of the Korean War, Truman had consciously starved the military in hopes of bringing federal budgets under control. The national debt in 1953 was nearing the fixed debt limit ceiling (then set at $275 billion).[30]

Additionally, despite the buildup of offensive nuclear arms by both the United States and the Soviet Union, America's continental defenses were woefully lacking. Even the ability to identify and track an incoming attack was largely absent. The Pine Tree Line, a U.S.-Canadian idea for a network of warning radars in southern Canada, was mostly incomplete.

The short-lived Mid-Canada line and the much more famous Distant Early Warning line were barely concepts when Truman left office.

Tailored information, used at some times to spur action and at others to allay worry, obscured both the country's strengths and its vulnerabilities. In 1951, the U.S. government released *Duck and Cover*, the now-iconic informational film teaching children to crouch under desks and cover their heads, mimicking a cartoon turtle named Bert, who told children: "We think that most of the time we will be warned before the bomb explodes, so there will be time for us to get into our homes, schools, or some other safe place. Our civil defense workers and our men in uniform will do everything they can to warn us before enemy planes can bring a bomb near us."[31]

With the Pine Tree Line incomplete and unable to track low-flying planes, "everything they can do to warn us" was likely to be less than what viewers might imagine. But *Duck and Cover* at least dispensed with claims that the attacks would be defeated. *Atomic Alert*, from the same year, told young audiences that "we have a warning system" and "we have the national defenses to intercept an enemy."[32] Neither was especially true when the film was released. Contemporary films for military personnel acknowledged that "World War II bombs were cream puffs compared to what you might be up against, but the defensive measures are about the same."[33]

National Security in a Turbulent World

A new posture seemed to be possible in the spring of 1953. In March, Soviet dictator Joseph Stalin died. The next two years saw second-tier apparatchiks jockey to become his successor, and Soviet policy would not be cemented until the leadership issue was settled. Six weeks after Stalin's death, Eisenhower delivered a globally lauded speech, "The Chance for Peace," in which he declared that "a nation's hope of lasting peace cannot be firmly based upon any race in armaments but rather upon just relations and honest understanding with all other nations." He noted that "the Soviet Union itself has shared and suffered the very fears it has fostered in the rest of the world" and that Western powers were ready to receive and reciprocate evidence of peaceful intentions. Segments of the speech sounded almost pacifistic; the president described how defense spending absorbed "the sweat of [the world's] laborers, the genius of its scientists, the hopes of its children" and that "every gun that is made, every warship launched, every rocket fired signifies, in the final sense, a theft from those who hunger and are not fed, those who are cold and are not clothed."[34] The speech, which won considerable praise in the Western press, served a dual purpose as propaganda and as earnest geopolitics. This was both cause and effect of the administration's understanding of Soviet peace rhetoric being meant to undercut anticommunist resolve in the West.[35]

Simultaneously, Eisenhower initiated Project Solarium, in which three groups of tasked advisers would analyze different potential Cold War postures. Three task forces examined alternatives: a continuation of the existing containment policy, selective protection of strategically important areas, and the "rollback" of communism, as some Republican campaign rhetoric had vowed. Whereas Task Force A advocated continued containment and saw the U.S. performance thus far as successful, Task Force B insisted that clear red lines be drawn to warn the Soviets away from potential aggression, and Task Force C sought to promote "maximum disruption and popular resistance throughout the Soviet Bloc" in the belief that destabilizing communism would help accelerate a victorious ending to the Cold War. Pleased with the task forces' work, Eisenhower intended the New Look to combine aspects of these approaches.[36] Nuclear-armed containment would continue, but rollback was dangerous policy, as was the drawing of extensive and explicit red lines that would extinguish the flexibility and credibility of effective nuclear deterrence. The U.S. Information Agency did wage energetic propaganda sorties to persuade the uncommitted corners of the world, which were often freshly free from the shackles of European colonialism, to embrace noncommunist philosophies and power blocs.[37]

Eisenhower remained a committed Cold Warrior, and his "Atoms for Peace" speech to the UN General Assembly in December 1953 reflected his concern about surprise attack. "The awful arithmetic of the atomic bomb" meant that building warheads and defensive systems would not "guarantee absolute safety" for any nation. Meanwhile, "an atomic attack" against the United States would bring "swift and resolute" retaliation.[38] Administration officials began speaking publicly about a "new concept" or a "new look" in the defense structure simultaneous with the popular reaction to the Atoms for Peace speech.[39] The New Look manifested the Eisenhower approach to national security. The "great equation" was to achieve security by combining spiritual, economic, and military force in a balanced way.[40] "Maximum effectiveness at minimum cost" was deemed "essential," because security required deterring Soviet aggression but without resorting to fiscal or social policies that would damage the country's national fiber.[41] Nuclear weapons were central in achieving "more bang for the buck."

Eisenhower understood that the Cold War had to be dealt with over the long haul with the support of a willing populace.[42] It was a new and potentially permanent reality, and a series of expedients to address emergencies would bring instability and ruin. Crash development programs for new technology wasted money and signaled panic. Abrupt shifts in policy reflected panic, too, and would tax the nation's emotions as well as its pockets. Neither Truman nor Kennedy fully appreciated the implications of this point. Psychological burdens existed as well; while civil defense films declared that "in this early and troubled stage of the atomic age, our very lives may depend on *always* being alert,"[43] constant wariness brings widespread anxiety and eventually apathy. Painfully silent about Senator McCarthy's political witch hunts, Eisenhower did know that "we can't defeat Communism by destroying the things in which we believe."[44] Fundamentally, the New Look sought sustainability. Truman had envisioned Soviet power growing toward a threshold point that would enable an unfettered attack against the United States, whereas Eisenhower interpreted the Soviet threat as "a long-term, tenacious, and unpredictable" danger.[45] Power projection (through Air Force and Navy planes capable of carrying nuclear weapons), coupled with the maintenance of allies and the use of propaganda and covert actions, defined Eisenhower's New Look.[46] Sustainable strength had to be the name of the game during the Cold War.

The Air Force was the principal beneficiary of the New Look; it controlled nearly half of the defense budget in the middle of the decade.[47] An independent military branch only since 1947, the service had a culture that exuded dynamism and confidence. Air Force commanders and their advocates in Congress expected the trajectory of their branch's relative position in national defense to continue to rise. In February 1954, a report in the trade journal *Aviation Week* gave the Air Force's friends in Congress (particularly the vocal Democratic Senator Stuart Symington of Missouri) a chance to deride Eisenhower as weak on defense and as starving the Air Force.[48] Symington, Truman's Air Force Secretary and a defense hawk into the 1960s, had earlier accused Eisenhower of dangerously "stretch[ing] out" the Air Force's schedules for successive reinforcements and modernizations of its strike force. As the highest civilian in the Air Force, he had happily, and frequently, said that his job was "not to develop strategy but to sell it." He redoubled his partisan attacks during the summer of 1954, and in doing so he won support from the influential journalistic Alsop brothers, Joseph and Stewart.[49]

In fact, the bomber gap was consistently one in favor of the United States. In the first nuclear years, the U.S. strategic force included B-29s and variants called B-50s, which were gradually phased out in favor of enormous ten-engine B-36 bombers, sleek B-47 jets, and ultimately the huge eight-engine B-52s. Soviet bomber aircraft, in contrast, were initially limited to the Tupolev 4, a reverse-engineered pirate copy of the B-29. The new aircraft of 1954, the Myasishchev M-4 "Bison," lacked both the range and the numbers that the Air Force and its sympathizers believed and feared. But lacking specific and reliable statistics about the new Soviet bomber, many commanders and analysts, and some citizens as well, believed that wisdom and caution favored ever more sobering estimates of Soviet capability. The logical response to counter this envisioned threat presumably was ever greater buildup. Although the bomber gap was never real, it carried cachet as long as actual Soviet strength was unknown.

Dissenters to the New Look within the military typically belonged to two different camps. Airpower advocates assumed that national security depended upon building large numbers of technologically marvelous weapons systems (capable of evading interception by flying higher, faster, and farther); ground power advocates worried that the U.S. defense structure was so atom-centric that no serious alternatives existed during crisis except appeasement and brinkmanship. These groups were not yet an existential political threat to Eisenhower during the mid-1950s because they disdained each other's ideas even more than they disagreed with the president.

In many ways, 1956 marked an apex in Eisenhower's presidency, but the period was nonetheless marked by several crises. The president's health was one concern; Eisenhower suffered a heart attack on September 24, 1955. Despite a smooth recovery, he did not deliver the 1956 State of the Union message in person. During his convalescence, he and press secretary James Hagerty pondered potential presidential successors in the event of his stepping aside after one term. Eisenhower knew that the political environment, especially in the wake of twenty years of unbroken presidential control by one party, could not afford a five-star-general-turned-president to appear to reach for power before being offered it. Lacking credible alternatives, Eisenhower decided that the country needed him to stay on.[50] He met with Vice President Richard Nixon the day after Christmas in an abortive attempt to remove him from the 1956 Republican ticket and galvanize a more robust, competitive, and moderate Republican Party. Nixon recognized the maneuver and refused to budge, and the president let him stay.[51]

New Soviet leader Nikita Khrushchev's ascension to power sent mixed signals about the direction his country might take. Khrushchev's "secret" address to the Soviet twentieth party congress in February included scathing criticisms of the previously sacrosanct Joseph Stalin, but de-Stalinization aimed to revitalize communism, not to reject it. As with Eisenhower's earlier Atoms for Peace speech, the condemnation of Stalin's "excesses" did not preclude sharp Cold War rhetoric. The old, and now marginalized, war hero Marshal Georgi Zhukov also spoke to the congress, blaming "the USA's 'position of strength' policy" for the arms race and international tension.[52]

The year also witnessed its share of quiet success. Eisenhower's ongoing concern about surprise attack led to reconnaissance projects that began to bear fruit. The U-2 spy plane provided inestimably valuable information, but its cameras could capture only thin strips of Soviet territory during any mission. Each mission was vetted and approved by the president, and only twenty-five missions were ever flown over Soviet territory.[53] Armed with photographs from U-2 missions, Eisenhower could quietly but confidently discount many of the alarmist claims of politicians like Senator Symington and of top Air Force generals clamoring for more resources. Since the U-2 violated Soviet airspace, however, Eisenhower declined to publicly counter the lingering bomber gap claims, which continued through 1956.[54] Disproving the bomber gap myth meant publicizing the U-2 missions, humiliating and potentially provoking the Soviet Union.

In addition to flights over the Soviet Union, two dozen other U-2 missions photographed Southeast Asia, communist China, and the Eastern Bloc satellite states.[55] In October 1956, U-2 photographs revealed military buildup in the eastern Mediterranean. Examined in retrospect, these images uncovered the international machinations of the United Kingdom, France, and Israel to retake the Suez Canal, which the autocratic Egyptian ruler Gamal Nasser had nationalized while attempting to play the superpowers against one another for arms deals and construction loans. This came in the wake of a rift between Eisenhower and British Prime Minister Anthony Eden, who, under political pressure, broke with the president in opposing a weapons test ban concept pushed by Democratic presidential candidate Adlai Stevenson.[56]

Just days before the U.S. election, the Anglo-French-Israeli Suez trap sprung. The action won antipathy in the nonaligned Third World, and it restricted U.S. options as Hungarian citizens rebelled against the communist puppet regime in that country. Eisenhower had no intention of aiding

Hungarians at the risk of global war, and U.S. assistance was limited to granting asylum to 30,000 dissidents, as Soviet tanks rolled into Budapest.[57]

As election day approached, columnist Arthur Krock blasted Eisenhower for the crisis in Suez and for the unfolding tragedy in Hungary as well. Critics, attributing the turmoil to Secretary of State John Foster Dulles, leveled an additional accusation at Eisenhower. The president, claimed Krock, "is not in charge of the store."[58]

Eisenhower spoke in Philadelphia on November 1, asserting that "the power of modern weapons makes war not only perilous—but preposterous." The timing of the speech, during the dual crises of Suez and Hungary and in the run-up to the U.S. election, indicated Eisenhower's interest in restraining the Anglo-French-Israeli assault against Egypt, and it implied that the United States would stand aside as the Hungarian uprising was crushed. Finally, it suggested that Eisenhower, who had rattled the nuclear sword to force a negotiated end to fighting in Korea in 1953 and who smothered a flare-up between communist China and nationalist Taiwan the following year, was loathe to again reach for nuclear brinkmanship in any but an existential crisis. By restating, in the midst of multiple international crises, that nuclear war was in fact unthinkable, the president dealt a blow to Stevenson's campaign.

The 1956 election demonstrated that the U.S. public continued to view Eisenhower in strongly positive terms, as a leader fit for the age, and as a man widely admired. He swept 41 of the nation's 48 states, with more than 57 percent of the popular vote. The United States was a place that still liked Ike because "he stood at the heart of American habits and values."[59] Liking Ike did not mean that the public gave Congress to the Republicans, however. Democrats retained narrow advantages in both the House and Senate.[60]

The political style that Eisenhower favored coincided with, and accentuated, a particular aspect of his personality. When he ran for president in 1952, Gallup polling found that he was the "most admired American." However, it was his "human qualities" such as "warmth and sincerity" rather than an appreciation of his expertise, accomplishments, and leadership skills that led to his spectacular popularity with Americans on the eve of his election. Four years on, Eisenhower remained extremely popular, but it was still the favorable perception of his approachable *personality*, rather than positive estimates of his leadership skill, that led Americans to admire and re-elect him. Americans generally felt the positive things in life had more or less occurred organically and without presidential input. Eisenhower insisted that the prosperity and security "didn't just happen, by God," but his approach to the presidency facilitated that mistaken impression.[61] Thus, confidence might falter quickly if catastrophe occurred and the public still considered Eisenhower to be a golfer-president.

Regarding national security policy, Eisenhower's style had been quietly successful in the period through 1956. The need for reconnaissance data, the obviously finite lifespan of the top secret U-2, the Soviet rejection of Open Skies, and the prospect of miniaturized thermonuclear weapons atop ballistic missiles made the establishment of a coherent space policy vital. The freedom of space concept, coupled with a scientific satellite project for the upcoming International Geophysical Year (IGY) of July 1957 through December 1958, answered Eisenhower's need. The low-key nature of the IGY satellite fit in this context. The secret interest in future reconnaissance satellite development did as well. But this policy and its components were vulnerable to buffeting by popular discomfiture and the ardent efforts of military officers, ambitious members of Congress, and other opinion leaders.

3

"What an Impressive Idea!"
Conceptualizing the Aerospace Bomber

A major industrial power can achieve almost any technological feat it can conceive of, provided only that it is willing to concentrate its energies and resources on that goal.
—JAMES KILLIAN, TECHNOLOGICAL CAPABILITIES PANEL CHAIRMAN

The hypersonic boost glide concept offers a major technological breakthrough in performance capabilities which should be exploited by future reconnaissance and bombardment aircraft weapons systems.
—AIR FORCE–NATIONAL ADVISORY COMMITTEE
FOR AERONAUTICS PAPER

The "New Look" Points Skyward

President Eisenhower saw space issues through a national security lens. Therefore, his administration gave early and deliberate thought to space policy. Because of the security issues involved, the administration declined to advertise its work or objectives. Since the public was aware of only the publicized, high-profile projects, it misinterpreted them as embodying space policy itself. Thus, the Eisenhower administration appeared to be a step behind in the space competition after the Soviet launch of satellites in 1957. This impression, although erroneous, threatened to upset U.S. space policy and redraw the boundaries of the arms race.

Before Sputnik launched wider public awareness of the issue, White House and Air Force space planning had proceeded on a quiet collision course. Both administration interest in a civilian-controlled reconnaissance satellite and Air Force research about a spaceflying hypersonic glide bomber began in earnest in 1954. Facing Cold War challenges, Air Force planners drew on their wartime experience, doctrinal legacies, and predictions about the pace of technological advance. Even before the end of the Second World War, the Air Force's highest ranking commander, Gen. "Hap" Arnold, directed Hungarian scientist Theodore von Kármán to consider trends of future technology.[1] In 1946, the newly minted Research and Development (RAND) Corporation, a pioneer think tank organized by the Air Force and the Douglas Aircraft Company, analyzed the feasibility of a reconnaissance satellite. Tight postwar budgets froze work for ten years after the Second World War, since many influential scientific figures judged missile and satellite projects to be hopelessly futuristic and therefore wastes of sparse cash.

To be accepted, doctrinal ideas about how to fight have to complement the policy goals and the culture of a given country. Advocates promised that precision bombing of the enemy's vital yet vulnerable assets precluded the ugly attrition of the First World War. Before, during, and in the wake of the Second World War, the mantra of higher, faster, farther flight continued to complement the perspective of a people loath to exist constantly in the shadow of war, although the Cold War made such existence unavoidable.

Like the United States, the United Kingdom and the Soviet Union sought to acquire Nazi scientists and hardware as Germany collapsed at the end of the war. However, German scientists tended to prefer the United States, which lacked the acutely earned vengefulness of the Soviet Union and possessed the economic heft (so vital for expensive research projects to advance) absent in the United Kingdom. The Soviet Union, which already had some domestic rocket researchers, made do with German scientists who had not been involved with missile development until their capture by the USSR. Granted higher pay and limited travel privileges and accompanied by their families, they were nonetheless kept within a stockade at Kapustin Yar, an old village on a flood plain of the Volga River in southern Russia.[2] Conditions were spartan, even by postwar Soviet standards, but work continued toward bold goals. Radar and rocket researchers reported to Georgi Malenkov, and Lavrenti Beria, chief of the fearsome secret police, controlled nuclear weapons development; each man reported directly to Stalin. Work proceeded in the immediate postwar years to produce missiles with twice the range and greater accuracy than the Nazi V-2 missile, which, with its short range and limited thrust, was unable to boost anything into orbit.[3]

Secretly, Soviet scientists envied the wildest of the Nazi technological dreams. One Soviet aviation design chief in 1945 marveled at Eugen Sänger's idea for an extremely long-range, high-altitude, hypersonic bomber: "These aircraft skip along the atmosphere; they dive down only after they have flown across the ocean in order to slice their way into New York! What an impressive idea!" However, Soviet experts also noted the innate difficulty in boosting a plane into space and building a craft that could bounce along the top of the atmosphere. "It's easier to make rockets than that airplane," a professor at Moscow's Aviation Institute aptly observed. Compartmentalization, secrecy, and professional suspicions complicated the work of Soviet and German scientists.[4] Rumors of Soviet weapons ambitions reached Western ears through Grigori Tokaev, who defected to the West and described Stalin's interest in the intercontinental Sänger aerospace bomber concept. Stalin was alleged to declare, "We need aircraft of the Sänger type. . . . If we have such an aircraft, it will be easier to talk to Truman. We may be able to quiet him down."[5] Such reports underscored the need for further information about Soviet weapons development, while rumors simultaneously strengthened Air Force technologists' conviction that deterrence and peace depended upon increasingly advanced U.S. weapons systems.

The U.S.-held ex-Nazi rocketeers continued to promote their technological visions to their new masters. Walter Dornberger, a major general when commanding the Nazi Peenemünde research facilities, tried to sell the idea of a vehicle capable of lifting into the high atmosphere and gliding to a target and back to a friendly base. The *Silverbird* (also called *Amerikabomber*) seemed adaptable to U.S. appetites for weapons that could destroy distant enemies and keep danger far away. Bell Aircraft's X-1 rocket plane broke the sound barrier in 1947, and the company began listening to Dornberger's boost-glide suggestions in 1952. In 1954, the Air Force contracted with Bell for a limited study of the "boost-glide" system, called the Hypersonic Weapons Research and Development Supporting System (HYWARDS). On May 12, 1955, the Air Force added a general operational requirement (GOR) for a "piloted very high altitude reconnaissance weapon system." Bell began work on this too, dubbing the project "Brass Bell." Finally, on June 12, 1956, the Air Force ordered another project, called BoMi (Bomber Missile) and later RoBo (Rocket Bomber).[6] HYWARDS, Brass Bell, and BoMi/RoBo each relied on the boost-glide concept for reaching the edge of space.

The Air Force's interest in boost-glide technology in the mid-1950s was kept low key. Space developer Walter C. Williams was then a young Air Force lieutenant and a graduate student in aeronautical engineering at Wright-Patterson Air Force Base in Dayton, Ohio. Interested in Bell's work, he attended a presentation at the Hayden Planetarium in New York City on May 4, 1954. "A bunch of old guys in civilian clothes" were already seated waiting for the talk to begin. Most of the other attendees turned out to be high-ranking officers from the Pentagon, but they wore civilian clothes to the lecture to avoid drawing attention and preventing "the god dam [sic] press [from] making a big thing out of this."[7] The military was more interested in long-range missiles than in space flight. Company head Larry Bell and Walter Dornberger explained their concept. They had abandoned Sänger's original "skip-glide" flight pattern because of the destructively intense heat it would bring to the vehicle; a smooth glide would provide nearly unlimited range to the craft. Still, they made little impression on the audience when they described their space weapons system that day in 1954. Angered by the jeering reaction, an irate Dornberger shouted, "I wish we had shot down more of your bombers in W[orld] W[ar] II!" and then sat down.[8]

Others, such as former Luftwaffe medical officer Hubertus Strughold, recognized the persuasive value of emphasizing continuity rather than novelty, writing that "today's manned rocket-powered craft have already advanced well into the area of partial space equivalence" and predicting further advances soon to come.[9] Aerospace weapons advocates continually struggled with the issue of whether continuity or revolutionary change best described—and most effectively sold—their technological vision.

Coincident with these preparations, policymakers sought to crack the problem of potential Soviet surprise attack. The collection of scientists and retired military figures who had constituted the Project Solarium panels had facilitated the New Look and impressed Eisenhower, who formed another ad hoc civilian committee, the Technological Capabilities Panel (referred to as the Killian Committee after its chairman, James Killian) in 1954. He encouraged dialogue, a team spirit, and regular meetings so that advisers could hash out competing ideas in his presence.[10]

Popular contemporary assumptions to the contrary, it was Eisenhower who called the shots in his administration. To preserve open discussion and debate, Eisenhower asked tough questions and reserved his own comment until others had advanced their views to the group. Having reached a decision, he expected debate to stop and his advisers to defend his policy.[11] A formalized system of regularly held meetings marked a contrast between Eisenhower's White House and those of Roosevelt and Truman before him and of Kennedy later. He wasted no time before establishing regular Thursday morning meetings for the National Security Council. A new Operations Coordinating Board assisted the president.[12] Ad hoc advisory panels complemented this system, addressing specifically defined study topics. In sum, the system permitted Eisenhower to develop what Raymond Millen of the U.S. Army War College identifies as the only comprehensive basic national security policy the United States has had in the modern era.[13]

The new committee, headed by James Killian of the Massachusetts Institute of Technology (MIT), built organizationally on the Project Solarium experience. Most members were scientists, and the committee was divided into three panels, to explore ballistic missile technology, antiballistic technology, and reconnaissance needs. The Killian Committee began presenting Eisenhower with updates as early as November 1954, and it delivered its final report on March 17, 1955.[14] The committee outlined four main phases of anticipated changes in power dynamics between the United States and the Soviet Union. These were presented as essentially inevitable and defined by technological developments, although policymaking might adjust the pace of development and the level of international tension.

The situation, the committee recognized, was dynamic and ambiguous. U.S. military power in the mid-1950s surpassed that of the USSR, but it was inadequate to deliver a crushing blow in a single strike. Both superpowers could be expected to further build their offensive and defensive power, which initially would result in a widening of the relative American advantage. "Our military superiority may never be so great again" as during a window expected to open around 1956 and close by 1960. American advantage would shrink through the early 1960s.[15] Beyond this lay the scenario later described as mutually assured destruction. Because of the mass deployment of multimegaton warheads and of defensive systems to defeat attack or mitigate damage, "both the U.S. and Russia will be in a position from which neither country can derive a winning advantage." This future would be "a period of instability . . . easily . . . upset by either side" and ending in potential "world catastrophe." Furthermore, Killian's group foresaw no end to this stalemate once it arrived.[16]

Technology did not offer a silver bullet escape, but the committee recommended development of specific new technologies, including intercontinental ballistic missiles (ICBMs) and intermediate range ballistic missiles (IRBMs).[17] Eisenhower acceded, accelerating the country's missile programs and building a continental defense system, both of which had languished under Truman and had been in deep freeze at the start of Eisenhower's own tenure.[18] Problematically, much more needed to be known about the USSR's deployed weapons, research programs, and potential targets. Whereas U.S. planners struggled to gain information about the closed Soviet system, their Soviet counterparts could develop target folders on U.S. sites by relying on publicly available information, buttressing intelligence gained from spies infiltrated into the West. Despite violating Soviet airspace and international law, reconnaissance overflight was "a rational response to Soviet secrecy."[19]

An advisory team headed by Ed Land favored, in addition to ballistic missiles, the prompt development of the U-2 reconnaissance plane and of reconnaissance satellites, as well as work toward the establishment of a diplomatic (rather than solely technological) safeguard that would protect them from the kinds of interception attempts that would inevitably challenge the U-2.[20] Following the outbreak of the Korean War in 1950, a secret U.S.-U.K. agreement committed the two nations to covert air reconnaissance of the Soviet Union, and combat aircraft modified to the reconnaissance role were used. The near-shootdown of a U.S. plane in April 1954 foreclosed further British participation.[21] Richard Leghorn, an MIT-trained physicist with experience in aerial photography, had posited as early as 1946 that "peacetime" reconnaissance flights would be "essential" in obtaining "prior knowledge of the possibility of an [enemy] attack."[22]

Land recommended "cute tricks . . . so close to the frontier of scientific knowledge that they remain unsuspected for months or even years," and in November 1954, Land and Killian presented the idea of a specialized, extremely long-range and high-altitude reconnaissance plane. Following Eisenhower's hard questions about the concept, it won approval, with the stipulation that the Central Intelligence Agency (CIA) and not the Air Force run the program.[23] Eisenhower's objective was strategic intelligence—information about what the Soviets were building and their overall posture. Under Air Force control, the program would tend to focus on a more tactical level of target selection, on the assumption that a general war was inevitable sooner or later. The plane would become known as the U-2. The finite lifespan of the "cute trick" that the U-2 represented forced policymakers and their advisers to consider further advances in reconnaissance, and Land's group therefore pressed also for the "immediate" start of a satellite program. They hoped that demonstrating the peaceful nature of a satellite before launching more sophisticated reconnaissance systems might minimize Soviet complaint or hostility.[24] Eisenhower's agreement marked the nation's first solid step toward establishing a space policy.

The administration and the Air Force disagreed about the proper objective of reconnaissance platforms, including satellites. Rather than searching for targets to win a war, Eisenhower wanted to hunt for answers that would avert one. Reconnaissance satellites depended on either diplomatic or technological means to preserve their safety. Air Force planners had little faith in any Soviet pledge (or a more ambiguous implicit acquiescence) to respect satellites. Institutional responsibilities drove this perceptual gap between the Air Force and the president. Eisenhower was responsible both for preventing war through wise policymaking and for presiding over the nation during crisis if prevention was not achieved, whereas the Air Force needed to be prepared for war and wage it if the time came. The president needed to know different things from reconnaissance than the Air Force would seek to discover, so he insisted that the U-2 be run by the civilian CIA.[25]

Avenues to Pursue: Open Skies and the International Geophysical Year

In 1950, scientists scheduled an International Geophysical Year to span from July 1, 1957, through December 31, 1958. The development and implications of nuclear weapons, and the trial of leading Manhattan Project researcher J. Robert Oppenheimer for his suspicious lack of enthusiasm for thermonuclear weapons, caused considerable rifts in the scientific community. These were more serious than Eisenhower initially understood, although his use of the Killian Committee helped recover trust between him and many scientists who were loyal Americans but who were discomforted by the arms race.[26] The international scientific community hoped for the IGY to help ease Cold War tension. Two International Polar Years had been held earlier, in 1882–83 and 1932–33. Planners now decided to radically expand the scope of the event by making a geophysical year. Highlighting scientific study disconnected from weapons development promised to point public interest in peaceful directions.

The IGY encompassed almost a dozen fields of earth science, but the prospect of an artificial satellite stood out as the potential show-stopper. By broadcasting a 108-megahertz signal, a satellite's passage could be tracked by science enthusiasts.[27] Soviet scientists intimated in April 1955 that the USSR might attempt an IGY satellite launch. The Air Force had meanwhile seen potential in building a revolutionary and extremely high-altitude vehicle. Although the Air Force rejected Lockheed's extreme altitude concept as inordinately vulnerable for meeting service needs, a boost-glide vehicle flying at hypersonic speed matched the Air Force's vision of its long-term reconnaissance requirements.

In July 1955, American, British, and Soviet leaders met at Geneva, Switzerland, the first meeting of the national heads of state since the Potsdam Conference at the end of the Second World War. Eisenhower proposed "Open Skies," a program wherein the two superpowers would agree to exchange data and establish a system of aerial inspection of one another's military facilities. By seeing that the other side was not preparing to launch a surprise attack, tension would ease. Initial Soviet reaction was divided, but the official whose opinion really counted was Nikita Khrushchev, and he rejected the idea outright. Khrushchev's refusal came just scant months after an essay in the Soviet journal *Military Thought* declared that the USSR needed a capacity to launch a pre-emptive nuclear blow.[28] If Open Skies had won Soviet accord, the U-2 program could have existed publicly and without the prospect of interception. Soviet refusal guaranteed that the U-2 mission would be flown covertly. Furthermore, Khrushchev's reason for refusal—a fear that Open Skies was a U.S. military ruse for accumulating targeting data needed for a surprise nuclear strike—ensured that when U-2 flights would begin, they would be interpreted as threatening.

On July 28, just days after Khrushchev rebuffed Open Skies, press secretary James Hagerty announced that the United States intended to lift a scientific satellite into Earth orbit during the upcoming International Geophysical Year.[29] The Soviet Union quickly formalized a parallel objective, which they had publicly flirted with two months earlier. The U.S. government explicitly declared interest in demonstrating "that space, outside our atmosphere, is open to all." The Eisenhower administration valued the freedom of space concept (crucial for pursuing national security objectives) far more than scientific discoveries deriving from the placement of a four-pound scientific satellite into orbit.[30] Some, including the president himself, noted that the first successful satellite launch would garner psychological advantage, but this was largely incidental in Eisenhower's eyes. By August, the Navy's Vanguard project was selected as the tool for securing the principle of "freedom of space." The rebuff of Open Skies in the atmospheric realm underscored the priority of establishing open access for unweaponized vehicles to the space above.

National Security Council document 5520, "U.S. Scientific Satellite Program," laid the foundation for space policy. It noted the "considerable prestige and psychological benefits" of launching the first satellite, the inferential relationship between satellite launch capacity and ballistic missile strength, and the fact that "a small scientific satellite will provide a test of the principle of 'Freedom of Space.'" The NSC predicted a total cost of $20 million for the Vanguard scientific satellite, its tracking instrumentation, and logistics costs.[31] Undertaken by the Naval Research Laboratory, the project would not hamper the ballistic missile work of the Army and the Air Force.[32] This arrangement also promised to avoid the embarrassment of having America's first satellite enter space on a rocket built by the Army's rocket engineering team, which was rife with ex-Nazi talent from Peenemünde.

The NSC recognized early on that weaponization of satellites made little technological sense. "Although a large satellite might conceivably serve to launch a guided missile at a ground target, it will always be a poor choice for the purpose" because "anything dropped from a satellite would simply continue alongside in the orbit"; deorbiting and aiming a weapon from a satellite at a specific target exceeded the technological challenges involved with a still-elusive earth-based intercontinental missile. "It should be emphasized that a satellite would constitute no active military offensive threat to any country over which it might pass."[33] Beyond technological hurdles, armed satellites would foreclose the freedom of space. With every foreign satellite potentially armed, none could be trusted, and technologists would turn to developing satellite interception methods.

The influential American Rocket Society (ARS) had already been lobbying for a scientific satellite. The society formed a sturdy rallying point for advocates of U.S. space science. It grew steadily in the 1950s, expanding more than three-fold between 1954 and 1959 to boast 13,000 members and constituting an active space science lobbying bloc.[34] Its opposition to weaponization coincided with administration decisions about the future of space policy.

Vanguard's cost steadily outpaced the initial estimate. In April 1956, special assistant for scientific liaison David Beckler told NSC member William Elliot that "it is not clear that this program would have been approved" had the contemporary cost estimates ($60 million for six satellite attempts or $90 million for twelve attempts) been known in 1955. Beckler nonetheless argued against cancellation, since "to slow down or cancel the program would involve a loss of prestige, would let down the scientific community and would invite a Soviet propaganda coup if she were able to come through with a satellite during the IGY time period." Beckler reiterated that "more than a scientific program is at stake," noting that a satellite launch was a "symbol of technological strength." Beckler repeated this before month's end, noting that "there is currently no emphasis in the U.S. program on the timing of the *first* satellite shot," but rather the focus is "to assure *a* successful satellite launching."[35] Enshrining the freedom of space meant more over the long run than being first in space, although the two accomplishments could coincide.

Vanguard's costs did not prompt endorsement of an Army alternative using the Redstone rocket built with the help of ex-Nazi scientists. The Army Ballistic Missile Agency boss, Maj. Gen. John Medaris, foresaw a possible satellite launch using a Redstone rocket in January 1957, though the possibility of success was deemed modest. If the sole purpose of lifting a satellite were to demonstrate missile technology, the chance would have seemed worthwhile. But DOD learned that the use of a military missile as a booster "may have the effect of disrupting our relations with the non-military scientific community and international elements of the IGY group."[36]

Advancing Technology and a Brewing Techno-Policy Clash

While the administration and its advisers deliberated about an unmanned scientific satellite launch in the upcoming IGY, the Air Force and the National Advisory Committee for Aeronautics (NACA) continued on a path toward new possibilities in flight. NACA engineers strove to transform the Air Force's technological dreams into reality. A higher, faster flying craft appealed to the Air Force, and at the end of 1956 the two organizations believed that "the hypersonic boost glide concept offers a major technological breakthrough in performance capabilities which should be exploited by future reconnaissance and bombardment aircraft weapons systems." While contemporary Air Force planes and the planned X-15 rocketplane would help pave the way in preparatory research, "a research aircraft system is required with performance characteristics and capability significantly beyond those of the X-15," because "the flight phenomena confronting hypersonic intercontinental weapon systems would not be encountered" by speeds approaching the Mach 7 anticipated limit of the X-15.[37]

A December 1956 report on hypersonic research held that "the hypersonic glide rocket concept offers a breakthrough in weapon system capability in terms of obtaining speed, range, and altitude simultaneously." For Air Force advocates of progressively capable systems that would fly faster, higher, and farther, such research promised a revolution in technology while retaining fidelity to what seemed a battle-tested tenet that technical superiority helped win (and ideally to deter) major wars. A secret section confided that "this research system should be able to generally support the Air Force research and development effort in the development of superior weapon systems for many years to come without having to plan and develop additional research systems with the consequent delay in progress." Authors pointed to Brass Bell and to BoMi/RoBo.[38] Hypersonic research, its advocates asserted, would pioneer a new age of flight and a new realm where the military could help preserve national security. And the two roles most identified, fittingly, were bombardment and reconnaissance.

Admittedly, opening the door to hypersonic flight meant encountering and mastering technical hurdles. NACA observed that "Brass Bell developed to the point when somebody sold the idea that there should be a reconnaissance vehicle to [observe] impact damage done by the ICBM." Air Force eyes looked to the future; an ICBM had not yet been built in February 1957. NACA officials noted that "although there are no basic obstacles to the realization of such a research airplane, there are many detail design problems," and the need for a heat-resistant, high-performance structure made the revised objective of a Mach 20 reconnaissance plane flying by 1959 "not realistic." Furthermore, the Air Force had run out of money for the BoMi/RoBo bomber, while Brass Bell and the HYWARDS research project seemed redundant. NACA sensed that the Air Force was "looking ahead to this as [a] bombing system for 1965–1970, but with no real expectation that it will meet these dates."[39] If the Air Force aimed at retaining a deterrent edge, then the pace of its technology needed to continually step ahead of rival Soviet offensive systems and defensive developments. Timelines remained fairly theoretical prior to Sputnik. Vigilance, for Air Force planners, equated to early study of fundamentally new technologies.

Determined to maintain Air Force-NACA cooperation, NACA director Hugh Dryden told his employees that "no vehicle as important as the boost-glide research vehicle . . . could proceed without the active participation and cooperation of the NACA." While NACA sought to assist Air Force aspirations in the hypersonic realm, Beckler urged Office of Defense Mobilization chief Arthur Flemming to increase the number of Vanguard satellite attempts.[40] Physicist Nate Gerson of the American Geophysical Union objected to the assumption that simply because upper air rocket research "is now imbedded in DOD, rocket research should remain there." He reminded the Technical Panel chairman for the U.S. National Commission for the IGY that "the primary function of DOD is the *immediate* preparedness of the nation for defensive, offensive or strategic type warfare." He suggested the bad precedent set by the National Science Foundation financing scientific study under the arm of DOD.[41] Months later, in discussion with the NSC, Eisenhower echoed Gerson's ideas.

NACA officials prepared to support the three projects that would soon constitute Dyna-Soar. Confusion about the separate but interrelated research projects posed enough complication that NACA designated a uniform, if uninspired, nomenclature for the project, "Round Three."[42] Rounds One and Two were research programs about high-altitude, high-speed flight, which Round Three research would advance. Within NACA, the engineers at Ames Laboratory and Langley Laboratory disagreed about the actual practicality of the Round Three concept; Alfred Eggers of the Ames group questioned estimates by Beckler's Langley team that the increasing vehicle temperatures could be kept at controllable levels during atmospheric hypersonic flight.[43] On April 30, 1957, the Air Force directed its Air Research and Development Command (ARDC) to consolidate boost-glide projects, pooling their limited funds. Williams, who had taken interest in hypersonic flight since the Bell Aircraft presentation in 1954, thought he had the perfect name for the project, whose dynamic shape would allow it to soar at the top of the atmosphere. It was something big, an idea whose time had finally come and that would inevitably rule over the earth. It would be called "Dyna-Soar."[44] It did not seem to occur to Williams that its name might one day evoke the image of a lumbering beast whose ultimate fate was oblivion. In 1957, there was reason to think that Dyna-Soar might indeed enjoy a day in the sun.

Air Force and NACA leaders met in June to discuss hypersonic objectives. The Air Force was represented by Lt. Gen. Thomas Power, commander of ARDC. He would be heading to the Strategic Air Command in less than two months to fill Curtis LeMay's vacancy as the latter became service vice chief of staff. Dryden spoke for NACA. Power indicated that a vehicle with global range enjoyed greater security against surprise attack, and Dryden said that development of an operational weapons system might lag behind the technological capability of global flight by fifteen to twenty years. Power, reminiscing about the B-52 Stratofortress at the forefront of the contemporary SAC, noted that "we started thinking about the B-52 about 10 years ago. So now is the right time to start thinking about

NACA saw the early Dyna-Soar concept as Round Three of an ongoing aviation effort. Round One broke the sound barrier, and Round Two extended flight to the edge of space. Although the Round Three Dyna-Soar challenged national policy, Round One (X-1, at the top right) and Round Two (X-15, rear view) are on display in the main hall of the Air and Space Museum in Washington, DC. *Author's collection*

a hypersonic glideweapons system." But Power demurred from the possibility of designating the vehicle a "prototype," suggesting that "the way to start is as a research vehicle." In contrast to his apparent enthusiasm in February, Dryden in his June meeting with Power indicated that "we would not want to start this project soon."[45] When the Air Force seemed interested in pushing ahead alone during February, NACA needed to keep pace or risk being left out. By June, awareness of technological challenges had dictated a more measured pace in pursuing hypersonic flight. In the run-up to Sputnik, the hypersonic bomber (by now named Dyna-Soar) was a long-range goal.

Statements by Deputy Chief of Staff for Development Lt. Gen. Donald Putt led NACA to reach four conclusions about the Air Force's position on Round Three: "(1) It is not yet time to start a project directed at a hypersonic boost-glide weapons system, (2) in any event such a project should be preceded by a hypersonic research airplane, (3) the hypersonic research airplane should be a joint project with NACA, (4) the Air Force should take some step to make the hypersonic research project a project in being, but should not apply any significant funds" to it at the expense of the X-15 rocket-powered research plane.[46] Dryden informed Ames Lab director Smith DeFrance that "the X-15 project is not out of the financial woods yet."[47] Pursuing both the bold X-15 and its even bolder Round Three successor seemed to risk the future of both. Technological progress was deemed essential, but only an acute emergency could warrant pursuit of programs at a pace that recklessly left them vulnerable to fiscally or scientifically founded skepticism.

As the Air Force slowly rolled toward its weaponized aspirations, the administration pressed ahead with its own plans. In keeping with the president's interest in placing limits on the increasing dimensions of the arms race between the Western and Soviet blocs, the State Department worked on the nation's position regarding potential disarmament. Although no satellite yet existed, space was addressed: "The parties will agree that within three months after the effective date of the agreement they will cooperate in the establishment of a technical committee to study the design of an inspection system that would make it possible to assure that the sending of objects through outer space should be exclusively for peaceful and scientific purposes."[48]

The need to maintain national security without damaging the national economy catalyzed internal debate in mid-1957 about a 5 percent reduction in basic research.[49] Simultaneously, the potential range of missiles sparked interservice frictions, prompting Eisenhower to rule in August that while he did not want the Army to be prevented from developing reliable missiles with longer ranges, "the technical angles, which include cost and procurement, are not matters for the President."[50] Nonetheless, "technical angles" carried potent policy implications and would thus creep into presidential policymaking areas.

Hypersonic study gained some traction in September. RAND's "missile-space" wish list grew apace, calling for "reconnaissance satellites; cislunar systems; interplanetary systems; navigation satellites; and communication satellites." In early September, NACA informed top Air Force military and civilian personnel, including Lieutenant General Putt, of NACA's engineering options in confronting challenges of hypersonic flight. Heat remained a stupendous problem. NACA preferred a complicated internal cooling system or an external insulation system to the use of heat-resistant ceramics. ARDC in turn informed NACA of its development program for a "hypersonic glide rocket weapon system (embracing HYWARDS, Brass Bell, and ROBO)." On

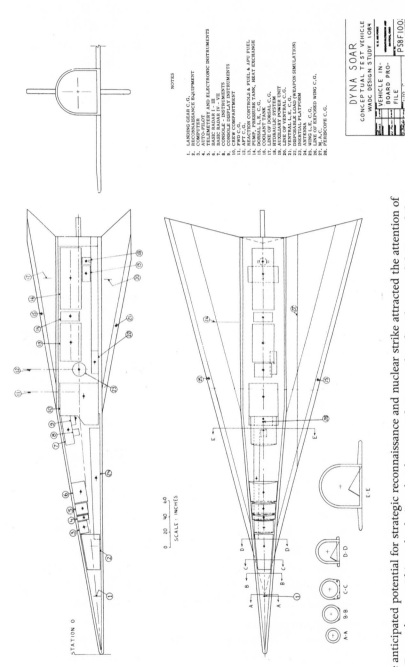

STATION 0

0 20 40 60
SCALE : INCHES

NOTES

1. LANDING GEAR C.G.
2. RECONNAISSANCE EQUIPMENT
3. COMPUTER
4. AUTO-PILOT
5. TELEMETRY AND ELECTRONIC INSTRUMENTS
6. BASIC RADAR I - III
7. BASIC RADAR IV - VII
8. CONSOLE - INSTRUMENTS
9. CONSOLE DISPLAY INSTRUMENTS
10. CREW COMPARTMENT
11. FWD C.G.
12. AFT C.G.
13. REACTION CONTROLS & FUEL & APU FUEL
14. PUMP, PRESSURE TANK, HEAT EXCHANGE
15. DORSAL L.E. C.G.
16. COOLANT TANK
17. LINE OF DORSAL C.G.
18. HYDRAULIC SYSTEM
19. AUXILIARY POWER UNIT
20. LINE OF VENTRAL C.G.
21. VENTRAL L.E. C.G.
22. DISPOSABLE LOAD (WEAPON SIMULATION)
23. INERTIAL PLATFORM
24. ANTENNA
25. WING L.E. C.G.
26. LINE OF EXPOSED WING C.G.
27. M.A.C.
28. PERISCOPE C.G.

A-A B-B C-C D-D E-E

DYNA SOAR
CONCEPTUAL TEST VEHICLE
WADC DESIGN STUDY 1084

VEHICLE IN-
BOARD PRO-
FILE
P58F100

The anticipated potential for strategic reconnaissance and nuclear strike attracted the attention of Air Force planners. This early design sketch appeared six months after the consolidation of the separate flight research, reconnaissance, and bomber design efforts. Weaponization is already a priority as designers conceptualize the layout of equipment within the craft (noted in item 22 in the key). *Air Force Historical Research Agency*

September 26, NACA sent comments on the project to ARDC. NACA agreed "that flight research with a vehicle having performance capability similar to that of the 'conceptual test vehicle' (Dyna Soar I), is indispensable prior to the development of an effective long-range reconnaissance or weapon system based on the hypersonic boost-glide concept." NACA forwarded its conclusions about the technical feasibility to the Air Force assistant secretary for research and development with the notation that it was "not being distributed beyond the military services at this time."[51]

By the fall of 1957, two different concepts for U.S. space policy were being aired. Eisenhower's concerns about tension and the arms race, and especially about the prospect of surprise attack, led him to believe that reconnaissance satellites moving unhindered and unchallenged through space marked the best way to gather crucial information about the Soviet Union in the long term. Toward that end, a modest U.S. nonweaponized satellite project had been undertaken under the auspices of the IGY. The administration aimed to demonstrate the project's independence from military application. This would deflect accusations that the satellite project diverted resources from top-priority ballistic missile development programs that the Killian Committee had recommended and that some airpower advocates also called for. However, the principal purpose in maintaining distance between the IGY satellite project and the ballistic missiles was to underscore the nonweaponized character of what was meant to be the world's first satellite. This, in turn, would facilitate the freedom of space message and help make satellite reconnaissance diplomatically feasible as soon as it was technologically possible. The Air Force continued to study the possibilities of hypersonic flight at the edge of space, expecting it to become a combat zone in the event of war. The level and intensity of top generals' interest in this long-term research fluctuated, and since NACA existed in an advisory and supporting capacity, it responded by keeping in step with Air Force interest.

Both potential policies—the one outlined by the NSC in 1955 and the one assumed by the Air Force in the mid-1950s—could represent a coherent and contextualized national security policy for space. Because freedom of space depended on the absence of weapons deployed in orbit, and because the Air Force assumed that air combat would someday extend into space, the two policy visions were incompatible; each implied a completely different use and future for that environment. But the secrecy and obscurity of the issues precluded discussion about them at this time. The obscure

Dyna-Soar had recently been consolidated into a single program, and even so, the action had been taken more to conserve disappearing research dollars than to adhere to doctrinal preferences. Dyna-Soar did not merit presidential attention in the late summer of 1957. Unexpected events and alarmist reactions to them would decisively alter the situation. The debate between the two visions was about to go unexpectedly public.

4

"Adjacent to the Abyss"
Leveraging Crisis to Promote the Aerospace Agenda

When the "beep" of Sputnik I was heard in October, 1957,
the true portents of air power became universally inescapable.
—EUGENE EMME, NASA HISTORIAN

That diversion of resources and energies is a political and not
a technological act.
—JAMES KILLIAN, PRESIDENT'S SCIENTIFIC ADVISORY
COMMITTEE HEAD REFLECTING ON THE SOVIET ATOMIC
PROJECT AND ON SPUTNIK

The Air Force and the Media

America's airpower paladins were not new to the world of advocacy in 1957, but that year marked an important shift in their quest for control over security policy. Thirty-two years earlier, Brig. Gen. Billy Mitchell had asserted that "should a nation . . . attain complete control of the air, it could more nearly master the earth than has ever been the case in the past." He and his contemporaries in and beyond the United States and its air service made intentionally controversial claims about airpower, aiming to spark debate and reshape policy. Alexander Graham Bell, inventor of the telephone, told naval personnel in 1916 that "the time . . . is almost now at hand, when sea power and land power will be secondary to air power, and that nation which gains control of the air will practically control the world." A brave new world was approaching. But in addition to this surprising new specter of warfare, Bell offered courage and hope: "The United States has a great advantage, so far as future progress is concerned. *We are an inventive people* [emphasis added]." The first "strategic" air raids using heavier-than-air vehicles occurred a few months after that

51

pronouncement was made. British Prime Minister Stanley Baldwin spoke for many when he predicted, during the interwar years, that "the bomber will always get through" and that "everything that has been used [in war] will be used again unless it is superseded by something more horrible."[1]

Airpower advocates used cultural vehicles including movies to drive their messages home. The year 1957 saw work on the blockbuster movie *Bombers B-52*, in which an Air Force ground crew sergeant (played by Karl Malden) is vindicated in his dedication to service. He declares that, "To me, it's important. More important than any other job I know." The appearance of the B-52 on screen is accompanied by triumphal music. Dialogue in the movie asserts that "we think that the way to prevent wars is to prevent major aggression" and that the B-52 is "the biggest jet bomber in the world. It can reach any target in the world." As if the audience could possibly have mistaken the airpower message and theme, the lieutenant colonel's character tells the sergeant that the B-52 is a "$9 million airplane, and we're going to prove it's worth every dollar. *If we can prove that, then we can keep it, and keep the peace* [emphasis added]."[2]

The Air Force promoted a public image of itself, in sync with cultural values and assumptions about technology and heroics. At a time when national defense enjoyed a lion's share of the federal budget, the Air Force in turn received the bulk of defense funds. To maintain its position, it worked to prove the value of its current weapons systems and the merits of the next generation, convinced that in preserving the developmental programs, the service would indeed preserve global peace and stave off Soviet attack.[3]

Bombers B-52 was part of a series of big-budget Hollywood films celebrating the role of American airpower. The Air Force actively cooperated with Hollywood efforts to make movies lauding its pilots, and Hollywood offered a movie star and bona fide war hero in Jimmy Stewart, who starred in *Strategic Air Command* in 1955. The same year, Gary Cooper portrayed a heroic title character in *The Court-Martial of Billy Mitchell*. Historian Steve Call correctly notes that these movies were "more than just Hollywood schmaltz." Movies were one avenue by which the Air Force and its airpower partners waged a metaphorical kind of strategic bombardment; an inundation of positive images appeared in *Life* magazine, and supportive articles were printed in large, general readership publications like the *Saturday Evening Post* and *Reader's Digest*.[4]

The popular media drumbeat was complemented by a growing professional presence as well. Trade journals, such as industry leader *Aviation Week*, catered to the cadre of true believers in airpower, both civil and military. The journal's circulation grew in the late 1950s to about 50,000. The popularity of this niche market was substantial enough to support the emergence of a new magazine, *Missiles and Rockets*, in October 1956.

Top Air Force officers added their voices to the chorus, as Chairman of the Joint Chiefs of Staff Gen. Nathan Twining, USAF, urged the country to mimic advanced Soviet scientific education methods by "weed[ing] out the less adept" students and "select[ing] the best for further education." The following year, *Bombers B-52* was Hollywood's attempt to glorify all of SAC instead of only its pilots, but the House Committee on Un-American Activities sympathetically published Curtis LeMay's declaration that "'protracted war' passed with the advent of the nuclear age," because "the determination of eventual victory or defeat . . . will have been reached in the first few days."[5] The combination of the impressive machine and its human master lay at the core of top Air Force thinking about the best means of wielding deterrent power in a credible and controlled manner. It was a view fitted to a prosperous society familiar with technology, one that agreed with Alexander Graham Bell's half-century-old characterization of Americans as an inventive people. And yet the world of airpower appeared in many eyes to fundamentally change in the autumn of 1957.

The Security Challenges of Late 1957

Eisenhower struggled during the second half of 1957 to control the national discourse, but it was an uphill battle as a phalanx of controversial subjects seemed to crowd in on the administration. The president encountered criticism from some people demanding even more money for the Air Force and from others insisting that the New Look's reliance on nuclear arms shackled the nation to a dangerous strategic binary of capitulation or oblivion.[6] Although committed to the maintenance of strong nuclear retaliatory forces, Eisenhower himself privately expected that a "limited" use of nuclear weapons would lead irresistibly to a general war. Chairman of the Joint Chiefs of Staff Adm. Arthur Radford agreed, saying that "today, atomic weapons have virtually achieved conventional status within our Armed Forces."[7]

Another distinct line of criticism came from a rising star at the Council on Foreign Relations, Henry Kissinger: "The more military strategy emphasizes the resort to an all-out nuclear strategy, the more responsible policy-makers will come to believe that no cause except a direct attack on the United States justifies the use or the threat of force." In Kissinger's estimation, the New Look implied an all-or-nothing approach to security that invited the "ambiguous challenges" of Soviet aggression. According to Kissinger, the United States would have the advantage in limited war because of its industrial base.[8] In 1957, Kissinger advocated the controlled and calibrated use of nuclear weapons in limited war, incorporating "'clean' bombs with minimal fall-out," defined zones of nuclear combat, and international inspectors inhabiting war zones to document nuclear effects.[9] Eisenhower, who felt certain that an exchange of nuclear attacks would inexorably spiral into general war, considered this line of thought perverse and absurd. Secret contemporary estimates concluded that about half of the U.S. population would be killed or maimed in such a war.

Absent accurate and timely intelligence, any policymaker could easily be buffeted by dissenting voices. The U-2 program had gone far toward providing that vital information since the summer of 1956. In mid-1957, the administration unveiled the *new* New Look, a further strengthening of the reliance on nuclear weapons. Deputy Defense Secretary Donald Quarles had helped lay groundwork for this by reiterating that it "is not necessarily valid" to assume that "sufficiency is a force *bigger* than that of any possible opponent" and that "the sufficiency of our own deterrent force is not necessarily invalidated by increases or decreases in Communist arms. . . . The power and policy of the Communist world leaves us no alternative but to move carefully along a path continuously adjacent to the abyss of total war."[10] This left Eisenhower's defense policy even more vulnerable to criticisms of inadequate flexibility.

In August, the Soviet Union tested a long-range missile, which blazed from the test site at Kapustin Yar across much of the broad expanse of the Soviet Union. U-2 missions occurred only infrequently and photographed narrow strips of Soviet territory, so much remained unproven and unknown: whether the USSR in fact possessed an ICBM, the size of the rocket's maximum payload, whether the USSR possessed a functional guidance system, or whether the capacity existed to recover items surviving the extreme heat of reentry. The August test definitively *proved* Soviet success regarding intercontinental range only. Furthermore, nothing in contemporary U.S. missile development suggested that a first-generation missile would be more reliable or accurate than a long-range aircraft piloted by seasoned personnel. News and rumors of Soviet tests appeared in air industry trade journals, but Eisenhower kept calm.

A far more serious problem for Eisenhower concerned segregation and the domestic racism that sustained it. Civil rights groups had sought for decades to roll back the crippling "separate but equal" racist policies that had won the Supreme Court's seal of approval in *Plessey v. Ferguson* in 1896. Until the 1950s, the National Association for the Advancement of Colored People, aware that it could not overthrow *Plessey* at a single stroke, brought several individual cases forward that demonstrated that "separate" conditions were often unequal. In many situations, such as voting rights, what was officially equal was dashed by local regulations, prejudice, and intimidation. However, the goal was to persuade the court that separation automatically constituted inequality. In 1954, with *Brown v. Board of Education of Topeka*, the court finally made a fundamental departure from the *Plessey* precedent.

The division grew in intensity, as communities and entire states resisted the integration ruling. The close of 1955 saw a charismatic Baptist preacher named Martin Luther King Jr. coordinate public transportation boycotts in Montgomery, Alabama, near the Air Force's graduate service school at Maxwell Air Force Base. The next year, former Dixiecrat Strom Thurmond of South Carolina organized a racist "Southern Manifesto," and most members of Congress from southern states signed on.[11]

In politics, Eisenhower accepted discussion, but he desired consensus. Regardless of his own views on segregation, Eisenhower believed in the court's role interpreting the Constitution, and the court had spoken. If consensus could not be adapted to the ruling, at least the ruling would be enforced. Early in 1957, the administration supported a civil rights bill, but the price of passage was the dilution of its strength. Meanwhile, Eisenhower struggled to approach segregation as a moderate facing a political issue rather than as a policymaker confronting a moral wrong.[12] Segregationist recalcitrance at Little Rock High School in Arkansas forced Eisenhower's hand in the fall. Governor Orval Faubus, aware of the state-level political dividends of upholding segregation, wrote himself into headlines by enforcing segregation before appearing to bow to Eisenhower's entreaties and then reneging on agreements made at a September meeting with the president. Eisenhower nationalized the Arkansas Guard on September 21 (removing the single biggest formal tool of coercion that Faubus might wield) and then ordered part of the 101st Airborne Division to Little Rock three days later. Angry southern state officeholders derided the presence as a military occupation and a presidential overreach.[13] As disappointing as later observers might find Eisenhower's reluctance to move with greater speed, the Little Rock episode demonstrates how he sought as president to work "gently in manner, strongly in deed."

Tension in Little Rock was an ongoing issue into early October, and a federal military presence remained there as a result. The country was in the early throes of a particularly bitter flu season by October. The harshest to hit the United States since the so-called Spanish Flu global pandemic of 1918–20, the 1957 flu killed 70,000 people. At NACA's Langley Research Center in Virginia, work included study of the effects of "flutter" (high speed-induced vibration) on control surfaces of the planned X-15 rocket research plane. On October 3, the Milwaukee Braves pulled an upset win over the New York Yankees in game two of the World Series.[14] The series, destined to go a full seven games and end with a Braves championship, took a one-day hiatus after the October 3 game, so that the teams could travel to County Stadium in Milwaukee for the coming three games. Columbia Broadcasting offered a new television program starting October 4, *Leave it to Beaver*. Remembered later as an iconic portrait of the idyllic 1950s, in fact it is an artifact of the time when control of space policy slipped from the grasp of Eisenhower's hidden hand and entered the unsettled arena of public debate and bureaucratic struggle. *Leave it to Beaver*'s first episode began at 7:30 p.m. Eastern Standard Time. Four hours, one minute, and twenty-six seconds earlier, a rocket had lifted from a launch pad in Kazakh SSR, carrying a 184-pound sphere bound for orbit.

Eisenhower was often skilled in dealing with the press and typically on fine terms with them, but his response to the Sputnik satellite launch backfired. Historian Stephen Ambrose observed that Eisenhower "had gotten through many a crisis simply by denying that a crisis existed," displaying calmness that the press and public eventually accepted.[15] After the orbit of Sputnik, this did not happen with regard to space policymaking. Eisenhower knew that "public relations can be stigmatized as mere propaganda," but "in a republic where the citizens must know the truth," communication remained necessary. In preparing for the Normandy landings in the Second World War, he had granted privileges to members of the media while simultaneously telling them that he considered them "quasi-staff officers." In working to bring the Korean War to a satisfactory end and the Quemoy and Matsu crisis to an acceptable solution in 1953 and 1954, Eisenhower included the media as tools for projecting calculated messages to overseas antagonists. The U.S. Information Agency (USIA) kicked its activity into high gear during the Eisenhower presidency because he deemed the promotion of friendly public affairs and propaganda to be of crucial importance during what he considered a "total cold war."[16]

In keeping with his overall presidential style, Eisenhower at first attempted to address Sputnik by minimizing it, hoping that its shock value would evaporate. He decided at a conference in the Oval Office on October 7 "not to shift from the present orderly procedure to produce an Earth Satellite."[17] But the president's calm, "low-key response to *Sputnik* completely failed to defuse the growing sense of public alarm." CIA Deputy Director of Plans Richard Bissell admitted that Sputnik I "created an atmosphere of crisis in the government, as well as in the country at large." Eisenhower's equanimity was based on U-2 evidence undercutting Soviet missile claims, but he decided against assuaging public concern, given the price (and risk) involved in disclosing the U-2 program's existence.[18]

Immediately after the Sputnik launch, scientist David Beckler informed the Office of Defense Management (ODM) of a "widespread belief among the scientists that our scarce scientific resources are not being husbanded in the most effective way; that there is lack of balance between long-term and short-term goals and inadequate focusing on important national objectives."[19] To sharpen the point, he forwarded a copy of Scientific Advisory Committee president Isador Isaac Rabi's year-old letter, which pled for resourcing for the Vanguard program, to Eisenhower's staff secretary Gen. Andrew Goodpaster, USA.[20] Outgoing defense secretary Charles Wilson conferred with Eisenhower on October 8, and the president personally conceded possible minor revisions to previous cost-cutting measures, but he preferred for reliable subordinates to lead these efforts.[21]

Sputnik helped establish the freedom of space objective that had heretofore been the driving force behind the Vanguard element of the IGY. Donald Quarles had pointed out to Eisenhower during the October 7 meeting "that the Russians having been the first with their Satellite to overfly *all* countries . . . have thereby established the international characteristic of orbital space. We believe that we can get a great deal more information out of free use of orbital space than they can."[22] A respected and experienced scientist in the private sector, Quarles had also served as Secretary of the Air Force, and his voice carried weight. Quarles repeated his points to the National Security Council on October 10. Sputnik in fact supported "our objective in the earth satellite program . . . to establish the principle of the freedom of space." Satellites, he said, "were of very great significance, especially in relation to the development of reconnaissance satellites." He also pointed to the conscious American distinction between the nonweaponized booster for Vanguard and the Soviet use of an ICBM booster to lift Sputnik.[23]

Still, the ambiguity of the situation gave rise to conflicting messages by administration officials. Charles Wilson's retirement as defense secretary, coincident to the Sputnik launch, brought Neil McElroy's succession on October 9. McElroy promptly declared at a press conference that "the [Vanguard] program *would* be speeded up," although staff noted that this "compet[ed] in effect with the President's press conference" statements.[24] Rumors, courtesy of Stewart Alsop, arose that Sputnik was itself a reconnaissance satellite and "is not blind," but Department of Defense (DOD) analysts assured Eisenhower that this was untrue.[25] Contemporary U.S. satellite plans, envisioning a U.S. reconnaissance satellite weighing three hundred pounds and carrying a one-hundred-foot resolution television camera that would either beam signals earthward or release film canisters for capture during reentry, did not foresee operational status before 1960. Still, editorial critics quickly characterized Eisenhower's response as inadequate and uninformed.[26] Walter Lippmann argued on October 10 that Sputnik "means that the United States and the Western world may be falling behind in the progress of science and technology." Lippmann compared prosperity to "a narcotic" that, combined with McCarthyism, had robbed the national stage of purposeful action. The president, Lippmann asserted, was "in a kind of partial retirement. . . . Thus we drift with no one to state our purposes and to make policy, into a chronic disaster like Little Rock."[27]

Mid-month, Eisenhower met with the ODM Science Advisory Committee to discuss the country's security. Urged to increase funds for basic research, in hopes of facilitating breakthroughs in the longer term, he asked the scientists, "How much is enough?" Despite the Soviet satellite, he doubted that the United States was truly outdistanced in science, citing the new U.S. nuclear submarine and judging it more strategically formidable. Rabi, who had urged expansion of the Vanguard program a year before Sputnik's launch, worried about "the quality and breadth of [the Soviet] educational system in scientific fields," and Ed Land said that the Soviets "are teaching their whole country." Eisenhower challenged the assumption that Soviet education churned out hordes of expert scientists. Land emphasized the need for greater "understand[ing of] the spiritual rewards of scientific life," and Rabi called for more continuous scientific representation in deliberations with the president, rather than "being called in after the fact."[28]

On this last point, Eisenhower agreed. The Office of the President was "crammed and inadequate," and the answer might involve "someone who can . . . bring in more specific ideas—a special assistant trained as a scientist." James Killian suggested that the Scientific Advisory Committee "could provide proper back-up for such an individual," and Eisenhower described having felt "a need for such assistance time and again." By late November, this would be accomplished, as the committee moved from ODM to the White House and Killian became the first head of the now-renamed President's Scientific Advisory Committee (PSAC). During the October 15 meeting, committee member Jerome Wiesner urged that U.S. missile programs be bolstered, noting that the problem in the missile realm was "that we got a late start."[29]

Facing Public Concern

Meanwhile, the administration sought to assure the public, and members of the press and the public sought to guide the president's hand. On October 15, Vice President Richard Nixon told an audience in San Francisco that "the Soviet Union is not one bit stronger today than it was before the satellite was launched." When Republican officeholders discreetly suggested establishing a new federal agency on science, the White House alluded to the imminent establishment of the PSAC and pointed to the larger problem of "attracting . . . an increasing number of individuals of quality and promise" to scientific study.[30]

Reporting international reaction to the Sputnik, USIA director Arthur Larson concluded "that the launching of the Soviet earth satellite has caused a greater psychological impact than any prior Soviet cold war action." Larson suggested projects "such as manned satellites, hitting the moon, or [building] a space platform" to demonstrate "the conscious intention to enhance America's prestige abroad."[31] He also pointed to the value of emphasizing technological power, given "the disproportionate impact that real or apparent scientific preeminence now seems to have on our military position and our diplomatic bargaining power." Air Force planners listening to Edward Teller, father of the hydrogen bomb and a consistent advocate for new and more potent weapons, demanded concerted space activity and believed that a "spectacular, but technically superficial demonstration" would allay concern and thereby do more harm than good.[32]

Also in the middle of October, Deputy Defense Secretary Donald Quarles, while accepting an award for merit from the American Institute of Consulting Engineers, declared that "to be effective from now on, our forces must be continually improved and modernized." The challenges of the Cold War had, therefore, to be met with "the same courage, determination, and single-mindedness that has always characterized [American actions] in wartime."[33] Bernard Baruch, who had dabbled in advising Woodrow Wilson, Franklin D. Roosevelt, and Harry Truman, confided to Eisenhower that "whatever the reason, the public is worried now and in a condition to see the needs on defense and our economy. I think they are in a frame of mind to give you the support and even the sacrifice to put both our defense and economy on a sound footing."[34]

Similar notes were sounded by the media, including an editorial in the *New York World-Telegram and Sun* demanding a national, and assumedly lunar, crusade. It called upon Eisenhower to "summon the industrial power of our country" and "give purpose, direction and determination to a program to get into outer space 'fustest with the mostest.'" Anything else would be to "falter" and cede valuable Cold War ground (whether literal or rhetorical, the essay did not specify) to Soviet ruler Nikita Khrushchev. "Your last three years in the White House, Mr. President, will not be pleasant for you, or for us, or for anyone in the free world," unless Eisenhower moved to "shoot the moon."[35] The demand meshed with the impulse of many people concerned that the Soviet accomplishment equated to U.S. technological and strategic vulnerability. "Shooting the moon," however, would be expansive and reactive policy. Seeking to deflect criticism by the formidable Missouri Democratic Senator Stuart Symington, Eisenhower emphasized that the United States had "leaned over backwards . . . *in order to avoid intensifying the armament race.*" Quarles added that "the problem is one of the long pull" and that Soviets' rejection of a bid to prohibit weapons from space indicated their "intention to exploit space for military purposes."[36] Such an outcome threatened the administration's reconnaissance satellite plans.

By the end of October, Quarles proposed tackling the ominous specter of Soviet weaponized space power head on, arguing that "we can play this game, too. We have other programs along these lines. The scientists have many ideas about what can be done in outer space. Some of these are actually being done. We will ask for any authority that may be required to do the things necessary to do in the national interest." The spirit of these assertions, echoed by advocates of what soon would be called "aerospace power," clashed with administration goals. Quarles was quickly warned off with a handwritten remonstration: *"Don't say this* [emphasis added]."[37]

November began with the Operations Coordinating Board (OCB) estimating the political damage done and with the president's team drafting speeches to refocus the public and alleviate fearful distraction. The OCB found Sputnik I distressing because "Soviet claims on scientific and technological superiority over the West and especially the U.S. have won greatly widened acceptance," harming U.S. prestige and perceptions of military superiority. "Discomfiture and intense interest" in the United States further unsettled allies.[38]

The November 1 draft of the first speech, dubbed "Science and Security," conceded that the Sputnik I accomplishment was impressive, but it contested the notion that the USSR actually stood ahead even in missile technology. U.S. ballistic missile tests, although not orbital, had reached altitudes in the thousands of miles. The draft added that "the standard of accuracy necessary to [place a satellite in orbit] is of a much lower order than that which would be needed to hit a particular target on earth," and the U.S. BOMARC and Snark cruise missiles already had that militarily relevant capability. The draft rejected the notion that Soviet citizens were all being trained as scientific geniuses, answering public parallels to the concerns Land had privately raised two weeks earlier. It emphasized that, although the United States had pioneered "nuclear energy, the cyclotron, the transistor, and the Salk vaccine," no nation had ever monopolized scientific discovery. "We have said, in all sincerity, that we did not consider the earth satellite project a race. But we may as well come right out and say that, so far as preeminence in the field of science is concerned, we *are* in a race, and a mighty important race it is." An honest appraisal of the situation "should give us confidence that, with hard work and self-denial, we must surely win."[39]

Refining the speech, Eisenhower and his staff soon added that the "most important thing on earth is *still* a start toward *disarmament*," declaring that "a race in missiles and satellites for military purposes" would be "tragic."[40] Eisenhower refused to leap voluntarily into a space arms race that would threaten both the future of reconnaissance satellites and the prospect of disarmament.

Then a second Sputnik—larger than the first and carrying the first creature into orbital space—entered orbit on November 3. Sputnik II also prompted far greater concern in the United States than its predecessor had done. Eisenhower's popularity dropped to just 57 percent from a high of 79 percent at the year's start.[41] Allegations of a "missile gap" and the Sputnik launch opened the door to increasingly overt challenges of the Eisenhower defense strategy. Major publications equated the Soviet achievement to a U.S. defeat. *Newsweek* called it a defeat "in pure science, in practical know-how, and in psychological warfare," and *Life* concurred. An assistant director of the Smithsonian Astrophysical Observatory predicted that the Russians would "reach the moon within a week."

Despite the mounting criticism, Dwight Eisenhower's conception of the presidential role remained fairly consistent.[42] But temptation grew for bold actions, and USIA director Larson noted that the "reaction of bewilderment is shifting to anger." The president's speech "has to have some simply understandable idea that *something has been changed.*" In the meantime, "DDE's prestige in his special area is slipping." It was felt critical that "reassurance in concrete terms" be provided to the country, assuring it "that this *setback can be overcome*"[43] and that Soviet earth satellites would not cause neutral nations to fall into the USSR's geopolitical orbit. Larson, who had already been urging the president to undertake some bold space project for the sake of international prestige and credibility, was sympathetic to such ideas.

As the first Sputnik had prompted editorial calls for Eisenhower to "shoot the moon," the next Sputnik brought more political fire. "You Needn't Be a Scientist to Understand Sputnick [*sic*], Mr President, But You Must Be a Leader," chided *Rockets and Missiles*'s executive editor, Erik Bergaust, in an open letter to Eisenhower. He asserted that Soviet satellites showed Eisenhower's inactivity regarding space to be "a dangerous mistake." The *New York Times* reprinted the letter on November 7.[44] ARS president Robert C. Truax worried that national planners sought excessive input from the armed forces and too little from the scientific community.[45]

As criticism mounted and the administration struggled to reassure friends and intimidate adversaries, yet another problem appeared. Earlier in the year, Eisenhower had organized an ad hoc committee to provide advice about methods for mitigating the effect of a Soviet nuclear attack. The Federal Civil Defense Administration, most widely remembered for its *Duck and Cover* film, had recommended spending $40 billion over ten years on passive shelters to protect Americans in the event of a nuclear war.[46] At the time, the entire federal budget was less than $80 billion annually, and the Civil Defense request made little headway with an incredulous Eisenhower, already convinced that no nuclear war between superpowers could be kept limited. Operation Alert, conducted as practice for evacuating key officials from likely target areas, included a calculation that a nuclear war would leave 60 million to 100 million Americans (one-third to two-thirds of the contemporary population) dead or injured.[47] Spending a fortune on shelters would encumber the U.S. economy while also diverting funds from active defenses (such as nuclear-armed air defense missiles) and the SAC's offensive force that provided deterrence and national security. Building the shelters thus meant relying more upon them. Survivors would be left to cope with the material and emotional debris of an apocalyptic landscape bereft of infrastructure, in desperate need of doctors, and with the war unfinished.

The issue became politically loaded. Democrats pressed for the shelter program, seeing an opportunity to skewer the president on a security issue. The House Committee on Government Operations had already reported the previous summer that "an effective nation-wide civil defense" would constitute a "strong deterrent to war."[48] Edward Teller judged the shelter plan to have "much merit," suggesting that appropriately equipped shelters could let the United States recover from a nuclear exchange and prevail over a devastated USSR. "We can be back in business within a few hours of any attack," Teller had ludicrously argued in January. Eisenhower had responded in April by establishing a committee under Rowan Gaither.[49]

Problems in chairmanship, the unwieldy size and organization of the committee, the perspectives of the report authors, and the group's self-designated autonomy combined to entirely contradict Eisenhower's desires and intentions. Earlier ad hoc committees such as Project Solarium and the Killian Committee had contributed ideas to execute Eisenhower's policy goals. With the Gaither Committee, much of the problem was hardwired into the group from the outset, starting with the chairmanship. Illness led to Gaither's replacement by Robert Sprague, who had earlier advised the Senate and the National Security Council and who expected Soviet nuclear arms to be a lethal and imminent threat.[50] The size of the Gaither group was detrimental as well, with eighty participants who were not subdivided into task forces as had been the case with the Solarium and Killian projects. A steering committee and an advisory panel had disproportionate input. Eisenhower held one of the two principal report authors, Col. George "Abe" Lincoln, USA, in high esteem, but the other, Paul Nitze, had been a staunch critic of the New Look and had been a leading author of NSC-68, the Truman model of containment that Eisenhower rejected and replaced upon taking office.[51]

Finally, before cancer forced his withdrawal, Gaither had been persuaded by a RAND colleague to radically widen the mandate of the study, presenting advice about the entire national security policy as a whole, rather than analyzing the civil defense issue alone. The outlooks of Gaither, Sprague, and Nitze ensured the report's alarmist tone and condemnation of the New Look. The report, called "Deterrence and Survival in the Nuclear

Age," urged further offensive nuclear buildup, strengthening of active defenses, and a nationwide passive defense shelter program. It was delivered on November 7, and it completely defied the president's intent. Leaks of the report's existence and stories of its purported contents wove themselves into the fabric of the public's restive concern that winter.[52]

Although the president still hoped to address the Sputnik issue in a single speech at Oklahoma City, revisions continued, and the result was a pair of major speeches. One point Eisenhower wanted inserted stressed that "we *can* have both a sound economy and a sound defense—in fact, the two must go together. The secret is determination to make the hard choices needed to find savings in other less essential areas. It can be done, and it must be done."[53] Disarmament continued to carry vital currency in policy and opinion realms, and the president cherished the message that a sound defense did not conflict with national economic health because national security required that preconditions for both be guaranteed simultaneously.

The final version made certain to note the potent capacity of U.S. intercontinental cruise missiles and of the "three [U.S.] rockets [fired] to heights between 2,000 and 4,000 miles" that collected "much valuable information on outer space." Speaking from the Oval Office on November 7, Eisenhower also showed the audience a nose cone retrieved from suborbital ballistic spaceflight; the Soviets had not yet demonstrated the ability to recover items from space. More strategically significant, their implied ICBM capability was silently discredited in observant eyes. In this speech, the president called for greater national attention to science and announced the formation of the PSAC.[54] Eisenhower aimed to show that the nation's security was not in doubt and that he retained policy initiative.

"Scientists were not taken wholly by surprise" the way the public had been by Sputnik II, National Science Foundation director Alan Waterman told listeners in San Antonio, Texas. Nonetheless, the Soviet triumph was "a major disappointment to American scientists." Waterman emphasized that domestic lack of understanding posed a greater real problem than did Soviet accomplishment: "The real truth of our present dilemma is the sobering fact that our people have neither fully understood nor appreciated the need for science and scientists in world affairs today."[55]

Drafting and redrafting progressed for the president's November 13 speech at Oklahoma City as writers toiled to explain his position. Eisenhower sought to dispel fears of imminent danger and to "call [young Americans] to a high and noble endeavor" of scientific study to prevent a real danger from emerging in the future.[56] President Eisenhower aimed to

After Sputnik I and II, President Dwight D. Eisenhower wanted to calm panic in the nation and to preserve his national security policy. The first object recovered intact from spaceflight, an American ballistic rocket nose cone, was a choice visual in this effort. *Library of Congress/photo by Warren Leffler*

forestall wasteful programs, adding to his desire to prevent the expansion of an arms race into the space realm. But top Air Force leaders did not share the opinion that such prevention could be secured by anything other than a U.S. military presence. And the public, its representatives, and the media grew restive at the prospect of repeated and ominous Soviet advances into space. In the next four years, a reckoning was due.

From Bad to Worse

The psychological pressure and the cacophony of voices during the fall of 1957 did more than harpoon Eisenhower's popularity rating; they taxed his health. During a press conference, he answered a reporter's question about potentially weaponized future Soviet satellites with some exasperation, saying, "Suddenly all America seems to [have] become scientist[s], and I am hearing many, many ideas." *Newsweek* had chimed in as well when it proclaimed Sputnik to represent a threefold defeat in scientific, technological, and psychological terms.[57] By the second half of November, Eisenhower was complaining to his lifelong friend Swede Hazlett about the pressures confronting him, and he lambasted his personal physician Howard Snyder for letting him play more golf than was good for him. On November 25, Eisenhower suffered a stroke.[58] Although he recovered with surprising speed, challenges continued to press in on the administration.

Having discovered an issue with press attention and now with popular recognition, Congress dove into space issues. A fierce partisan and an ardent airpower advocate, Senator Stuart Symington relished the opportunity to initiate hearings critical of Eisenhower's policy. Although this kind of spectacle could damage the Republicans and benefit Democrats, it also had the potential to bog space issues down in partisan rancor. As a result, Georgia Democrat Richard Russell asked fellow Democrat (and Senate majority leader) Lyndon Johnson of Texas to begin an inquiry on space,[59] thus short-circuiting Symington's capacity to take the same action. This put a rift in what had been a congenial dynamic between Symington and the majority leader.[60] However, with his eyes set on gaining the White House in 1960, Johnson seized the opportunity to further build his already imposing political stature. Johnson's hearings started the same day that Eisenhower suffered his stroke.

Public pressure and Eisenhower's interpretation of the problem helped guide federal money into public schools. Insistent voices called for immediate funding of science and mathematics education and urged that federal money be added. Eisenhower remained uncomfortable with introducing a permanent federal presence into education, while science adviser Killian favored a broad educational model that fed the sciences without spurning the humanities. Eventually Killian and the president's brother Milton persuaded Eisenhower of the need for a holistic approach to, and federal support for, education. The final version of the National Defense Education Act nevertheless favored areas that were identified as strategic, such as science, math, and foreign languages.[61]

The first U.S. reach toward orbit ended with a bang, but not the kind that policymakers had intended. The administration had painstakingly steered the Vanguard project so that the satellite booster was lifted by a scientifically oriented rocket rather than by a converted missile booster, although this did not always prevent reporters from confusing the terms and negating what the White House deemed an important distinction. Also, in contrast to the Soviet launch of Sputnik, which had occurred in secret to conceal any potential failure, Vanguard's progress would be witnessed by the world. But instead of a launch, the world watched the booster lift a few feet before exploding in a fireball.

Controlling the message was an especially difficult challenge for Eisenhower from the fall of 1957 onward. The Gaither Committee's defiance of Eisenhower's guidance irked the president, and leaks about the report would cause lingering headaches. Determining exactly how much political damage is directly attributable to Sputnik is impossible. A contemporary Gallup poll found that half of respondents concluded that Sputnik had damaged U.S. prestige, but Gallup had focused its survey on a small sliver of the population deemed professional or middle to upper class. A *Baltimore Sun* study found that only one in eight people in its surveyed population could identify the Martin Company as the contractor for Vanguard.[62] The public's opinion was especially open to influence through the concerted efforts of opinion leaders in the government, media, or the military. Public commitment regarding an issue, however, depended on more than mere advertising; policy needed to mesh with enduring cultural tenets.

The combination of the Sputnik launch and Eisenhower's leadership style led the American public to imagine that a coherent national space policy had not been developed. The image of American progress was a constant theme in the internationally oriented messages of USIA, and throughout the Western world the perception prior to Sputnik had been that because communism was antithetical to independent thought, communist countries could succeed in S&T only by stealing secrets or poaching scientists. The launch of Sputnik underscored the presumptive ties between technological achievement with perceptions of national identity, and the Soviet satellites disturbed some Americans.[63] The calmness in Eisenhower's reaction, first to one satellite and then to two, was misunderstood as being a shallow lack of scientific understanding. Sectors of the media and Congress had interests in nurturing this concern, but the White House itself enabled concern to take hold. The shortcoming, as well as the strength, of Eisenhower's presidential style was that it set the country to underestimate Eisenhower's actions. Many, like Walter Lippmann, believed that the country would "drift with no one to state our purposes and to make policy." This false perception handed an enormous opportunity to anyone with a seemingly credible alternative policy for space that would assuage concern in the public, press, and Congress and that could guarantee national safety. Into this environment, the Dyna-Soar space bomber project was born.

5

"The First of a New Generation"
Dyna-Soar Spreads the Air Force's Wings into Aerospace

Looking further into the future . . . the Dyna-Soar, the first
"space bomber" . . . will be capable of circumnavigating the earth
at extreme altitudes and attacking its target with space-to-surface
missiles.
—Gen. Curtis LeMay, USAF

This program represents the first of a new generation of strategic
weapons systems.
—Gen. Thomas White, USAF

The Effect of Sputnik

Strategic Air Command adopted "Peace is our profession" as its motto in 1958. The command was a product of America's earlier strategic bombing thought and a creature of the Cold War. The U.S. Air Force's highest ranking officers in the 1950s had been bomber pilots and unit commanders in the 1940s. They were seasoned veterans who had seen the devastation of a world war, and the experience had convinced them of the potential of airpower in general and the efficacy of strategic bombing in particular. Peace was preserved through strength, and strength would be maintained through constant technological progress.

Gen. Thomas Power was not alone in believing that "the requirement for an effective deterrent force will remain with us for an indefinite number of years."[1] Technological progress and recent history suggested that the tools of destruction were discarded only when more fearsome ones were developed and that such developments were not only constant but also occurring with accelerating frequency. From this perspective, wisdom demanded staying ahead of the veritable shock wave of technological development. Anything less appeared to invite calamity, even oblivion.

A crucial conclusion stemming from this airpower perspective was that the key driver in the arms race was technology rather than policy. Since airpower advocates witnessed a spiraling expansion of technology, they often resisted terminology that would come naturally to airpower-friendly politicians, who spoke in terms of an "ultimate weapon." General Power, SAC's new commander, explained that it was "highly unlikely that there will ever be such a thing as an 'ultimate' strategic weapon." No technological monopoly could remain indefinitely, and countermeasures would arise to challenge technological breakthroughs. Rather than thinking in terms of the ultimate weapon, revolutionary new weapons and methods "only raise [technological strengths] to a higher plateau."[2]

Even prior to the launch of Sputnik, the Air Force had entertained some modest interest in the hypersonic flight concept that epitomized "higher, faster, farther." As ARDC commander in 1957, Power had signaled an intensifying interest in hypersonic flight, and NACA's researchers cooperated with the Air Force interest in extreme high-speed and high-altitude flight. Tension and competition between operational needs and research opportunities had prompted the separation of ARDC from Air Materiel Command (AMC) in 1950. Although internal frictions continued, the service's interest in constantly improving tools, its organizational decisions, and its adjustments to contract policies and development methods reflected a "culture of innovation" in the Air Force.[3]

Air Force officers were generally committed to piloted systems as a means of exerting "combat power," the ability to influence the battle zone through the potential or application of destructive force. The Air Force is popularly assumed to have clung to bomber aircraft and to have resisted ballistic missiles and the space age. Although many resisted the idea of bomber pilots being supplanted by "silo sitters" in the missile age, the Air Force was actually highly interested in space systems—but it wanted the flexibility and potency of *piloted* systems.

Good history resists notions of inevitability. Missile development was fraught with complications and difficulty, and building a reliable, accurate, powerful rocket was a mammoth task in itself. Until 1954, it was not clear whether a practical, miniaturized thermonuclear warhead could be constructed; without that, a practical ballistic missile would still not be a practical *weapon*. Early missiles were vulnerable technologies; corrosive liquid fuels had to be stored separately and missiles were fueled just prior to firing, making the missile and its ground crew profoundly vulnerable in the event of an enemy surprise attack. Vannevar Bush's advice against missile research in the mid-1940s had been in earnest, and it had been given with reason. Even in 1957, as the Soviet Union first tested an ICBM, Henry Kissinger predicted that such devices would not be possessed by both superpowers until about 1963.[4]

Power argued at the end of 1957 that "the manned aircraft offer[s] certain advantages that would be difficult, if not impossible, to achieve with missiles," such as a pilot's ability to cope with an unexpected contingency, to observe the results of an attack, to decide future courses of action, and to send nonnuclear warheads onto targets.[5] Power's support for hypersonic research indicates that top-ranking Air Force commanders thought in terms of crewed flight systems. At the time, those were airplanes in the atmospheric arena, but this realm would continue to expand upward and outward. After all, aviation history (which, for heavier-than-air vehicles, still covered well under 60 years) showed continual growth in maximum altitudes reached by new craft.

"Aerospace" referred to the various altitudes within—and beyond—the atmosphere, comprising a continuous environment. Control of the air had been contested during the wars of the twentieth century, so it was logical to expect that extra-atmospheric realms would eventually witness military struggles in future wars. Some futurist and optimistic advocates would ultimately claim that conflicts in space might spare the Earth from the experience of general warfare. The Air Force identified the development of aerospace forces as a necessity and an inevitability, and it envisioned Dyna-Soar as an integral component of that aerospace force.

Sputnik did not create Dyna-Soar or aerospace thought. In February 1957, nearly eight months before Sputnik's launch, Maj. Gen. Bernard Schriever addressed an astronautics symposium in San Diego, California. A tireless advocate of the ballistic missile, Schriever told his audience, "In the long run, our safety as a nation may depend upon our having space superiority." Furthermore, "our prestige as world leader might well dictate that we undertake lunar expeditions and even interplanetary flight when the appropriate technological advances have been made and the time is ripe."[6] Sputnik merely seemed to dictate a hastened stride into space.

Extreme ambition characterized Dyna-Soar's intended development schedule. The ARDC presented its consolidation plan for Dyna-Soar to the Air Staff on October 17, 1957. HYWARDS, now redesignated Dyna-Soar I, would focus on a first flight in 1963 to test the project's overall feasibility.

The ARDC hoped that the reconnaissance vehicle Brass Bell, renamed Dyna-Soar II, would make its first flight in 1966 and would have an initial operational capability set for 1969. Dyna-Soar II was to have a range of five thousand miles. The considerably more complicated task of perfecting the strategic bombardment vehicle RoBo, now called Dyna-Soar III, aimed at a first flight in 1970. Planners intended Dyna-Soar III to fly faster than its predecessors and almost twice as high as the operational Dyna-Soar II. A booster rocket would send the bomber variant to an altitude of over 50 miles, giving it an orbital earth-circumnavigation capability as it cruised at 25,000 feet per second.[7] The ARDC aimed for an initial operational capability date of 1974 for the Dyna-Soar III bomber.[8]

ARDC's "Abbreviated System Development Plan" for the Dyna-Soar program reported that Dyna-Soar offered "tremendous capability and potential . . . in the accomplishment of Air Force missions of the 1967 through 1980 time period and beyond. . . . It is believed that the boost-glide concept is feasible, and that the classic sustained flight air breathing engine approach to strategic mission accomplishment is very limited and has very little growth potential." Dyna-Soar I was to provide the technical information required before operational extensions could be built. "A Dyna-Soar II type vehicle [for reconnaissance] appears to be a very realistic design goal for the first weapon system" because "there is a very great need to provide 'eyes' for the ballistic missile and the boost-glide weapon systems," Dyna-Soar III.[9] ARDC saw Dyna-Soar as a future successor to the still-new B-52.

Sputnik's orbit heightened the Air Force leaders' sense of urgency regarding the Dyna-Soar program and aerospace philosophy. Reeling from the news of Sputnik I, which was still less than two weeks old, the Air Staff directed that the ARDC compress the already bold development schedule, accelerating Dyna-Soar I's first flight from 1963 to 1962. The Air Staff wanted Dyna-Soar II's first flight in 1964 and an initial operational capability three years later. These goals were incredible enough. Even more staggeringly optimistic, the Air Staff wanted the final bombardment version, Dyna-Soar III, to fly in 1965 and to be operational starting in 1968.[10] To that point, the United States had not successfully sent into space and recovered any creature larger than a mouse, and even then, the vehicle had traveled on a simple ballistic, nonorbital trajectory—up and down in a parabola. However, as Gen. Nathan Twining remembered afterward, Sputnik "didn't change the strategy any. We knew the Russians were working like the devil. . . . It . . . just spurred us on."[11]

Air Force intentions were undeniably ambitious, but the Apollo manned lunar landing in the 1960s would soon show that bold and spectacular accomplishments in compressed timeframes were not necessarily impossible. Reaching the moon and exploring it, and providing a means of return, certainly represented an enormous challenge, which the nine-year deadline suggested by President Kennedy in 1961 accentuated. Apollo and Dyna-Soar were vastly different in purpose and technology. Apollo was not intended to become a weapons platform, and it was not meant to be an operationally regular means of traversing space. Nevertheless, portions of the U.S. media and eventually members of the Kennedy administration came to compare Dyna-Soar with its capsule contemporaries as if they were apples and apples.

Air Force commanders insisted that technical supremacy was a sacrosanct prerequisite for national security. Addressing NATO's Advisory Group for Aeronautical Research and Development on November 25, Twining said that "technical supremacy is clearly essential to our freedom, progress, and survival."[12] Defense publications echoed this sentiment. In May 1957, LeMay declared that "the objective of our national defense policy is deterrence," which depended on rejuvenated technical superiority. New weapons would constantly be needed to maintain a margin of deterrence, and LeMay reminded his audience that "any operational commander . . . must expect that any new weapon system will have very low reliability." This factor warranted caution, but it was not deemed a reason to spurn innovation.[13]

Twining revisited this issue six months later, in the wake of Sputnik I and II, in an address to the American Ordnance Association in Cleveland, Ohio. The missile "is the latest weapon which is having a tremendous impact on military thinking," and it required responsiveness: "We must not be deluded into thinking that the machine will replace the man." Security demanded military capabilities that could "meet every possible contingency," and human beings possessed judgment and mental flexibility lacking in machines.[14]

The Air Force Scientific Advisory Board (SAB) declared in December that "Sputnik and the Russian ICBM capacity have created a national emergency." The board urged "a massive first generation IRBM and ICBM capability as soon as possible . . . a vigorous program to develop second generation IRBM's and ICBM's," acceleration of military reconnaissance and satellite programs, and, presumably under Air Force auspices, creation of "a vigorous space program with an immediate goal of landings on the

moon." Another priority was to develop systems for detecting and counter-
ing incoming ICBMs.[15] Edward Teller, who also advocated a $40 billion
investment in nuclear air raid shelters, was a member of the SAB when it
urged massive expenditures to accelerate what were already top-priority
missile development programs.

In broad terms, the proposals would be fulfilled. The United States
would develop successive generations of ballistic missiles. Reconnaissance
satellites were indeed a priority, and their importance would come to
impact the course of the nation's larger space program. Work to warn of
incoming ICBMs would yield the Missile Defense Alarm System (MIDAS) in
the coming decade, although an anti-ICBM system would remain an elusive
and controversial goal. These issues would eventually have an impact on
the Dyna-Soar program.

Staking Out an Aerospace Organization

A week following the first Sputnik launch, NACA endorsed boost-glide
flight. Its Committee on Aircraft Construction unanimously endorsed
a "manned hypersonic boost-glide rocket-propelled research aircraft
and recommend[ed] that this program be continued by the NACA at the
maximum possible rate." NACA and its successor, NASA, supported but
would never control the Dyna-Soar program, and Hugh Dryden acknowledged that "new technology and changing defense requirements result in a
need for continual reappraisal of flight propulsion systems."[16]

The aerospace vision was an ardently held belief of leading military
and civilian officials in the aviation sector. An extension of the airpower
views that had already taken root in the United States, it had been nurtured
by deliberate Air Force and media efforts that were in keeping with
American cultural legacies. Air Force Chief of Staff Gen. Thomas White
explained that "in the future whoever has the capability to control space
will likewise possess the capability to exert control of the surface of the
earth. . . . There is no division, *per se*, between air and space. Air and space
are an indivisible field of operations." To airpower advocates, this was not
a bold overreach. White explained that "the Air Force has been penetrat-
ing 'space' for several years with manned aircraft" such as the Bell X-2 and
the North American X-15 rocket research planes. His statement echoed that
of an Air Force officer who in 1954 had alluded to the X-2 flights as being
"part[ially in] space."[17]

Unity of command is frequently identified as one of the "principles of warfare." In keeping with this, White asserted that "the Air Force's missions and combat technology, manned aircraft, unmanned systems and manned spacecraft join together in compatible and complementary roles to form a *functionally complete system.*" As a result, the Air Force objected to Navy proposals for a separate military space branch: "Operational employment should definitely be vested within the appropriate service or services,"[18] and the "appropriate service" was the Air Force. Aerospace would be the term through which to express this idea.

On November 20, 1957, new Defense Secretary Neil McElroy told congressmen his idea to establish an Advanced Research Projects Agency (ARPA). It would work on behalf of the president's intent that space projects be consolidated, for the time being, above the level of the individual military branches. This proposal clearly threatened the Air Force's vision of its own role in space, and two days later, the Air Force assistant director of the office of legislative liaison offered a solution to the potential problem posed by ARPA and the "freedom of space" posture. Major General Schriever, already a renowned missile advocate, told the Senate Armed Services Committee that "if you ever separate research and development of military weapons from the users," the development of weapons systems would be unnecessarily delayed. By "jump[ing] the gun on the problem of astronautics by appointing either a Director or Assistant Chief of Staff for Astronautics," hostile entities such as ARPA might be foiled.[19]

While Air Force leaders considered a possible astronautics agency, they simultaneously sought to nudge national policy into alignment with their own ideas. Deputy Chief of Staff for Development Lt. Gen. Donald Putt sent White "a policy statement that affirmed the loyalty of the Air Force to national objectives and asserted that the control of space was essential to national security." Unsurprisingly, Putt's statement identified the Air Force as the appropriate organization through which space-related national security could be guaranteed. His reasoning echoed White's assertion that air and space could not be divided: "[there is] only one indivisible field of operations above the surface of the Earth.'"[20] White's words were gaining traction in the Air Force because they reflected a widely held idea that was in keeping with the perspective of airmen. The perceived need for space-oriented defense projects and the presumption the Air Force should be responsible for them reinforced the conviction that the establishment of an astronautics office would be needed.

That office would need a home. Irritation in Air Force headquarters with the proliferation of assistant chiefs of staff precluded the creation of a new astronautics office at that level. White decided to place the office under the deputy chief of staff for development. On December 10, 1957, Putt named his deputy director for research and development, Brig. Gen. Homer A. Boushey, to head the new Air Force directorate of astronautics. The civilian bosses quickly snuffed out the Air Force's initiative. The service continued to insist that it needed an agency ready to liaison with the pending ARPA; the civilians relented after months of negotiation and on the condition that the word "astronautics" not appear in the office's title. The agency, which would be established on July 29, 1958, would be called the Directorate of Advanced Technology.[21]

Although civilian authorities had proclaimed in December 1957 that "astronautics" was a politically intolerable term for the nascent Air Force agency, the Directorate of Research and Development continued to think in astronautical terms. A statement on the service's space program, dated January 24, 1958, listed the missions deemed "essential to the maintenance of our nation and prestige" under the heading of "Air Force Astronautical Program." The programs listed were even more telling.

The Air Force subdivided five space programs responsible for fulfilling a variety of missions into twenty-one discrete categories. The first program was "ballistic test and related systems," aimed at space research; the second was a "manned hypersonic research system," which was to start with the X-15 rocket-powered research plane. Dyna-Soar, third on the program list, accounted for fully one-third of the USAF space subdivisions. Its first subdivision was a "manned capsule test" to demonstrate manned space flight. The remaining subdivisions turned more specifically to the ambitious winged vehicle: a "conceptual test" to build on the capsule, a "boost glide tactical" system for "weapon delivery," a "boost glide interceptor" and a "satellite interceptor" for "countermeasures," a "global reconnaissance" system for "reconnaissance," and finally a "global bomber," identified as having a "reconnaissance" mission. The list's last two items were the WS-117 reconnaissance satellite program and a "lunar system," culminating in a "manned lunar base," that would also be tasked with "weapon delivery reconnaissance."[22] Air Force intent demonstrated the wide gap between its interpretation of "strategic reconnaissance" and what Eisenhower and his confidants saw it to be.

The Air Force did not shy away from declaring its interest in developing space vehicles for reconnaissance and for combat, both offensive and defensive. Concepts such as "armed reconnaissance," by which fighter aircraft scanned enemy territory looking for targets and attacking them, indicated the marriage (for the Air Force) of the reconnaissance role with an inherent connection to combat preparation or even to combat itself. In this context, the "reconnaissance" requirement for the "global bomber" version of Dyna-Soar becomes unsurprising.

Space seemed an inevitable arena for future conflict and an invaluable "high ground" from which to dominate the Earth below. "Obviously," Power argued, "the Soviets would use satellites for similar purposes" of reconnaissance and bombing, and this in turn raised the need for developing anti-satellite systems.[23] Air Force planners assumed that the superpowers' arms technologies would grow ever more potent and sophisticated. Sound policymaking, from this perspective, meant providing ample funding for an unyielding cycle of weapons research, development, and acquisition.

The requirement for Dyna-Soar as a "hypersonic conceptual test vehicle system" emerged simultaneously with the service's astronautics program. Air Force development requirement documents explained that "the ultimate objective [of Dyna-Soar] is to have a manned or unmanned global recoverable bombardment/reconnaissance system available at the earliest possible date." An eye would be kept on the development of "future boost-glide systems," controllable reentry, orbital test flights, and potential mating with a future nuclear rocket. Spin-off applications were welcome, but the primary Air Force objective was "rapid accumulation of data for boost-glide weapon system development." Boeing representatives had recently learned "about the NACA's desires in a research vehicle apart from the requirements that the Air Force might establish for Dyna Soar I. . . . They were told that the NACA is now looking at the effect on the design of trying for higher speeds, and that NACA is very much interested in having prospective designers examine this same question. The Boeing visitors agreed."[24] The Air Force and NACA were beginning to march in diverging directions with respect to Dyna-Soar, just as some people in and outside the administration were considering the transformation of NACA away from serving as an aviation research assistant to the Air Force and instead standing as its civilian rival in space.

The astronautics program, the Dyna-Soar development requirement, and the record of the NACA-Boeing discussion bring significant points into sharp relief. Dyna-Soar was the basis of a broad aerospace concept, filling a crucial niche and providing a foundation upon which even more grandiose

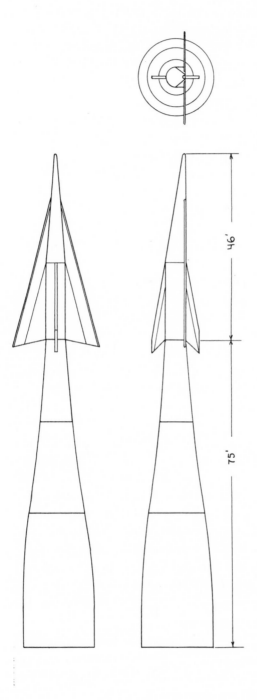

DYNA SOAR

CONCEPTUAL TEST VEHICLE

WADC DESIGN STUDY. 1084

LAUNCH

CONFIGURATION

DRAFTSMAN O. Leon5	DATE MARCH 1958	NAME	U. S. AIR FORCE
CHECKER			WRIGHT-PATTERSON AFB DAYTON, OHIO
ENGINEER	march 1958		
EXAMINER			DWG. NO.
APPROVAL	march 1958		P58B1001

| SCALE 1:120 | 58WCLS4117 |

The contours of Dyna-Soar remained in flux as of March 1958, barely six months after the project's consolidation. This sketch shows the designation of a "conceptual test vehicle," as planners worked to thread the needle between NACA/NASA's research priorities and the Air Force's operational aspiration. *Air Force Historical Research Agency* systems could be built. Dyna-Soar was to fill the very wide gaps: complementing the weapons delivery and reconnaissance functions and carrying the responsibility for countermeasures against hostile space activity. The fate of the Air Force's space aspirations would factor crucially into the fate of the Dyna-Soar program. NACA's engineering tradition of mastering aerodynamic challenges led it to focus on the flight research aspects of Dyna-Soar. Policymakers quickly saw NACA as raw material for building a space administration beyond military control, and this would substantially alter the relationship that had existed between the Air Force and the civilian aeronautics researchers.

NACA continued to seek a role, asserting that the Air Force's team "cannot adequately handle such a project" and pointing to ARDC's interest in NACA involvement. Still, NACA intended to split authority (with the advisory committee overseeing applied research and the Air Force being responsible for weapons aspects) while concentrating program finance and administration. Air Force representatives were intent on seeing the program first and foremost as a weapons system, meaning that flight research was a necessary precondition but not a focus.[25] The definition of a "conceptual test vehicle" straddled somewhat conspicuously between the NACA preference for research vehicles and the Air Force desire for something short of a prototype system leading to an operational weapons system.

NACA recognized the Air Force's mounting determination to see Dyna-Soar as a state-of-the-art weapons system rather than simply as a revolutionary flying craft. An internal NACA memo noted that the Air Force had "invented" the idea of the "conceptual test vehicle . . . to assure that the requirements for both bombing and reconnaissance would be considered. Since the initiation of the DynaSoar project, the account has been continually shifting toward the weapon system and away from the exploratory aspects."[26] NACA officials were concerned that the aerodynamic research elements of HYWARDS might be forgotten by an Air Force focused on Brass Bell and RoBo, now embodied by Steps II and III of Dyna-Soar.

An accord came in May, when Air Force chief of staff White and NACA director Dryden signed a joint memorandum. In keeping with Air Force priorities, Dyna-Soar existed to "develop the military potential of hypersonic boost glide" flight and pave the way for future usable refinements. "Overall technical control," as well as financing of the vehicle's design, construction, and testing would be Air Force responsibilities. NACA would provide its "advice and assistance" to the Air Force.[27] A month after the agreement, Air

Force personnel worried that the pending National Aeronautics and Space Act establishing a civilian space agency would muddy the waters. "The lack of clear cut distinctions" in authority between military and civilian institutions "may well inhibit sound planning, programming and operations of NASA, DOD, or both," and some in the Air Force saw NASA encroachment as an unwelcome possibility. However, language in the memoranda of understanding changed only to reflect the new NASA acronym.[28]

The suspicions of the technological determinists were confirmed in the spring of 1958. U.S. scientists had eliminated 95 percent of the fallout from large thermonuclear weapons, leading advocates to believe (and nuclear arms critics to fear) that a resort to nuclear arms in the midst of crisis would be judged practicable. In May, Sputnik III (weighing 1.5 tons and carrying a dozen scientific instruments for measuring radiation, electromagnetism, and meteoric particles in low Earth orbit) was, according to Symington, a "warning for all but the blind to see—a ton and a half of Soviet technology circling our country at will."[29] Those who were converts to the aerospace perspective interpreted current events and trends as unquestionable proof of the wisdom in moving forward with projects including Dyna-Soar.

In December 1957, the Army Ballistic Missile Agency had proposed "consolidation of military and scientific space objectives," with Army rockets predictably playing important roles as satellite boosters. Putt quickly decided to rush a test phase of Dyna-Soar, demonstrating human survival in space by placing a man in a specialized canister shot into a short ballistic space journey.[30] LeMay was soon briefed on three alternative courses for putting a human into space: a vehicle developed from the still-not-flown X-15 research plane, an accelerated Dyna-Soar program, or a manned ballistic capsule identified as a demonstration step toward making Dyna-Soar.[31]

NACA representatives discussed the situation in March, agreeing that "a recoverable manned vehicle in orbit is of primary national importance and that the ballistic type of vehicle . . . offers the most promise for an early solution to the problem." Consistent with their earlier interest in the Dyna-Soar program, NACA's Hartley Soule and his associates noted a glider's long-term advantages: "The use of lift offers a possible means for reducing the severity of the heating and deceleration problems" as well as operational flexibility. However, without criticizing Dyna-Soar, NACA favored a simpler system using a ballistic trajectory (lifting from a rocket booster and simply falling back earthward) as the best "early solution." In bilateral talks, NACA representatives focused on "operational and engineering aspects," whereas Air Force planners saw both a man in space demonstration and the hypersonic flight regime as prerequisites supporting an armed Dyna-Soar project.[32]

This "Man in Space Soonest" (MISS) project would spin off to become the Mercury program, the unlikely military utility of which would permit its being shifted from the Air Force's control to that of NASA in October 1958. The genesis of the Mercury capsule, recognizable in the Air Force's January 1958 astronautics program, was a capsule preceding Dyna-Soar intended to demonstrate human capacity to survive in space. The Air Force's interest in and use for the program were therefore confined to the aspect that least impressed NACA personnel. Though astronauts would be the face of NACA's successor agency, engineering remained NACA's concentration for the time being. The Air Force's space interest remained fixed on the actions that could be undertaken to strengthen the nation's deterrent posture. Many space advocates presumed that humans would probably have a role to play, in person, in securing the nation's defense.

Giving Voice to the Aerospace Concept

Influential service voices strove throughout 1958 to describe the Air Force they envisioned. "Soon we may have missiles launching space vehicles and space vehicles launching missiles," Twining told an audience at the National War College in March.[33] In an article for the journal *Ordnance*, LeMay placed Dyna-Soar in the context of the Air Force's vision of the present and future. Noting that the Soviet Union unveiled a new jet bomber two months after Khrushchev had claimed that "fighter and bomber planes can now be put into museums," LeMay explained that "before manned bombers can be completely replaced with unpiloted systems, unpiloted systems must possess" the ability to react to enemy action, deliver firepower, penetrate enemy defenses, and be selective in finding and striking targets. Precision was another important factor. "The speed and range of manned bombers and the wide range of tactics which can be employed permit great flexibility of operations," and manned aircraft would not be obsolete until unmanned systems could match such capabilities. LeMay pointed to the B-70 chemical-fuel bomber as something to increase the capability of manned systems.

LeMay then shifted his outlook to the more distant horizon: "Looking further into the future, the Air Force has under development, Dyna-Soar, the first 'space bomber' which will be capable of circumnavigating the

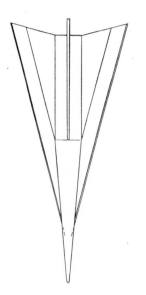

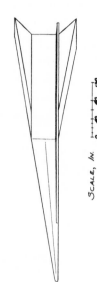

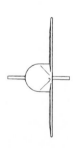

LENGTH 77 IN.
DIAMETER 67 IN.
WING
  AREA 421 SQ. FT.
  SPAN 264 IN.
  ASPECT RATIO 1.15
  ROOT CHORD 45 IN.
  SWEEP—L.E. 75 DEGREES
       —T.E. 14 DEGREES
  RADIUS—L.E. 2 IN.
  THICKNESS 6 IN.
DORSAL
  AREA 37 SQ. FT.
  HEIGHT 39 IN.
  ROOT CHORD 290 IN.
  TIP CHORD 43 IN.
  SWEEP—L.E. 63 DEGREES
       —T.E. 43 DEGREES
  RADIUS—L.E. 2 IN.
  THICKNESS 2 IN.
VENTRAL
  AREA 39.5 SQ. FT.
  HEIGHT 39 IN.
  ROOT CHORD 290 IN.
  TIP CHORD 63 IN.
  SWEEP—L.E. 43 DEGREES
       —T.E. 43 DEGREES
  RADIUS—L.E. 2 IN.
  THICKNESS 6 IN.
M.A.C. 306 IN.
TOTAL LENGTH 580 IN.

SCALE, IN. 0 40 80 120

DRAFTSMAN
CHECKER
ENGINEER
EXAMINER
APPROVAL
DATE   NAME

DYNA SOAR
GLIDER
THREE VIEW

U. S. AIR FORCE
WRIGHT-PATTERSON AFB
DAYTON, OHIO
DWG. NO. D578 B 1002

A "three view," from ahead, above, and beside, of the Air Force's intended pioneer space bomber. Many aspects of the vehicle shape would change substantially, as the project progressed and encountered technological critiques (notably Phase Alpha) between this sketch from March 1958 and Boeing winning the contract in December 1959. *Air Force Historical Research Agency* earth at extreme altitudes and attacking its target with space-to-surface missiles."[34] Although stereotyped as an obstinate advocate of the old bomber over the new missile, LeMay was instead a defender of the idea that piloted systems outperformed wholly automated ones. Eventually the B-52 and even the Mach 3 B-70 bomber would be in air museums. But "it will not be because the man was eliminated" from weapons systems; rather, "new equipment [will have been] perfected to propel him higher, drive him faster, keep him in motion longer and so enable him to perform more tasks better."[35] Dyna-Soar seemed to promise a continuing, productive role for man in the "missile age." Like LeMay, White told the National Press Club that "if manned systems can perform some tasks better, then we want manned systems; if unmanned systems can do the job, then we will be the first to accept and use them."[36] White's advocacy of Dyna-Soar reflected a belief in the potential of a crewed vehicle with the speed of the missile.

Air Force officials used the *Air University Quarterly Review* as a vehicle to explain the relationship of piloted vehicles and missiles. One colonel cautioned readers early in 1958 that "we must agree that missiles are welcome and compatible additions to the Air Force family of weapons, but it is incumbent on all airmen to keep them in perspective." Another author expected decisive impact from missiles but fretted that "boredom and monotony will be a constant companion" of missile base crews during peacetime.[37] Schriever addressed the National Geographic Society in November, and while he predicted "for the foreseeable future a missile and bomber 'mix,'" he meant a force including both missiles and aircraft.[38] LeMay, among others, would vocally support an appropriate "mix" rather than the abrogation of missile development or the wholesale abandonment of aircraft. When coming online, Dyna-Soar would nest with other Air Force systems.

Isolated thinkers in the Air Force questioned the top leadership's interpretation of national security requiring a continuous aerospace force. One colonel, Martin Schofield, argued for the United States to establish international control of space from a position of strength as a leading aerospace pioneer.[39] However, the current within the service favored the aerospace vision enunciated by its chief of staff. Schofield's article encountered prompt refutations in the next issue by a pair of two-star generals, one of whom cited Chief of Staff White's assertions about the "high ground" of space and claimed that "in our time this vantage point may become the surface of the moon itself."[40] While officers scuffled on the pages of the

Air University Quarterly, Boushey pontificated on the presumed military potential of the moon, first in a speech to the Aero Club in Washington, DC, and then in an article for the *Army-Navy-Air Force Register* a week later. His article, titled "Lunar Base Vital," proclaimed that "the moon provides a retaliation base of unequalled advantage" for either the United States or the Soviet Union, depending on how they paced their space exploits.[41]

Boushey repeated this view in April, telling congressmen that "the moon could be used as a launching site for deeper penetration of space, as a supply base for earth satellites, as an astronomical and meteorological observatory, and as a means of world-wide surveillance that could be a deterrence to aggression."[42] Boushey's claim would encounter some detractors, but the idea of a moon missile base—a literal manifestation of "lunacy," as it in essence was—found its defenders as well. Before the moon could be used, there needed to be a way to get there; lunar ambitions and Dyna-Soar plans fit into the larger context of using and defending the continuum of air and space.

Boushey's office unveiled the Air Force Astronautical Program in late January, and the ARDC spent the following seven months developing a cluster of study requirements for space defense. Six of these seven projects fell within three overall systems: a "strategic orbital system," a "strategic lunar system," and a "strategic interplanetary system." A seventh study requirement was for a reconnaissance satellite serving "as a possible support system—along with the photographic satellite of WS-117L, the meteorological satellite, man-in-space, and Dyna Soar—for the strategic orbital system."[43] This presumed application for Dyna-Soar indicates the divergent directions in which the program was heading during the late 1950s. In keeping with the ARDC's study requirements, Dyna-Soar would be a multifaceted vehicle capable of helping support a more permanent orbital system.

Yet Air Force approaches to its funding issues simultaneously pulled Dyna-Soar in another direction. In February 1958, Air Force and NACA representatives worked to draft a memorandum of understanding about the Dyna-Soar I project. Dyna-Soar I was the test bed, providing a foundation from which development of Dyna-Soar II (reconnaissance) and III (bomber) would follow. A February draft noted that "the conceptual test vehicle [Dyna-Soar I] has to be related to a weapon system to expend weapon system funds." Planners wanted NACA as an "active partner" in addressing the technological challenges.[44]

NACA's focus on flight research and the Air Force's prioritizing of the weapons phase of development placed NACA in a strange position. In February 1958, its staff urged director Hugh Dryden to insist that, despite the assistance that the ARDC had recommended NACA provide, "the Dyna Soar I must be more closely related to the Weapons Systems which will follow it than were the X-15 and the preceding research airplanes."[45] A memorandum of understanding the following month confirmed this relationship: "Overall technical control of the project will rest with the Air Force, acting with the advice and assistance of the NACA."[46]

Air Force control of Dyna-Soar was understandably important to its exponents. Over the course of 1958, ARPA emerged, and NACA transformed into NASA. Authority for a number of military service space projects began shifting toward ARPA and eventually to NASA. Not all of these transfers were permanent, and a contemporary Air Force history noted that "the Air Force was pleased to have" authority over its research on nuclear propulsion in space, high-energy fuels, effects of weapons on electronics, the 117L satellite, and Project SCORE (the orbit of an Atlas missile fuselage and the first voice broadcast bounced from space). "Nevertheless, the Air Force was seriously disturbed," the history explained, "by ARPA's persistent splintering of projects into components."[47]

Dyna-Soar's funds needed to be secured. On September 8, White wrote to his deputy for development, General Putt. Having learned that ARPA entertained transferring $10 million of its own money invested in MISS to instead fund Dyna-Soar, White concluded that NASA aimed to take control of the "manned capsule test" project that the Air Force in January 1958 had identified as a subdivision of the Dyna-Soar program.[48] Through MISS, the Air Force aimed to demonstrate a human's ability to *survive* in the new environment as a precondition for his *working* in it.

Dyna-Soar won special attention when White submitted his list of priority programs to Air Force Secretary James Douglas on October 3. Several systems, including the solid-fuel Minuteman ICBM, the KC-135 tanker, the B-52/GAR-77 bomber/missile combination, and strategic airlift resources each got prominent attention in turn. Then White turned to Dyna-Soar: "*This program represents the first of a new generation of strategic weapons systems* [emphasis added]. It will enable the United States to maintain its military posture of major war deterrence by having in-being a known capacity to nullify air defense systems designed to combat air-breathing manned vehicles and predictable ICBM trajectories."[49] The chief of staff's description indicates that, in contrast to later criticism, the Air Force held a clear and bold vision of what Dyna-Soar would do: serve as the successor of the modern manned bomber planes and mark a point from which a new era in military aerospace would be inaugurated.

And yet when civilian authorities seemed critical of the program, the Air Force would deliberately seek to obscure the envisioned technological potential of the craft. This explains the inaccurate interpretation—that the Air Force lacked clear ideas about Dyna-Soar's role—that later took hold when actually the Air Force intended to make it a space-traversing offensive weapons system capable of pioneering reconnaissance and other duties in the expanding fringe of navigable sky. Convinced that this national security objective would win acceptance but that an isolated flight research project would be poached away from the military as MISS had been, the Air Force married the Dyna-Soar I conceptual test craft to the Dyna-Soar II and III weapons systems.

The Air Force did not want to lose its space glider to ARPA or NASA, and this goal prompted the Air Force's first major adjustment of Dyna-Soar's definition. The idea originated with Secretary Douglas, the man who had ordered the directorate of astronautics (a direct and deliberate Air Force competitor to ARPA) dismantled. He replied to White in mid-October, noting that ARPA was claiming authority over projects "only . . . which are designed to attain or exceed velocities necessary for stabilized orbits." Presumably, therefore, "this might, in effect, constitute an arbitrary but clearly defined boundary on one side of which development projects are clearly the responsibility of the Air Force whereas, on the other side, the policy and program direction responsibility becomes the responsibility of the ARPA." Therefore, he continued, it "becomes advantageous to clarify the development objectives of the Dyna-Soar Program."[50]

Douglas' suggestion went as follows: "Since the purpose of the Phase I is to obtain hypersonic research data and accumulate research and development experience in this kind of vehicle, it should clearly be designated a military research vehicle programmed for *sub-orbital* velocities." By temporarily ceding the space arena to ARPA, Dyna-Soar could proceed. Air Force officers might quickly object, however, that a sub-orbital Dyna-Soar I answered NACA's area of interest while scorning that of the Air Force. Anticipating the reservations, Douglas insisted that "this would not be a handicap to the project, and modest growth could at some future date meet any military purpose requiring continuous orbiting flight.[51] Ostensibly, this course of action would ensure that Dyna-Soar I was not "splintered" away from Dyna-Soar II and III as the MISS human demonstration had been from the Dyna-Soar program. The Air Force's highest ranking officers were keen on avoiding what they identified as a very real danger. Splintering was bringing the 117L satellite to grief, and the human demonstration phase of Dyna-Soar had just been lost to NASA. If the same happened to the hypersonic flight research elements of the program, what would be the future of the Phase III bomber?

However, "safety" had been bought at an extremely high price. First, the intimate linkage of Dyna-Soar I to its weaponized successors meant that any national policy precluding space weapons systems would pose an existential threat to the *entire* program. This risk might have seemed minimal at a time when a potent Soviet Union seemed capable of surprises in space and might be secretly at work on more sinister capabilities. Certainly, President Eisenhower had expressed hope that space might be preserved for peace. But in cold war, as in war itself, policy is not fixed but rather is a product of interaction.

The second part of the price for keeping Dyna-Soar I under Air Force auspices was that the Air Force declared it to focus on research of hypersonic *suborbital* travel. Douglas' letter to White assured the chief of staff that this definition "would not be a handicap" because it could easily be reversed at a future date. But the purpose of Dyna-Soar I seemed abruptly less ambitious, and it suddenly seemed less distinctly different from the pending X-15 project. Artificial ambiguity about the program's purpose caused nagging problems in the face of skeptical inquisition. In short, the Dyna-Soar program's wings were clipped in 1958; they were clipped by the Air Force itself, at the behest of its secretary, on the premise that the act would protect their program from abduction by new civilian space research agencies.

Whether Douglas' October 1958 suggestions secretly aimed to trap Dyna-Soar cannot be definitively known. However, his prompt destruction of the Directorate of Astronautics lends strong reason to believe he aimed to damage the program. Suborbital redefinition temporarily contained Dyna-Soar's development, buying time for Eisenhower's space policy but without forcing the president to cancel Air Force work outright.

Although the Air Force accepted Douglas' suggestion about couching Dyna-Soar I in the suborbital realm in the short term, its officers were determined to promote the aerospace vision that they were confident would ultimately define security issues in the skies. Therefore, Air Force rhetoric emphasized expansive continuity not only between air and space, but also among the vehicles designed to traverse them. LeMay, speaking at a public information seminar banquet in November 1958, described "the X-15 rocket research plane" as "our first 'spacecraft.'" Noting that it was a continuing joint effort with the Navy and NACA (now NASA), he added that "the experience of the United States Air Force in working with science and industry for the development of faster, higher-flying aircraft with longer range gives us confidence in the future of our equipment in space."[52]

LeMay told television audiences that "we use the term aerospace" because "space is actually just a continuation of the air in which we now operate. Since there is no definite boundary between the two, it is very difficult to determine where the functions of an aircraft stop and those of a spacecraft begin." Noting that the X-15 was "a research vehicle . . . never intended to have a combat capability," it gathered data to support "the design of our future combat aerospace vehicles."[53]

The sticky question of continuity bedeviled civilians, too. NASA Assistant Director of Research Ira Abbott attempted to explain to a House subcommittee the fuzzy line separating "air" from "space." He told them that "advancing technology combined with intensive international competition for faster and longer range aircraft and for missiles to cause revolutionary developments in aeronautics" early in the Cold War. "Gradually," Abbott explained, "the 'frontiers of aeronautics' became the 'threshold of space.' From the point of view of research no clear cut distinction can be drawn between aeronautics and space[,] as may be illustrated by the experimental X-15."[54]

In ironic parallel, a review conducted by the *Air University Quarterly Review* and published in their fall 1958 issue focused on the X-15 as it explored "the spiral toward space." Its closing section, titled "The Future," focused exclusively on Dyna-Soar, which was described as "the project which is perhaps closest to the continuation of the X-15." The Air Force journal's contributors and editors were clearly struggling (and failing) to remember NACA's new name and structure. "The Air Force and the NACA [sic] are now engaged in developing this advanced test vehicle, thus continuing the productive partnership that resulted in the earlier research aircraft and the forthcoming X-15."[55] Significantly, NASA had been structured for the purpose of answering public outcry for space activity and for determining that space efforts, at least for the time being, were *not* dominated by the military services. As a result, misidentifying NASA by its predecessor's name was much more than a slip of the tongue. It implied a lag in understanding that NASA was the Air Force's competitor in a way that had been neither the case nor the intention with NACA.[56]

The staff review highlighted Dyna-Soar coming into its own, noting that "preliminary investigations indicate that it will be possible to vary the original thrust and thus the velocity of the Dyna-Soar, enabling the pilot to complete one or more orbits around the earth and make a normal landing. . . . It will utilize both centrifugal effect and aerodynamic lift."[57] This orbital potential lay far beyond anything that the X-15 might approach and belied the disingenuous nature of the Air Force's "suborbital" Dyna-Soar. The staff review explained that the Air Force saw the vehicle's capability as a function of its booster rocket.

The review also looked to Dyna-Soar's future, declaring that "possible operational craft following the test version of Dyna-Soar will be capable of various missions, including bombing and reconnaissance. As a weapon this vehicle promises to have some of the best features of both guided missiles and aircraft." Specifically, the authors pointed to "long range and great destructive capability, especially with nuclear payloads," galvanized by the pilot's human judgment. "Such a weapon would be particularly useful in case of loss of our overseas bases." The essay finished by acknowledging that the future is an unknowable realm, but "we can say with some assurance, however, that the information gained from the flights of the X-15 and similar aircraft will be extremely valuable. In fact, later craft in this same line of development may well play the predominant role in the conquest of space."[58]

Meanwhile, space issues remained significant to Air Force leaders. In February 1959, White reasserted to the House Committee on Science and Astronautics an idea that he had been enunciating forcefully since the winter of 1957–58: "Since there is no dividing line, no natural barrier separating these two areas [air and space], there can be no operational boundary between them. Thus air and space comprise a single continuous operational field in which the Air Force must continue to function. The area is aerospace."[59] In March, White was pleased by indications from ARPA director Roy Johnson that that agency desired "to farm out military space programs to the services as soon as feasible in order to reduce the budgetary impact on ARPA. . . . I want to make certain that the Air Force is in these programs to the full extent that its missions demand."[60]

White recognized that ARPA's motivation meant "transferring the [financial] impact to the services," and in this light he thought "it might be wise to sacrifice in some areas in order to be able to take on space projects."[61] At the end of May, contemplating the "military use of space" in the context of the fiscal year 1960 budget, White reiterated that "we consider that space is contiguous and an undivided medium with the atmosphere. The requirement to operate above the earth's surface is indistinguishable in our eyes, whether it be in the atmosphere or above the atmosphere. It is one medium as far as we see it from a military point of view."[62] White's statements show a definite consistency in their allegiance to the idea of "aerospace" and to the belief that the Air Force's expanding defense responsibility was simply a "continuing" one.

LeMay told an Air Force Association convention that "aerospace power by its very nature is global" and that "the speed, range, mobility and flexibility of aerospace power—which are constantly being improved by modern technology—allows [sic] us to take full advantage of this worldwide medium and to expand the scope of Air Force operations." Early the following week, responding to a question about aircraft during a radio interview from the Pentagon, the vice chief of staff reminded the interviewer and listeners that "rather than manned aircraft, I think we should refer to them as manned vehicles. Our advanced aerospace vehicles don't look much like aircraft as we now know them. But whatever these machines are called, I feel that their future is unlimited."[63] LeMay did not single out Dyna-Soar by name, but the piloted craft was clearly on his mind.

The same September day that LeMay continued to speak about aerospace power, White moved to accelerate progress on Dyna-Soar. But at just this point, the Dyna-Soar program faced another round of challenges. "We have a contract pending on Dynasoar," White noted, and "industry is anxious to get settled" whether Martin would work in combination with Bell or with Boeing on the project. "Let's get [the] Sec[retary of the] A[ir] F[orce] to get on with it." White's assistant executive informed the deputy chief of staff for development that "General White desires that the pending contract on Dyna Soar be expedited," a point that the chief of staff reiterated on September 5.[64]

Dyna-Soar was a priority for General White, but it was not the only program vying for the service chief's attention, and it was not even the only space project on the horizon. Schriever rushed separate letters to White that he was "deeply concerned" about the "severe administrative and funding restrictions on the MIDAS development program" and "the events which are besetting the SAMOS Program [the Satellite and Missile Observation System, a planned reconnaissance satellite]."[65] Such calls did not preclude White from returning his attention to Dyna-Soar's progress the following month, however.

Vehicle weight posed a crucial issue, especially because of the weakness of contemporary rockets. Dyna-Soar was expected to weigh about ten thousand pounds, more than triple the weight of the heaviest craft yet put into orbit (Sputnik III). What if Dyna-Soar could be sliced into two pieces: a test vehicle weighing five thousand pounds, and then the full-weight vehicle along the lines the Air Force envisioned? Air Force Assistant Secretary for Research and Development Victor Charyk, saying on September 25 that the Titan C booster would not be developed, "suggested a 'two-phase' program, but with a different glider."[66]

Major General Victor Haugen, deputy chief of staff for development, met with several colonels in early October and declared Charyk's proposal to be "a wasteful and very undesirable" option. Waste was one problem, and dead ends were another; civilian bosses had already indicated hesitancy when director for defense research and engineering (DDR&E) Herb York minimized the military subsystems' testing that April, and the officers may well have recognized that the larger operational vehicle would be in danger if not married to the test vehicle. The colonels urged that Dyna-Soar remain at its envisioned size, and Haugen agreed, also noting that he "did not believe that management would be changed sufficiently to require a new competition," since this too would slow the program and risked offering another opportunity to stymie it altogether. It was noted that "General [Roscoe] Wilson said he had given ARDC and AMC ten days to get together on a management plan for Dyna Soar."[67] The alternative to building a scaled-down "5,000–6,000 lb. airplane" in keeping with DDR&E objectives was to start with "a 9,000–10,000 lb. design vehicle off loaded to weigh 5,000–6,000 lbs. at launch and boosted by a modified ICBM." Thus, "the test phase of this step could evolve into an initial operational capacity" as Dyna-Soar II and ultimately "a military system based on Dyna Soar technology," Step III.[68]

Even if necessity demanded that the full-scale, full-weight Dyna-Soar could not be lifted imminently, the Air Force determined to hollow out its vehicle rather than construct a smaller test variant. Charyk wanted to authorize a contract for the fiscal year 1960 part of only Dyna-Soar I, and this action would certainly have served to corroborate Air Force suspicions.[69] In the eyes of its key thinkers, Air Force planning—and U.S. sovereignty—depended on the development of operational aerospace systems. According to this perspective, the Air Force had to operate across the aerospace realm. Vehicles needed to be operational weapons systems rather than to end simply as test platforms. Dyna-Soar stood to be the first in a line of new aerospace defense systems.

Between the Sputnik orbit and the latter months of 1959, high-ranking Air Force officers strove to express and elaborate on their aerospace vision. Although space-traversing vehicles were revolutionary technologies, the aerospace idea was consistent with earlier doctrinal thought espousing higher, faster, farther-ranging vehicles. This facilitated support within the service. Dyna-Soar was the flagship vehicle for the new age of flight. As a winged, piloted vehicle, its appearance also helped make the extension of airpower thinking seem as contiguous as the aerospace realm. Vice Chief of Staff LeMay's support for Dyna-Soar and what it stood for did not always prompt him to identify the project by name, but his boss's rhetoric was even more consistent and emphatic. White would struggle in the coming years to help make Dyna-Soar fulfill the promise that he and others saw in it. The culmination of both his directives and his rhetoric would lead to a December 1960 showdown between Chief of Staff White and the president during a National Security Council meeting, with both arguing for very different conceptions of America's aerospace future.

6

"The Air Force Must *Not* Lose Dynasoar"
Air Force Reaction to Eisenhower Policy

The aerospace is an operationally indivisible medium consisting of the total expanse beyond the earth's surface.
—Basic Doctrine AFM 1–2, December 1, 1959

Just as the Wright Brothers in their time could not imagine planes like the B-70 or modern jet transports, so it seems that man today cannot imagine the extent to which space flight will be used. . . . Dyna-Soar is one of the steps toward that future.
—Air Force booklet on Dyna-Soar, April 1960

The Suborbital Dyna-Soar

In the two years following the Sputnik launch, Air Force officers ranging from articulate colonels to the chief of staff himself had propounded the idea that the sky was a single operational medium and that space therefore belonged in the Air Force bailiwick. By the end of 1959, the Air Force would officially unveil an "aerospace doctrine," borrowing the term Chief of Staff Gen. Thomas White had coined late in 1957 to emphasize the continuity of air and space. Aerospace advocates looked to Dyna-Soar as the pioneer of the new realm taking "steps toward [the] future."[1] Thus, it was expected to serve as the first generation of piloted military technology in space—offensive military power in particular, but providing defensive and reconnaissance capabilities as well. Technological hurdles and problems directed by civilian officials caused Gen. Curtis LeMay to question Dyna-Soar's competitiveness relative to other futuristic weapons systems. However, despite obstacles, General White remained determined to uphold Dyna-Soar in the face of all challenges.

"The A[ir] F[orce] must *not* lose Dynasoar," White unambiguously said in a memo fired off to Assistant Secretary for Research and Development Victor Charyk.[2] Other military space programs had been pulled into ARPA in 1958. ARPA had in fact been created as a repository of the space projects of the military services, and the agency had reviewed them, returning some to their originating services, detaching valuable parts of others (like the future Corona program from the WS-117 satellite) first, and handing others over wholesale to NASA. The latter fate had befallen the Army's missile team in Alabama, although its researchers (Wernher von Braun prominent among them) did not mind the move and were pleased at the prospect of pursuing their vivid space designs.

So far, unlike other military space projects, Dyna-Soar had been spared. The preliminary test to use a ballistic capsule to demonstrate human capacity to survive in space had been shifted to NASA and renamed Project Mercury. But ARPA never gobbled up the glider program itself, and the civilian aeronautics organization remained generally cooperative with respect to hypersonic flight research. The Air Force had followed Secretary Douglas' advice to tie the Dyna-Soar I research vehicle to the armed follow-on variants in order to prevent further splintering of the program. Then, in the fall of 1959, the chronic paucity of rocket booster thrust raised the issue of lightening or shrinking an initial test vehicle. Since a scaled-down vehicle might easily be splintered away from the main program, planners decided to gut the inside of the initial vehicle to reduce its weight. White made clear his determination to prevent splintering of the project in his October 27 memo to Charyk.[3]

Charyk was among the few people privy to the existence of the Corona reconnaissance satellite program. It was Corona that gave the president's freedom of space policy coherence, but because it was such a secret initiative, the president's critics were convinced that the freedom of space goal was a sign of Eisenhower's inability to perceive the promise of space. Aware of Corona, Charyk recognized the complications that a progressing Dyna-Soar program might eventually pose. Technological debates also simmered, as scientists disagreed about whether flat-bottomed winged vehicles (such as Dyna-Soar) or other shapes (such as a tub-shaped "lifting body") were best suited to space missions. Experts debated the merits of the different approaches, giving Charyk cover for impeding Dyna-Soar development.

White was also concerned about turf feuds within the Air Force that were costing valuable time. The day after communicating his priorities to Charyk, White wrote to Lt. Gen. Roscoe Wilson, deputy chief of staff for development since July 1958. White complained that inefficient work on Dyna-Soar might yield long-lasting ill effects: "I am exceedingly dissatisfied with the progress made on Dynasoar." A significant factor was competition within the service: "If the Air Force expects to get into the space business we have got to get over these internal hassles which have already delayed this program an unconscionable amount. I want immediate action, and if there are impediments that can't be settled at your level let me know."[4] When Wilson cited the project's management structure as a sticking point, White's reply was immediate and emphatic: "I want: Dynasoar expeditiously settled on a realistic basis. Beyond that I expect the staff and the commanders to get the thing *settled* and implemented."[5]

By October 30, planners concluded that the most workable solution was to establish an office at Wright-Patterson Air Force Base for joint management by AMC and ARDC. The Air Force moved energetically in early November to present its refinements to Air Force Secretary Douglas, but his schedule prevented a meeting until the afternoon of November 6. At that point, "Mr. Douglas approved the Dyna Soar plan substantially as presented. However, he allowed that he wasn't committed by that approval to the exact sum of $96.0 million in FY [fiscal year] [19]61" that the Air Force planners thought necessary.[6]

Despite White's insistence that the Air Force retain control of the glider program, an internal report stated that "the loss of the Dyna Soar project to NASA appears imminent." A DOD budget review tentatively eliminated all Dyna-Soar funding for fiscal year 1961, and Eisenhower's new scientific adviser George Kistiakowsky and DDR&E Herbert York "discussed (and may have decided upon) cancellation of the Dyna Soar, with NASA to pick up the pieces as experimental in-house work." NASA, it was believed, was considering a "winged Dyna Soar-like vehicle" as part of their Mercury program.[7]

Space vehicles could not exist without a means of reaching beyond the atmosphere, so booster capability represented a significant issue. "In a 29 October 1959 high level O[ffice of the] S[ecretary of] D[efense]-NASA presentation on the SATURN (now scheduled for transfer to NASA) Von Braun justified the choice of 220 [inches] as the diameter of the second stage SATURN booster (rather than 160 [inches]) entirely on the basis of assumed DYNA SOAR requirements."[8] Rumors of NASA takeover of large booster development were true: OSD and NASA reached agreement on October 30 that there was "a definite need for super boosters for civilian space exploration purposes, both manned and unmanned," and President

Eisenhower signed this agreement on November 2.[9] The Air Force would be confined to using relatively small boosters converted from ICBMs.[10] Air Force planners toyed with the idea of potentially employing the service's solid-fuel Minuteman missile as a booster for Dyna-Soar.

The implications of losing big boosters would be staggering for Dyna-Soar. To prevent Dyna-Soar I's loss to ARPA, the Air Force had emphasized its suborbital characteristics to prevent its being stolen away as a "space research vehicle" instead of being a test platform for weapons systems. As contributors to *Air University Quarterly Review* had recognized, the power of the booster would do much to determine the range, altitude, and capability of a Dyna-Soar vehicle. Now an agreement for NASA to monopolize big booster development threatened to prevent the Air Force from being able to develop a booster capable of lifting Dyna-Soar to the heights for which it had always been intended.

Relations between the Air Force and NASA had, in fact, been deteriorating for some time. NACA had served as an extremely useful instrument for the Air Force and aviation for four decades, assisting in aerodynamics and flight research, but NASA fast appeared to stand as a competitor rather than as a helpmate. In the words of a contemporary Air Force history, "A background of unhappy incidents in NASA-USAF relations built up."[11]

The moon was a serious locus of contention. The Air Force had, in 1958, developed a series of study requirements envisioning a strategic orbital system, a strategic lunar system, and a strategic interplanetary system; prior to this, the Air Force's astronautics program had envisioned five major categories, including Dyna-Soar and a Lunar System. Throughout that year, Deputy Director of Research and Development Boushey had publicly advocated for military bases on the moon. In March 1959, ARDC had invited NASA "to participate in contractor midpoint briefings" related to the study requirement for a lunar observatory within the context of the Air Force strategic lunar system. NASA sent a single staffer, and the "response was markedly unenthusiastic." A month later, NASA created its own Lunar Exploration Group, and although the Army and Navy had taken part in its formation, the Air Force had not. Thus, the service was surprised by NASA's abrupt announcement, on April 17, 1959, of "plans for long-range scientific exploration of the moon."[12]

The *Air University Quarterly Review* fired back with space-oriented articles. The Spring 1959 issue included another article by Boushey that touted the value of piloted space vehicles able to conduct bombardment and thereby "serve as a deterrent to armed aggression, just as our Strategic

Air Command bombers safeguard the peace of the free world today." While conceding that satellites as yet lacked much payload capacity, he hastened to note that his opponents offered arguments "much like those faced by Wilbur and Orville Wright."[13]

Boushey predicted "both military and civilian" space applications, but "our very security depends upon our military capabilities." He added that "unless these men can do something useful, there is no reason for landing them on the moon in the first place. So we must give them a goal, a job to do. And it must be a job that will contribute to scientific and military colonization of the moon." The immense costs of such a program would be dwarfed by the problems of a preemptive Soviet lunar presence.[14] In an essay supporting Boushey, Lt. Col. S. E. Singer declared that in the two years since Sputnik, "the American public" had "settled back into a complacency only somewhat less rigid than that of the pre-Sputnik era. . . . It is hard to escape the conclusion that there is military sense to General Boushey's concept of a lunar-based missile force."[15] Singer hoped that a Martian atmosphere might include oxygen "that one only needs to compress . . . for use by man" and predicted that "water is probably present in sufficient quantity" as well. Singer saw the moon's absence of an atmosphere as advantageous, minimizing the destructive force of a Soviet nuclear warhead aimed at a future U.S. lunar base. Furthermore, "the topographic features of the moon, its many craters and clefts" might "provide a very large number of potential [missile launch] sites requiring little or no additional construction." Lunar mineral wealth, he hypothesized, might even help make a moon missile base feasible from an economic standpoint.[16]

"There must be a somewhat visionary or even fanciful approach to the future as well as a conventional one," Singer wrote, aiming to justify his wild expectations. The optimism in each of these assertions is striking, and their cumulative effect is almost overpowering. These authors, and their audiences to whom they repeatedly spoke and wrote, exuded a remarkable amount of both technological optimism and darkly determinist pessimism about security policy. Utilizing space was an "overwhelming urgency" because "for each new offensive weapon, a military opponent will attempt to provide a defensive or counterweapon. . . . Each new fantastic space weapon will probably generate a requirement for an equally fantastic defensive weapon."[17] There would be no ultimate weapon but rather a technological spiral. LeMay would make similar points in addresses, indicating that he did not expect nuclear weapons to be "ultimate"; the process of innovation continued, just as security needs did.

Aerospace proponents added a note to the emerging debate about limited warfare. Critics of President Eisenhower's New Look and its reliance on nuclear retaliatory power had voiced concerns throughout his tenure, but these criticisms gained traction in his second term. Critics from the services, such as newly retired Army Chief of Staff Gen. Maxwell Taylor, advocated a "flexible response" with a more even balance between atomic and conventional forces as an alternative to the rigidity implicit in Eisenhower's policy of massive retaliation. Eisenhower's policy emphasis on nuclear weapons and deterrence, combined with his awareness of the dangers connected with federal deficits, had bitten deeply into conventional forces' budgets in general and the Army's in particular. By preparing for a range of conflicts, Taylor argued, the United States could more effectively counter aggression; revitalizing the conventional forces stood prominently in this outlook. Taylor retired in July 1959 and published *The Uncertain Trumpet* outlining this plan in January 1960. Kissinger's *Nuclear Weapons and Foreign Policy* had already advocated U.S. acceptance of attritional limited nuclear war on the dubious assumption that Soviet policymakers would watch their country crumble but refuse to raise the ante by escalating a limited nuclear war. Some in the Air Force similarly sought to find ways for effectively using "nominal-yield nuclear weapons" since "we must not deprive ourselves of the unique advantages offered by imaginative employment of nuclear weapons."[18]

A very different, but still fantastical, vision of limited war appeared from some corners in the Air Force. Singer's 1959 article in the *Air University Quarterly Review* had proclaimed that "since Korea the term 'limited war' has enjoyed great popularity." Like Kissinger, Singer explained that limited wars had in fact been a mainstay in history: "David and Goliath fought a limited war, though larger forces were available to settle the issue at stake. . . . The history of warfare is a history of limited wars. Whether this has owed more to man's inability to wage truly total war than to his desire not to is surely debatable. Modern nuclear warfare makes this issue irrelevant," since nuclear weapons introduced that capacity. Additionally, "the possibility of miscalculation make[s] limitation a doubtful and dangerous concept."[19]

But Singer did not banish the thought; indeed, he aimed to salvage it. "At first blush the use of the moon as a locale for conducting limited war is surely somewhat fantastic," he explained. However, "space warfare as a form of limited warfare is a possible solution to this dilemma," since "the moon and its environs are close enough to the prospective combatants to be convenient but far enough from the inhabitants of earth to ensure their safety. . . . The possible annihilation of mankind in a total nuclear war is also fantastic. Perhaps a fantastic problem requires a fantastic solution."[20] A contemporary Air Force history of its space efforts envisioned "a self-contained, manned military space force with multimission capabilities" as "an ideal deterrent. . . . Elimination of this force by the enemy would be prerequisite to attacking the nation, and the outcome of the aerospace battle could well be decisive without involving surface forces."[21]

Civilian agencies challenged Air Force ambition. In June 1959, ARPA informed the Air Force that some funds previously slated for SAMOS and MIDAS would be transferred instead to other space projects, and ARPA simultaneously "directed a major technical reorientation of Samos" while rejecting Air Force plans for the next phase of development of MIDAS. The Air Force had to work to regain managerial control over these projects.[22] Eisenhower's reorganization of the defense structure in 1958 prompted the creation of the position of DDR&E. Henceforth, military space projects would be compared to earth-based alternatives, invalidating by implication the idea of a diversified space-based deterrent. Although "the Air Force retained responsibility for determining requirements to satisfy its assigned mission," it faced challenges in persuading "higher echelons to approve, fund, and assign the projects necessary to meet these requirements." Lt. Gen. Roscoe Wilson, deputy chief of staff for development, feared that "all services will bog down in red tape."[23]

Air Force planners had sought late in 1958 to prevent the outright seizure of Dyna-Soar by ARPA by characterizing Dyna-Soar I as a hypersonic, high-altitude, *suborbital* research vehicle preceding the Dyna-Soar II and Dyna-Soar III weapons systems. Perhaps deliberately, Secretary Douglas had led his service to be too clever by half. In mid-April 1959, York directed what the Air Force recognized as "certain fundamental changes." Dyna-Soar I would now focus on hypersonic flight at a velocity up to 22,000 feet per second (15,000 miles per hour), and work toward an orbital capability and development of military subsystems could occur only if these activities did not interfere with its redefined focus on manned, hypersonic, maneuverable research.[24] The Air Force could keep Dyna-Soar, but York's directive had transformed the Air Force's raison d'etre for Dyna-Soar into an incidental goal.

Air Force Magazine editor Claude Witze provided the Air Force with NASA's reassuring answers to his recent questions. "There is no desire, or intention, on the part of NASA management to expand the civilian space agency's role" to the extreme detriment of the military, and "it is necessary for NASA to demonstrate . . . and for the Military Services to recognize . . . that the civilian agency is not competing with the Military Services. There needs to be the same interchange of information, the same harmony, between DOD and NASA, as existed between DOD and NACA."[25] Reality contradicted the comforting words.

The Air Force developed doctrinal thought in space largely by adapting existing airpower doctrine and projecting it into the space field. But assertions and suppositions outpaced technological accomplishments; Dyna-Soar itself developed little during 1959, and much of the 1959–60 period "was spent in resolving technical and managerial questions." Civilian officials created stumbling blocks to prolong this situation. A small-scale study by ARPA implied its authority to intercede with a larger footprint if it chose to do so. Nevertheless, the announcement of a contractor (Boeing) appeared to signal that the project was on solid enough footing.[26]

Aerospace Doctrine

Charyk, whom White had charged with expediting Dyna-Soar, instead slowed the program by establishing a new study of the boost-glider in mid-November 1959. Rather than unilaterally divulge top-secret information on Corona, Charyk pointed to technological questions regarding Dyna-Soar when initiating the Project Alpha study. Ordered on November 23, it "prohibited the obligation of more than three or four million dollars for Dyna Soar until [Charyk] had approved detailed financial plans and work statements," starting with the Phase Alpha study.[27]

Within the SAB's Aero and Space Vehicles Panel, Alfred J. Eggers continued to critique the project's contemporary configuration for the weight penalty that came with winged design. Eggers touted a semi-ballistic model as more efficient for delivering payloads into orbit. The panel wanted Phase Alpha to address the aerodynamic concerns. Panel chairman Courtland D. Perkins (who was also chairman of the Princeton University aeronautical engineering department) confided to the panel "that 'Dyna-Soar at this point was easily killable,' but, because Air Force leaders wanted Dyna-Soar, 'the [USAF] SAB should help [the] USAF [retain it].'"[28] Air Force planners stated that "Phase Alpha will delay the Dyna-Soar development by not less than three months," and White directed LeMay to "talk to Charyk" about how to proceed.[29]

New "aerospace" doctrine rolled out a week later, on December 1, 1959. Air Force Manual (AFM) 1–2, "United States Air Force Basic Doctrine," enshrined the term with fanfare rather than in true elucidative detail for contemporary personnel. Aerospace was decreed to constitute "an operationally indivisible medium consisting of the total expanse beyond the earth's surface." Air Force weapons platforms were deemed to "comprise a family of operating systems," utilizing the six features of flight in this extended realm (range, mobility, flexibility, speed, penetrative ability, and firepower delivery). The manual emphasized the linkage between the continuity of aerospace and the flexibility offered by the extended realm. The manual closed by asserting that the "nation, or group of nations, which maintains predominance in the aerospace—not only in its military forces but also in its sciences and technologies—will have the means to prevail in conflict."[30]

Apart from a definition of "aerospace" as "the total expanse beyond the earth's surface," AFM 1–2 did not address space. It did, however, fold space and air realms together.[31] Embracing atmospheric flight in the definition meant that, in rhetorical terms, every plane in the military arsenal had already demonstrated the value of aerospace power. Given this premise, aerospace power could be described as an existing fact and not as a revolutionary assertion or a service power grab.

Key Air Force leaders, in the closing years of the Eisenhower administration, believed that Dyna-Soar would hold the key to U.S. security by establishing aerospace defense and by maintaining and extending the deterrent power that seemed crucial in preventing Soviet aggression and general war. Along with projects such as the WS-117L satellite, the X-15 research rocket plane, and the Skybolt standoff missile, Dyna-Soar was listed as a Category I development project, considered "vital to the survival of the nation and . . . provid[ing] technological capabilities vital to the development of future Air Force systems."[32]

Retired Air Force generals such as Ennis Whitehead supported the airspace push. In March 1960, he wrote a personal letter to Chief of Staff White: "I believe that I shall live to see the time when Strategic Air Command is renamed Strategic Air-Space Command. A name is important if one is to identify his service and its mission clearly to laymen."[33]

Dyna-Soar's importance to top-ranking generals was reflected in the fact that, even as Phase Alpha stalled the glider program, commanders did not rush to reprioritize other armed space concepts above the Dyna-Soar project. Lt. Gen. Dean Strother, acting chairman of the Air Force Council, reported to White in March 1960 about the satellite interceptor project known as SAINT. "An orbital based tracking capability, and probably an offensive capability, are prerequisites to Air Force achievement of an effective active ballistic missile defense capability. The Satellite Intercept and Inspection (SAINT) demonstration is an advanced research exploration that would furnish considerable information to assist in making future space decisions," Strother wrote. He indicated that the Air Force should present its views as cooperative with ARPA so that the Air Force might be confirmed "as the agency to conduct the demonstration" of the project. SAINT would be orbital, able to inspect and rendezvous with satellites, and was to possess a "kill-on-command capability." Even so, Strother recommended to White that "the SAINT program should be placed in Category II of the Advanced Systems Importance Category List" below the status of Dyna-Soar.[34] Dyna-Soar aimed to have an antisatellite capacity while focusing on an offensive role that complemented prevailing deterrence attitudes.

Strother confirmed this in another memo, anticipating the future Cold War threat: "The offensive aerospace threat in the 1960–63 period will include a mixed force of ICBMs (250 in mid-1961), and bombers (1250), some armed with air-to-surface missiles." This expectation, like other Air Force predictions of Soviet strategic strength, far exceeded Soviet strategic buildup. "It is expected during the 1964–70 period that Soviet reconnaissance and warning satellites will be fully operational, as well as manned maneuverable space vehicles, manned lunar satellites *and possibly boost-glide vehicles* [emphasis added]."[35]

The U.S. Air Force expected to need Dyna-Soar partly because the Soviets were feared to be at work on one of their own. Since the "bomber gap" in 1954, the Air Force repeatedly overestimated Soviet advanced bomber development and strength. The Spring 1959 issue of *Air University Quarterly Review* had voiced concern of Soviet interest in a Dyna-Soar–type weapons system. In his article "Past and Present Soviet Military Doctrine," Dr. Kenneth R. Whiting pointed to a second-hand, but chilling, story by the Soviet aeronautical engineer Lieutenant Colonel Grigori Alexandrovich Tokaev.

Tokaev abandoned the Soviet Union for the West in 1947, telling his hosts that he had been put in control of the Soviet counterpart to Operation Paperclip, the effort to capture, collect, and utilize Nazi rocket scientists. Shortly before his defection, "Stalin interrogated him very closely on the

Sänger Project . . . a German engineer's plan for an aircraft with better than intercontinental range," Whiting explained. This project had also triggered Walter Dornberger's advocacy for boost-glide vehicles, leading eventually to Dyna-Soar.[36] With dictators running a superpower as a closed system, with that superpower espousing a hostile sociopolitical and economic outlook, and with expatriates attesting to sinister Soviet intentions and fantastic technologies, aerospace advocates felt quite justified in demanding that the United States hurry to develop its own presence in space.

Phase Alpha concluded in the spring of 1960, permitting work to resume as the project was again authorized to draw on its funds. A "full scale, but lightened" and unmanned vehicle would be tested first, and then progressively larger boosters would be employed to lift Dyna-Soar to orbital altitudes and allow piloted, maneuverable flight.[37] The Air Staff reviewed Phase Alpha in early April and then passed it to the under secretary of the Air Force. It next went to DDR&E York, who approved the program on April 22.[38] The year 1960 was shaping up to be eventful.

For the most part, the Air Force's Dyna-Soar advocates emerged from Phase Alpha confident that the program remained just as vital to national security and that it stood even more vindicated than it had when the study began. This did not suggest that its advocates had seen Phase Alpha as a blessing, however, because it certainly seemed in their eyes to have been an unfortunate loss of valuable time. Strother synopsized the Phase Alpha results for White: it had addressed Eggers' questions about Dyna-Soar's efficiency and "confirmed that all of the vehicles involved will require additional development effort" because of formidable heat problems.[39]

Contemplating the technological challenge caused a recalculation of Dyna-Soar costs. Dyna-Soar I was set to cost $493.6 million, a nearly 25 percent increase from the earlier estimate of $397 million. Strother noted that the cost estimate for Dyna-Soar II had remained steady.[40] But Dyna-Soar program head engineer Bill Lamar and program director Col. Walter Moore had also altered the development plan. Step II would be split into two portions. Step IIA would garner research data about orbital velocity flight, and with IIB, the Air Force would derive an interim reconnaissance and satellite inspection capacity. Step III remained the operational weapons system, which would be capable of orbital bombardment missions. If not fully cognizant of Charyk's motivations in ordering Phase Alpha, Moore did perhaps sense that it would be best to only outline the longer term armed derivatives of Dyna-Soar.[41] In the wake of earlier civilian actions, the

Air Force had growing reason to be vaguely suspicious of obstacles thrown in the glider's path.

Moore was likely right to be sensitive. In mid-June, York gave his formal approval of feasibility work on the SAINT program, but in mid-July, Charyk "directed that all references to a 'kill' capability in the system be eliminated, restricting technical effort to inspection functions only." Contemporary Air Force historian Max Rosenberg recognized this as being "related to the President's 'Space for Peace' program." SAINT now stood for "Satellite *Inspector*"—not "Satellite *Interceptor*."[42]

The inadequate thrust of contemporary U.S. booster rockets also posed a technological challenge. Weight estimates for Dyna-Soar tended to stand around ten thousand pounds, and talks between the Office of the Secretary of Defense and NASA threatened to preclude the Air Force's developing big boosters capable of lifting such heavy loads with the force required to achieve orbital speed. Air Force leaders recognized the booster problem, in terms of not only insufficient thrust but also daunting cost; they noted that "there should be increased effort to reduce booster costs, since those costs will be a large item of expense in any active space defense system."[43]

Defense Department officials formally endorsed Dyna-Soar's new development plan in late April, shortly after the conclusion of Phase Alpha. But York, not convinced of a military need for a boost-glide weapons system, reiterated his now-year-old guidance that Dyna-Soar I focus on hypersonic flight research and that military subsystems testing not be allowed to hamper the priorities that he had set for it.[44] Although the Air Force would continue to control Dyna-Soar, the pace of its development could be impacted by civilian-ordered studies, and the parameters of the project could be established along lines that fit NASA interests rather than Air Force priorities.

The Air Force noted York's disposition regarding space policy. NASA historian Eugene M. Emme, an Air Force Reserve officer and a former member of the Air War College faculty, briefed Chief of Staff White about York's outlook in mid-1960, noting York's concentration on reconnaissance systems rather than on defensive or offensive weapons.[45] In short, York was staying in touch with the president's policy. He and the administration as a whole did not favor space-based weapons systems that would soak up research money, expand the arms race, and provide less reliability than other (Earth-based) systems that already existed.

In Pursuit of New Capabilities

Schriever had been explaining to Congress the different needs of military and civilian systems. The military needed large numbers of simple, rugged, robust vehicles, with relatively long operational lives. Time was a vital element in defense, so military vehicles needed to be available with little notice. Scientists, Schriever noted, faced different challenges: vehicles would carry differing payloads from one mission to another, but fewer vehicles would be needed, their systems could afford to be more complicated, and vehicle life did not need to be very long if one-use capsules were used (as Mercury and the nascent Apollo programs indeed envisioned). And of course the time element in science answered to completely different imperatives than it did in war.[46]

The loquacious Boushey enunciated some of this in 1960 as well, and he had connected his points specifically to Dyna-Soar: "An essential ingredient of these systems will be some device that can return man from orbital flight in a routine, nonadventurous manner that can be supported operationally. . . . It is this essential capability that is sought in Dyna-Soar." It "must be maneuverable not only for providing flexibility of operation but also as a corollary to its data acquisition of capability. Dyna-Soar must be able to test military equipment and the machine relationship. *Dyna-Soar must achieve orbital capability.*"[47] Boushey spoke directly to the Air Force's idea for Dyna-Soar, and his description contrasted with the profile York had outlined since April 1959. "Just as the Wright Brothers in their time could not imagine planes like the B-70 or modern jet transports, so it seems that man today cannot imagine the extent to which space flight will be used." The future was irresistible, and "Dyna-Soar is one of the steps toward that future."[48]

Aerospace power remained of great interest to the Air Force's highest ranking commanders. Moreover, Chief of Staff White resolutely saw Dyna-Soar as the vehicle, both in the literal and the figurative sense, for bringing the Air Force into the space portions of the aerospace realm. His deputy, LeMay, grew somewhat less certain after the delays imposed by Phase Alpha that Dyna-Soar could maintain the development schedule that the Air Force believed crucial. Program Review Exercise 62-E aimed to determine the Air Force's fiscal year 1961 and 1962 budget needs and to identify specific programs meriting emphasis. Three months of work went into the review. The Air Force Council deliberated the findings on May 23 and 24, and then Vice Chief of Staff LeMay reported to White.

Deterrent capacity retained priority, but in the review, LeMay did acknowledge funds to be finite, and therefore specific priorities were necessary. The review emphasized "1965 as the key year when it can be anticipated that heavy funding will begin for those operational systems such as Dyna-Soar, Space Counter Weapon System, SAINT, B-70, and control systems which are initiated in the early sixties." As such, programs expected to offer operational capability by 1965 would receive priority. The study's paperwork explored a variety of systems ranging from strategic bombers to missiles, air-refuel craft to nuclear capable fighters, radars to weather planes. Projections for Dyna-Soar were included as well, but the study predicted that it would not yield an operational system by fiscal year 1967; the same was the case for projections of the troubled ANP (aircraft, nuclear powered). The study anticipated some operational capacity for B-70, SAINT, and MIDAS by that time, however.[49]

Although popularly characterized as a staunch and fossilized advocate of the manned bomber plane, LeMay would be more accurately understood as a manned aerospace system advocate. His statements repeatedly acknowledged the need for a mixed force of manned systems and missiles, and his report on Exercise 62E stood in keeping with this. "For the foreseeable future, a mixed force of missiles and manned aircraft is required. There must be a follow-on capability to the B-52 for finding and eliminating residual targets. However, the Air Force cannot afford a package of B-70, ANP, Phase III Dyna Soar, and a Long Endurance Boundary Layer Control aircraft, all as follow-on systems to the B-52." It was time to prioritize. "The B-70 appears to be the only weapon system that offers the capability at the time required. Therefore, the B-70 should be reinstated on a Weapon System basis."

A second conclusion was that "at this time, manned orbital flight within a weapon system, as represented by Dyna Soar III, must be considered of lesser priority than the B-70 as a manned strategic weapon system," a product of the project's technological complication and its development delays. A further factor strengthening the B-70's position in LeMay's eyes was the prospect of converting it into an ANP at some time in the future. "All efforts on a military nuclear powered aircraft should be directed toward a significant improvement of the B-70. This, therefore, means a concentrated, reoriented ANP development effort phased toward a weapon system at a date not later than 1970."[50] With boost-glide technology so long stalled, wisdom appeared to point to hedging Air Force bets by developing a vastly speedier bomber plane and then modifying it with nuclear jet engines to exponentially extend its flight duration.

LeMay's system for prioritization followed three major strands of logic: first, resources spread in all directions would constitute waste that civilian bosses would disallow; second, systems incapable of operational status by 1965 claimed less priority; third, hope could be preserved for elements that might be folded into existing systems. As a result, the wish for the ANP could be preserved on the notion that the nuclear engine might someday be dropped into the back of a then-existing B-70 aircraft.

Although delays had led LeMay to rank Dyna-Soar's priority behind that of the B-70, the Dyna-Soar program proceeded, and it retained its status as a Category I priority for *development*. Dyna-Soar's place in LeMay's mid-1960 report then comes into clearer focus, as the vice chief of staff recognized the program's slow progress as hindering its near-term operational future. But the service did not discount Dyna-Soar's future significance, and research continued for years. Air Force planners felt certain that Charyk's ordering of Phase Alpha had thrown Dyna-Soar irrevocably behind schedule by three to six months. LeMay prioritized the B-70 in the middle of an election year in which Eisenhower had resuscitated the program in an effort to deflect candidate John F. Kennedy's criticism of him and attacks on Vice President Richard Nixon. Exotic defense technologies seemed likely to be reprieved and supported, especially in the event of a Democratic victory in November, since that party cried out for increased defense spending.

On May 1, 1960, a former Air Force pilot climbed into his U-2 spyplane and flew from his base in Turkey. A Soviet surface-to-air missile cut his mission short, destroyed his plane, and led to Francis Gary Powers' capture by the Soviets.[51] It was the first time that a surface-to-air missile had brought down a piloted aircraft. As with other moments significant in space history, military history, and national policymaking, Powers' shootdown and its aftermath made up part of a chain of events that would come to impact the Dyna-Soar program as well.

Satellites got a boost in the wake of the U-2 affair. Corona, of course, was already deep in its development and testing. Discoverer satellites had been launched as early as February 1959 as researchers struggled to find and defeat each of the many technological problems that stood between the state of the art and the operation of a recoverable-capsule reconnaissance satellite. Congress, unaware of Corona, pressed for more rapid development of the MIDAS and SAINT satellite systems, even voting sums for the programs in excess of the amounts the administration had requested.[52]

By the time of Powers' flight, ten Corona tests had been conducted. Most tests in 1959 had successfully achieved orbit, but the task of reliable recovery introduced daunting complications. Until mastered, cameras could not be fitted to Corona payloads, and satellite reconnaissance would not be accomplished. Recovery attempts during much of 1960 went poorly. Early in the year, recovery efforts for a satellite launched in November 1959 failed. Discoverer IX was launched on February 4 but failed to orbit, and Discoverer X was destroyed seconds after its launch on February 19. Attempts after Powers' interception also proved disappointing. Discoverer XI in May could not be recovered, and Discoverer XII in June failed to orbit.[53]

Discoverer XIII launched, orbited, and was successfully recovered at sea August 10–11, 1960. The Air Force's public affairs fact sheet from the day of the launch characterized it as "the most completely instrumented satellite vehicle of this series yet launched into orbit by the U.S. Air Force. The shot has been termed a 'Diagnostic' test." Although acknowledging the recovery objective, the sheet shows a hesitance to anticipate success: "The additional telemetering instrumentation placed aboard Discoverer XIII may lead the way to solving this very difficult undertaking and should bring us much closer to recovery than we have ever been before." Although recovery attempts would be made, "emphasis during the separation and re-entry sequence . . . will be on diagnosing, through telemetry, exactly what is happening to the capsule."[54]

In his own remarks after the satellite's successful recovery, White's description carried a more boastful note: "All of us vividly recall specula- tion with which the free world met the news of the Russian Sputnik overhead less than three years ago. Let me assure you that the tremen- dous and complex test successfully made by Discoverer XIII is by far a greater single achievement in the technical realm than that of the relatively crude early satellites." The United States now had a clear, spectacular first. Retrieval of the Discoverer XIII capsule from the waters off the coast of Hawaii was the first recovery of a man-made orbital, and Discoverer XIV was the first midair recovery during reentry.[55] This encouraged even those unaware of Corona's existence. It gave renewed reason for hope, which had been damaged in May both by the shootdown of Gary Francis Powers and the five-ton Sputnik IV, which beamed prerecorded voice signals by radio and whose cabin was designed to sustain a human being and was occupied by a dummy cosmonaut. The USSR appeared to be pressing ahead steadily and successfully with its plans for space.

Seeing space as a boundless frontier, the Air Force expected space accomplishments to feed one another, extending the service's capability and the nation's security. It greeted the Discoverer XIII success with enthusiasm, predicting that "while many of the results of the Discoverer program are directed primarily toward applications in the SAMOS and MIDAS programs, the long term benefits of Discoverer are serving to increase significantly man's knowledge of the spatial environment, to enhance his skills and techniques in space operations, and to provide experience in the development and operation of a reliable space vehicle."[56] Air Force leaders saw Discoverer as a stepping stone toward greater accomplishments and operations. This opinion was in keeping with the larger approach to technology to which Air Force leaders were committed. The Air Force, intent on deploying more sophisticated satellites of its own, suggested that "from this viewpoint DISCOVERER is a 'test bed' for [future Air Force] space programs." Many of the "components and techniques" used by Discoverer "are directly applicable to SAMOS, MIDAS, and programs with a man-in-space application."[57]

But the Air Force was speeding toward collision with a rising wall of secrecy. In a special National Security Council meeting on August 25, 1960, Eisenhower's scientific adviser George Kistiakowsky discussed the newly successful Corona and the Air Force's contemporary SAMOS with the president. Six days later, the Office of the Secretary of the Air Force for Missile and Satellite Systems (SAFMS) was created. The office quickly cut the Air Force out from knowledge of the national reconnaissance projects. "Effective immediately, the satellite reconnaissance program will be managed within [a new] structure," Air Force Secretary Dudley Sharp directed White on September 13. The secretary of the Air Force also held authority over the Air Force's SAMOS project. The Satellite Reconnaissance Technical Advisory Group and the Satellite Reconnaissance Advisory Council provided input.[58]

SAFMS was linked directly to the secretary of the Air Force, without review channels "between the Director of the SAMOS Project and the Secretary of the Air Force." This meant more direct civilian control, and it also threatened to exclude Air Force personnel from awareness of their own service's satellite work. Information would be limited: "Need-to-know briefings . . . will be given by the program management staff."[59] When kept uninformed, those outside the information loop are likely to act separately or even divergently from policy.[60]

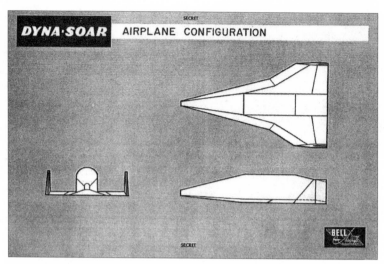

Dyna-Soar underwent five iterations between early 1959 and the end of 1961. Each involved minor adjustments to the dimensions of the craft and the wing sweep. By December 1961, Dyna-Soar was set to be 35 feet, 4.14 inches long, with a wingspan of 20 feet, 10 inches, and a wing sweep of 72 degrees, 48 minutes. *Niagara Aerospace Museum*

This policy of information exclusion was not accidental. President Eisenhower believed that awareness within the Air Force of the secret Corona program could well lead to a leak to the public and the world stage. After all, the Air Force had an extensive track record of using information to promote its image and role to the broader public. The Air Force's own satellite reconnaissance program SAMOS could attract national attention to satellite reconnaissance issues. The Air Force propensity for open and frank discussion of its technological futurism suggested the likelihood that work on Air Force reconnaissance satellites would throw a spotlight on the reconnaissance satellite concept.

Perhaps, given Eisenhower's perception of his needs as a policymaker and the Air Force's past actions, the decision to establish the SAFMS was unavoidable. At first, former Deputy Chief of Staff for Research and Development Brig. Gen. Richard D. Curtin directed the SAFMS.[61] By the time the office was rechristened the National Reconnaissance Office, its co-directors were Victor Charyk and Richard Bissell. Charyk had ordered Phase Alpha, and Bissell had run the U-2 program.[62]

Fences were coming to space. But those fences were not completely understood by the Air Force's aerospace advocates, and in the late summer of 1960, the Air Force retained Dyna-Soar on its "Category I list for priority development." In October, a development directive outlined work on Step I and Step II vehicles and stated that "a concurrent aggressive study effort is directed in the area of military application of the Dyna Soar concept as a basis for possible future Step III weapon system development programs."[63]

Lieutenant General Wilson noted in late September 1960 that "Dyna-Soar . . . offers an enormous potential for future maneuverable capability in space and the atmosphere." More specifically, he declared, "The Air Force considers Dyna-Soar the most important research and development project it has. . . . The Dyna-Soar will open a new era. . . . It is the first step towards practical man-in-space flights." Schriever agreed, proclaiming in *Air Force Magazine*, "Dyna-Soar has important operational potentialities . . . as the first piloted military space system planned by the United States."[64] In the words of Brig. Gen. Homer Boushey that election year, "The Air Force believes that the Dyna-Soar development can now be begun with ample confidence."[65]

7

"A Capacity . . . Adequate for Our Own U.S. Purposes"
Space Security Policymaking in the Eisenhower White House

Actually nothing can be "dropped" from a satellite. . . .
The earth itself is the best and most effective weapons carrier.
  —President's Scientific Advisory Committee,
    March 12, 1958

No reasons have been advanced which indicate that this
research and development activity is "necessary to make effective
provision for the defense of the U.S." other than the "feeling"
that the military ultimately will require manned satellites or other
space vehicles.
  —Robert Piland, July 23, 1958

Weighing Security Alternatives

Eisenhower had firmly established a popular persona as a genial and nonpolitical public servant residing in the White House. Scholars have since recognized political aims in his actions and postures, including the public image he constructed. Determined to govern as a moderate, grafting the right wing of the Republican Party onto a viable mainstream, he remained committed to preserving national security through nuclear deterrent power, maintaining friendly relations with allies overseas, and keeping defense spending from spinning out of control.

An affable five-star general could pursue these goals through much of the 1950s by relying on capable subordinates to take action (and perhaps take the lead) in promoting the president's agenda. Eisenhower's eminent reputation offered a measure of insulation against critics of his defense policy, especially during his first term. When security issues offered opportunity for critics to rally and grow more vocal, his methods became as much of a trap as a tool for President Eisenhower.

By the standards of the Cold War, the United States was not vulnerable. Imperfect U.S. interception capability notwithstanding, actual Soviet power to deliver coordinated and accurate nuclear strikes fell far short of the worst fears. In contrast, the United States already possessed an extensive arsenal as well as large numbers of nuclear-capable aircraft. To supplement the Air Force's strategic capability, some Army projects involved tactical nuclear weapons. Work was under way to develop reliable IRBMs and ICBMs, and by the end of 1957 the Navy had ordered a submarine capable of launching nuclear weapons. Sufficiency was Eisenhower's objective, and sufficiency meant deterring Soviet attack. But ranking officers such as SAC commander General Power argued that "sufficient" retaliatory capability was a difficult factor to quantify with certainty since Soviet dictators cared little for the suffering and casualties of their people.[1] The *amount* and *scope* of deterrent strength needed were important issues of speculation. Were free-fall and air-delivered nuclear bombs, as well as land- and sea-based ballistic missiles, the proper parameters of the U.S. deterrent force? Was an aerospace option appropriate? Was aerospace an option or a necessity?

The challenge and the context drove many people toward technological determinism. For example, one Army briefing on the satellite program declared that Sputnik II "has materially increased the threat to our national security posed by Sputnik-I. . . . As we stand on the threshold of space travel we must truly visualize applications heretofore relegated to the 'Buck Rogers' category."[2] High-technology advocates critical of the administration accused it of a lack of vision bordering on incompetence.

Coherent policies depend on consistency between overarching philosophical outlooks and specific policy decisions. Eisenhower's difficulty emerged because his New Look security posture, endorsed in aggregate by the public in the 1952 and 1956 elections, conflicted with increasingly numerous and vocal cries for a new direction. Eisenhower defined U.S. security in terms of a balance between sufficient national security and internal strength and freedom. Strategic deterrent forces embodied the former, and economic health and stability were crucial to the latter. Embracing "applications heretofore relegated to the 'Buck Rogers' category" would undermine U.S. security, according to this outlook. At the same time, however, some alterations would have to be made regarding space issues. Eisenhower recognized that "we must be selective or we will be broke,"[3] and an appropriate organizational structure could coordinate controlled space projects that complemented national policy while placating the presumed public demand for space activity.

The foundation of national space policy had been laid, on advice from the Killian Committee, by the NSC in its document 5520 back in 1955; organizational questions still remained, however. The ARS advocated civilian oversight of space, arguing that "the strictly military aspects of space flight are probably receiving adequate support at the present time" but that "many scientific, commercial, and politico-military applications of even greater long-range importance" were "in the opinion of the committee . . . being neglected." In the aftermath of the first Sputnik launch, the ARS recommended that "a national space-flight program be initiated" under an agency with "independent status similar to that of the Atomic Energy Commission [AEC] or the National Advisory Committee for Aeronautics."[4] Indeed, policymakers considered granting responsibility for space projects to AEC, NACA, or a planned Defense Department arm overseeing military research and development.

The National Science Foundation (NSF) agreed, and its members discretely emphasized the value, "particularly in terms of U.S. and world public opinion . . . of a strictly scientific, civilian-managed program which could not be suspected of military purposes," urging that U.S. space science efforts "should be initiated without delay."[5] NSF member Alan Waterman wrote directly to James Killian urging that a space agency be established and that it be "independent of existing agencies."[6] The *Harvard Business Review* asserted that "the most critical question facing the United States today is how it can regain undisputed technological leadership and weapons superiority," faulting the military's ponderous research system and proposing "a new weapons planning system . . . under civilian control to assure a scientifically directed, long-range weapons research program such as will permit the prompt and wise selection of weapon systems for development and production."[7] While concerned about Soviet achievements, the journal professed confidence in American research and development capacity.[8]

Noting the vocal opposition to military oversight of "the more purely scientific and non-military aspects of space research," Killian declared late in 1957 that "we must have far more than a program which appeals to the 'space cadets.' . . . Other nations will continue to hold the leadership" in space science unless this was accomplished.[9]

Establishing the Advanced Research Projects Agency would buy the administration time as it pondered the exact shape of the organization for space projects that it deemed desirable. Within DOD but beyond the control of any of the individual services, it would pool the various military space projects and prevent service ambitions from getting quickly out of hand. Defense Secretary Neil McElroy had announced in November 1957 that he "intended to establish an overall agency to direct work on missile defenses and other space projects." The Air Force, claiming to think McElroy was "only . . . opposed to any formal announcement about the new Air Force office," recognized ARPA's potential to limit the services' prerogatives in technical development of space projects.[10]

Therefore, Deputy Chief of Staff for Development Lt. Gen. Donald Putt announced the establishment of a directorate of astronautics on December 12. Putt named Brig. Gen. Homer Boushey as director and empowered him to "plan, organize, and manage the air programs in astronautics." Boushey's office would oversee ballistic missile defense, boost-glide vehicles, the ambitious WS-117L satellite, and other projects. Described as "a program aimed at developing an earth satellite as a space platform," the Discoverer satellite project would later be detached from the Air Force project and would provide a cover for the top-secret Corona surveillance satellite program.[11]

The Air Force Directorate of Astronautics would have coordinated the boost-glide Dynamic Soarer as well. Consolidated that summer to conserve the evaporating research dollars of three parallel programs, Dyna-Soar would enter space atop a booster rocket and at the end of a mission would harness a lift-to-drag ratio that would allow a controlled aircraft-like landing method. To operationally inclined airmen, this was distinctly preferable to the alternative approach to space flight and recovery: a ballistic trajectory for a small capsule occupied by a crewman with only the slightest influence over his landing. Four years earlier, NACA engineers had already theorized ballistic methods to be more immediately practicable for spaceflight recovery.[12] However, NACA had been considering the technical challenges of recovery, rather than the operational challenges of combat. A ballistic capsule skirted many technological challenges of the boost-glide alternative, but it almost dictated that space flight would be episodic rather than regular and that "landing" would entail extensive search and recovery efforts for every mission. While the tiny Dyna-Soar effort was too small to yet merit White House attention in the fall of 1957, Putt's announcement creating the Directorate of Astronautics drew Secretary Douglas' swift opposition. He had already "made it clear that no action would be taken to establish any office of astronautics," and now the Air Force's top civilian reproached Chief of Staff General White and ordered the directorate's immediate dissolution.[13]

The Air Force's ill-fated Directorate of Astronautics reflected the service's aerospace vision of technological development driving flight ever higher, faster, and farther, toward the "ultimate high ground" of space. White himself would declare that "Air Force goals have changed in degree only; the basics have been constant—greater speed, longer range, and higher altitude."[14] The ideas were beginning to rub off on the influential and ambitious Senate Majority Leader Lyndon Johnson.

Definitions of space impacted organizational distinctions. The ARS's outlook sharply contrasted with the Air Force idea of a continuous realm of sky, declaring that "*Astro*nautics can no longer be considered as an appendage of the Science of *Aero*nautics." The ARS did, however, concede the need for "*Chemospheric* superiority," which it described as "the successful operation of hypersonic gliders for bombing and reconnaissance."[15] This was, in fact, the purpose and role for which Dyna-Soar had been undertaken. In fact, the ARS report, dated August 23 and revised October 10, coincided with the consolidation of Dyna-Soar from its three component research projects: the HYWARDS research vehicle, the Brass Bell reconnaissance craft, and the BoMi/RoBo bomber. Above the chemospheric realm, the ARS also delineated "ionospheric superiority" as the "capability of operating satelloids and satellites for reconnaissance" and stated that "exosphereic and free-space operations, up to altitudes of several thousand miles, are of potential politico-military usefulness."

This did not equate into an ARS endorsement of military employment of space. "*Politically*," they wrote, "space flight can not but make still more apparent the impracticality of war as a means of solving differences between nations." The committee looked to "freedom of the seas" as a precedent for a "freedom of space" and to the "establishment of this condition in international law [as] a prerequisite to any national space flight program."[16] Arranging agreement of an international law would prove a challenging task. With so much about space still unknown, different ideas competed about what should constitute the U.S. position about those laws and even whether deriving formal laws and definitions about space would be beneficial. NASA historian Eugene Emme would soon note that "international law remains extremely impotent and unenforced; while law in air space is a primitive jungle."[17]

Uncertainty can give way to fear, which, in its turn, can bring instability. The 1957 Gaither Report's brazen expansion of its own mandate had irritated Eisenhower when the group delivered its report on nuclear shelters—and all the other deterrence issues it saw fit to address—in November. When the president failed to adopt the panic of the committee's key participants, members leaked parts of their report to the press, further irking the administration.[18] Early in January 1958, Robert Sprague, who had chaired the committee since Rowan Gaither's illness, urged Secretary of State John Foster Dulles to have the president form a study group to consider "possibilities for fruitful use of the near-term future during which the U.S. will have a margin of strategic bombardment capability over the Soviet Union." Sprague advocated waging preventive nuclear war against the USSR over brinkmanship, containment, or what he described as "plac[ing a] reliance in God to find a solution." A few weeks later, Dulles confided to his disarmament adviser Gen. Alfred M. Gruenther a "concern" about Sprague "because he seemed to be emotional about certain aspects."[19] The Soviets' space accomplishments, despite not proving a true military superiority, greatly jarred the confidence of individuals and much of society. In a January 6 NSC meeting, Vice President Richard Nixon and Eisenhower's assistant Gordon Gray suggested that the government release the report by the committee Sprague had chaired; rumors about the report had been "fantastically worse than what the Gaither report actually said."[20]

During that same meeting, however, Eisenhower indicated more interest in "improving the early warning system and the dispersal of SAC bases," since "our main reliance would still be placed on manned aircraft" for "a long period of time in the future."[21] The following week, Defense Secretary McElroy told the House Armed Services Committee that the Soviet satellites had demonstrated "that the Russians are farther advanced scientifically than many had realized and are in fact challenging the scientific supremacy of the United States" and that "the weapons of the future may be a great deal closer upon us than we had thought, and therefore, the ultimate survival of the nation depends more on the speed and skill with which we can pursue the development of advanced weapons . . . than ever before." One part of the solution, he told the House members, should be "increased effort behind space programs," including anti-missile missiles and satellites.[22]

Internal and External Influencers

In the meantime, Congress invited Air Force testimony and parroted Air Force ideas, which found their way into newspapers. On January 7, Lyndon Johnson declared to fellow Democrats that "control of space means control of the world. . . . Whoever gains that ultimate position gains total control over the earth for purposes of tyranny or for the service of freedom."[23] The *New York Times* released a story citing Maj. Gen. Bernard Schriever's testimony to the Senate Preparedness Subcommittee, claiming "that the Air Force hopes to launch a 'military reconnaissance satellite' by the spring of 1959."[24] This intention was in line with an optimistic appraisal, and Schriever repeated his prediction of a "limited reconnaissance capability" by mid-1959 to Senator Symington and to Connecticut Republican Senator Prescott Bush. Although the first of the Discoverer test series was launched in early 1959, a functioning reconnaissance satellite was not accomplished for another year and a half.

White's testimony urged prompt and vigorous action, and Schriever backed his chief. "I think the number one thing is to get some things in outer space" soon, White argued, while boasting of space-capable Air Force flight suits. Waxing ambitious, White then unfurled the colors of the aerospace perspective: "I actually foresee the use of weapons in space, both on offensive and defensive. I can imagine a satellite being a missile launching platform." He also hypothesized about the development of countersatellite weapons. The technological spiral would be set in full swing. Schriever told Johnson that "there is a great similarity in operating in the air, in the atmosphere above the earth, and in operating in space." Without endorsing the statements of Army missile chief Gen. John Medaris, Schriever also recommended the development of large booster rockets by about 1965, "for the next generation of space vehicles."[25]

Air Force statements were all the more potent because a series of would-be satellites disintegrated in embarrassing, spectacular, and public launches during the winter of 1957–58. Whether space boosters or ballistic missiles, the technology left much to be desired. One general, despairing of contemporary unreliable missile technology, called the Atlas missile "a plain out and out bastard. You cannot even tell where these things are going to go after you push the button, apparently."[26] As unfortunate as the problems were, and as embarrassing as the rocket explosions continued to be, the problems were in the technical engineering realm; the scientific problems with first-generation missiles essentially had been worked out.[27]

Dyna-Soar's considerable weight underlined the handicaps of weak and unreliable rocket boosters.

Organizational considerations posed even graver problems for Dyna-Soar's long-term future. Eisenhower conferred with scientists James Killian, George Kistiakowsky, and Herbert York on February 4, 1958, three days before the official establishment of ARPA. Eisenhower regretted that missile programs had not been centralized in the Office of the Secretary of Defense, beyond the reach of individual military services.[28] On February 7, the day of ARPA's creation, Killian told the president that "mil[itary] uses of outer space" were "extraordinarily limited," although they could entail warning against strategic attack, reconnaissance, and missile travel as worthwhile.[29] Since ballistic missiles traversed space, some consideration was given to whether ICBMs might potentially be defined as "space weapons," explaining Killian's inclusion of missiles in his list of military uses.

PSAC prepared to address organizational issues directly in a February 21 memo. The committee emphatically declared that U.S. space efforts should be civilian-managed "both at the policy and at the operating levels."[30] It also differentiated between space "exploration" done by civilians and "control" exercised by the military. Although conceding that "certainly, ICBM's will transit portions of outer space in performing their missions . . . for the moment the chief military interest lies in better methods of surveillance, communications and long-range weather forecasting." Significantly, however, "the *control* of outer space, basically a military matter, involves many troublesome questions of international law."

The report noted suggestions "that whatever form of organization is agreed upon to initiate the space exploration program it should be attached temporarily to ARPA." Indeed, a number of space projects, held by various military services, would be transferred to ARPA and subsequently to the civilian space authority. Three candidates emerged as to which civilian organization that would be. The Atomic Energy Commission was "organizationally competent" but lacked experience in flight technology. The newly established ARPA appeared to be able to "take on the job with a minimum of additional legislation." NACA had "been in the space exploration field for a long time." However, an addendum to PSAC's preliminary observations report noted the precedence attached to whatever decision might emerge: "it would seem highly probably [sic], however, that even if a *temporary* arrangement of this sort were made it would be extremely difficult, if not impossible, to detach and to reassign the space exploitation project at a future date."[31]

A March 12 revision of PSAC's report honed the message. A "civilian-oriented, civilian-managed organization for space" would be better able to engage with scientists and non-governmental organizations than would an organization connected to the military.[32] Commenting on a satellite's use as a weapons platform (already predicted by White and Johnson), PSAC pointed out that "actually nothing can be 'dropped' from a satellite. All objects aboard are travelling at the same speed, and if pushed overboard will simply continue in orbit. A bomb would have to be 'fired' back toward the earth. The problem then becomes the same as launching a full scale ICBM from earth." Thus, "the earth itself is the best and most effective weapons carrier."[33] This argument seriously undercut the case for Dyna-Soar and a diversified deterrent extended into space.

The report went on to observe that "in contrast to the somewhat limited practical military use of its outer space currently envisioned, there are many important civilian and scientific uses."[34] Separately, Killian and other PSAC members wrote Eisenhower that "because of the importance of civilian interest in space exploration, the long term organization of Federal programs in this area should be under civilian control." The letter to Eisenhower contained, almost verbatim, the March 7 draft's assertion that space exploration should be done under civilian auspices because space "has a relatively limited military significance, at least for the foreseeable future."[35] This did not preclude the use of space as a medium through which ballistic missiles would travel if general war were to come. The same day that he had written of the limited military significance of space, Killian notified Eisenhower that the Ballistic Missiles Panel had reported that "technological progress to date indicates the feasibility of greatly improved programs for a second generation of ballistic missiles."[36]

Technological advance brought mixed results. Unreliable and expensive first-generation missiles left Eisenhower cold, and although the feasibility of second-generation weapons meant that first-generation systems could safely be kept within limits, the steep costs of successor missiles helped temper Eisenhower's enthusiasm as well.[37] Sufficiency was the watchword. In late April, Eisenhower told advisers "that he still had more faith in the delivery capabilities of the aircraft than he had in all these missiles at the present time." Defense Secretary McElroy concurred.[38]

The year 1958 saw the transformation of NACA into the National Aeronautics and Space Administration. During the spring and summer, while his scientists urged that space activities be organized under civilian-controlled auspices, Eisenhower likewise expressed that he wanted space responsibilities centralized. Speaking to Killian and York in early March, "he pointed out . . . the need for central direction of our space activities."[39] Three weeks later, press secretary Jim Hagerty suggested that Eisenhower give another speech about space, but Eisenhower demurred, explaining "that he thought he should not spend his time on a speech that did not deal with peace, disarmament, mutual aid, freer trade, or the like. Other things he considers other people can do. . . . His immediate job is peace and the things that go into the making of peace."[40] In mid-June, Eisenhower conferred with a group of scientists that included his PSAC chair Killian. Both the president and his trusted science adviser agreed that "the great problem is to achieve a public understanding of the basic considerations [regarding space policy], since otherwise they [the public] are expected to take positions on policies which they cannot understand."[41] But an enunciation of that policy in a clear, emphatic, compelling way to the public remained incomplete.

Issues of space and issues of peace intersected at the boundary between "space" and "air"—or would do so, if indeed the advocates of separation were to prevail over aerospace theorists. White again promoted the aerospace term and its underlying worldview in an article that August.[42] Definitions had interservice and policy implications as well as scientific meaning. A year prior to Sputnik, one book on space had noted that "that old question, 'how high is the sky?' is still not answered." Various scientists delineated convincing—and competing—definitions by focusing on various scientific aspects of the atmosphere.[43]

Charting the boundary of the atmospheric sky carried significant strategic and political implications. As sovereign political entities, nation-states have internationally respected rights to control their own airspace. Violating these rights constitutes serious offenses and implied hostility. Both the United States and the Soviet Union had been dragged into the last world war through surprise attack, and both were understandably anxious and suspicious of any apparent preparation for a future surprise. Careful secrecy characterized Eisenhower's use of U-2 spy planes in undertaking vital reconnaissance missions between 1956 and 1960. Despite airspace sovereignty issues, intelligence information was desperately needed, and the extremely high-altitude U-2 had been built as a stop-gap to provide that information, violating Soviet airspace in the process. Planners hoped that its extreme altitude would protect it from Soviet interception and possibly even from detection, but Soviet radar operators saw the plane on their scopes from the time of the first flight in July 1956. Secretary Dulles believed that their inability to bring the planes down might preclude the Soviets' publicizing U.S. overflights and denouncing them in front of the world. The Soviets did discreetly notify U.S. officials of their knowledge of, and displeasure with, U-2 flights. The USSR sternly but quietly insisted "that the Government of the United States of America will undertake further investigation of this act of violation of the airspace of the USSR by an American military aircraft and will punish severely those guilty of this violation" and that it also "take the necessary steps to prevent violations of Soviet airspace by American aircraft in the future."[44]

Richard Bissell felt that Eisenhower took such protests seriously, as after one event the president slowed the project down for "quite a few months" before approving another flight.[45] In total, twenty-five U-2 flights passed over the USSR between July 1, 1956, and May 1, 1960.[46] Each was soberly considered and carefully approved by Eisenhower, and each was a violation of Soviet airspace. Determining the boundary between "air" and "space" meant determining how future intelligence might be gathered. While the U-2 was indisputably an airplane (capable of extremely high altitude) and thus a violator of Soviet *air*space, defining the threshold of space stood to affect whether a reconnaissance-equipped Dyna-Soar or a camera-equipped satellite such as Corona might one day be considered in similar violation.

Deputy Defense Secretary Quarles had recognized this when the Soviets first launched Sputnik. The Soviet metal sphere overflew many foreign countries during its orbits, and the absence of a wave of international diplomatic protest implied international acceptance of the freedom of space philosophy.[47] Eisenhower was happy to capitalize on this. If freedom of space for peaceful purposes could be formalized, reconnaissance satellites could securely assume the role that the U-2 was precariously filling.[48]

Such views appalled an Air Force convinced that American security rested on its deterrent power rather than reliance on international agreements and hopes of Soviet sincerity. And such international agreements could be completely devastating for the Dyna-Soar program. A defined boundary between "air" and "space" would overturn the aerospace concept and wreck the future of Dyna-Soar and follow-on systems. And in the eyes of its personnel, dedicated to defending the country by keeping the Air Force preeminent over Soviet rivals, a stunted Air Force meant an unnecessarily vulnerable nation.[49]

Defining Objectives in Space

National Security Council document 5814/1, developed in the summer of 1958, marked a compromise between the space interest of the joint chiefs and the fiscal consciousness of the Bureau of the Budget. The document called for the United States to become a "recognized leader" in space.[50] The NSC noted the "critical importance" of reconnaissance satellites "to U.S. national security" and the need for studies "to determine the most favorable political framework in which such satellites would operate." The document deemed that the sky "is divided into two regions," where "'outer space' is considered as contiguous to 'air space,' with the lower limit of 'outer space' being the upper limit of 'air space.'" It outlined three uses for outer space: for "vehicles or other objects that achieve their primary purpose in outer space," for "transmission of electromagnetic energy for . . . communications," and for "vehicles which traverse outer space, but which achieve their primary purpose upon their return to air space or earth." Whereas Dyna-Soar would operate in space as a reconnaissance or bombardment platform capable of maneuverable reentry on an airfield, in-flight ballistic missiles were vehicles that "traverse outer space."[51]

NSC 5814/1 aspired to two potentially complementary goals: preserving space for peace while simultaneously securing an opportunity to use "space for security" through the development of reconnaissance satellites.[52] International respect for reconnaissance satellites, as a means of verifying the developments of rival nations, would certainly have depended on an atmosphere of lessened tension; once operating, they might potentially help ease tension and fear of surprise attack. "Military" satellites and "weapon" satellites might still have sounded like synonyms, and an internationally acknowledged distinction between "military" (as relating to a defense or national security institution) and "weaponized" (as a tool for destruction) character would be required. Many issues remained contested in 1958. Discord in NSC meetings emerged in July debates about whether ballistic missiles constituted "space vehicles" and what being a "recognized leader" in space entailed.

Quarles claimed not to have chosen a side about whether ballistic missiles ought to be included in U.S. policy on outer space, but he noted that because some practicable ballistic paths did not exceed an altitude of one hundred miles, "we must not let ourselves be trapped into the feeling that all ballistic missiles must necessarily traverse outer space." Eisenhower agreed that "we could and should differentiate policy guidance on these two classes [ballistic missiles and everything else] of outer space vehicles."[53]

Two powerful figures, Secretary of State Dulles and Budget Director Maurice Stans, questioned the wisdom of the nation's proposed status in space. Dulles noted a passage in 5814/1 calling for the United States to achieve "either superiority over or parity with the USSR in all outer space activities." Rather than a policy inviting "expenditures and efforts . . . almost without limit," the country instead needed "a capacity in outer space adequate for our own U.S. purposes, but not necessarily superior or equal to the capability of any other nation." Alan Waterman countered that the national interest demanded bold commitments to research and development in areas where "the success or ultimate utility . . . could not be definitely foreseen. . . . A really determined effort" would be required to achieve parity with the Soviet Union, which "had such a long head-start in outer space activities."[54]

Stans supported Dulles, partly by turning Waterman's argument on its head. Dulles had doubted "the wisdom of a policy which committed the United States to ape whatever we imagine any other nation is doing or is going to do." Stans echoed that U.S. policymakers had traditionally set policy according to "our own national objectives rather than the objectives of some other country." Furthermore, he added, U.S. efforts *couldn't* mirror Soviet efforts unless their objectives were somehow discovered. The Soviet head start guaranteed that they "will maintain their lead over the United States in outer space activities for at least two years [and] accordingly, we could not possibly achieve parity with them within this time limit . . . no matter what we do." A mantra of competition against the USSR highlighted that the military aspects of space would also undercut "all the programs for the peaceful use of outer space."[55] NSC 5814/1 identified the eventual need for manned space exploration as "undoubted," partly because "to the layman, manned exploration will represent the true conquest of outer space."[56] However, this did not need to mean military exploitation of space beyond reconnaissance purposes by satellites and ballistic missile trajectories in the event of war.

The aerospace perspective provided the groundwork for Dyna-Soar's raison d'être. Eisenhower's determination to distinguish missiles' travel *through* space distinctly from vehicles' travel *in* space compartmentalized the air and space realms, in keeping with his security policy. Dulles and Stans reminded an already alert Eisenhower of the need to defend national security from both external antagonists and economic dangers. Proof, let

Dyna-Soar never flew, but unweaponized projects such as the M2-F3 (a lifting body craft based on competing aerodynamic theories posed during Phase Alpha) were tested later. *NASA*

alone mere rumors, of a Soviet boost-glide bomber would not alone justify crash efforts to build Dyna-Soar. It had to stand on its own merits, as measured by national policymakers.

The residue of Air Force priorities threatened to seep into civilian space developments. Representatives of NACA's Lewis facility in Cleveland, Ohio, had met with planners from several leading aviation companies to explore man-in-space vehicle concepts. Many proposals reflected the conclusions of NACA researcher Alfred Eggers that a blunt-bottomed cone was the most expeditious approach to putting a human into space. However, NACA personnel noted that "Bell, Northrup, and Republic Aviation set this ideas [*sic*] aside as a stunt and consequently these contractors stressed the more elaborate recoverable hypersonic glider vehicle as the practical approach to the problems of flight in space." Northrup in particular "stressed . . . the military advantages and capabilities of the reconnaissance-bomber glider vehicle" when presenting their adapted concept of Dyna-Soar to NACA. This was expected to weigh a colossal ten thousand pounds, and Northrup representatives said that a "near-orbital manned flight" could be anticipated in 1964. "Much of the allotted discussion time was spent in reviewing [the] weapon system capabilities," reflecting that significant elements of the aviation industry expected the Air Force to prevail in its policy race against Eisenhower administration constraints.[57]

The size, parameters, and organizational control over space projects continued to prompt discord. The Bureau of the Budget balked at the extent of space spending. Stans, who pointed to a $10 billion deficit in 1959 "and the prospect of deficits continuing into the 1960s," wanted to slash the $500 million–$600 million projected space budget in half.[58] In the years to come, space budgets would instead blossom to sizes ten times larger than the one that had seemed so upsettingly large in 1958. Stans also observed DOD's reluctance to relinquish its manned space projects to a new agency. On June 10, conversations between General Schriever, ARPA representatives, Nobel-winning physicist Edward Purcell, and NASA's acting administrator Hugh Dryden concluded that extending national sovereignty endlessly skyward was "absurd." It was impossible, they thought, "to try to apply national sovereignty to an ascending cone of outer space, to stay within which it is necessary at sufficient altitude to travel faster than the speed of light." In practical terms, national airspace sovereignty would have to end somewhere, or space exploration would not be possible. Purcell suggested a freedom of space travel principle to secure "passive transit" in space, specifically describing "satellites which take photos and are recoverable or [which] telemeter them [the images] back to home territory" as an example of passive transit.[59]

Purcell suggested a "delineation of responsibilities," granting ARPA "manned maneuverable space flight" and "manned maneuverable orbital flight" to challenge Air Force plans. The group did believe that "man can do some things better than a machine in an unknown environment," and it listed alternative projects to give a human occupant of a spacecraft with a "circular orbit at 300 mi[les]–7 times a day covers the whole Soviet bloc in 8 days." Possible craft listed were an "ejectable capsule," the X-15 rocket plane, and Dyna-Soar, described as a maneuverable vehicle to be potentially ready in 1963.[60]

Killian's assistant, Robert O. Piland, warned him that the Defense Department was gaining control of the nation's space program. Despite ARPA's authority over "man-in-space and communications projects," Piland feared resurgence by the military branches to "set a precedent for future, more extensive manned projects" under military control. York, ARPA's chief scientist and soon to become the DDR&E, informed Killian that although "consideration of space weapon systems, such as bombardment satellites, would be limited to studies in the 1959 budget . . . ARPA had an obligation to conduct the studies to insure that military opportunities were not overlooked."[61]

Quarles and York agreed that the "man-in-space program" was neither "primarily" military nor "primarily" civilian and that long-range objectives in space would determine its ultimate position.[62] Piland warned that Defense authority in a man-in-space project would "set a precedent" regarding future projects, regardless of whether they had a military application. He indicated that man-in-space efforts were no longer on a "crash" basis, insisting that "responsibility should be assigned on more basic grounds than 'who has' or 'can do what' at the moment." The administration aimed to follow its ideas about centralized organization of space projects, and authority over many space projects passed to ARPA and then on to NASA. This transfer was not inevitable and was not taken for granted. In the meantime, "for all intents and purposes, ARPA has, of necessity, become the Nation's first space agency."[63]

Piland explained that the "man-in-space project would still be the responsibility of the NASA since the military operational system is not yet defined." Meanwhile, "DOD would be permitted to not only develop and operate systems of a military nature and conduct research necessary to develop or determine feasibility of a particular system but would also be permitted to conduct any space activities that might have military application."[64] The Air Force had an expansive view of what constituted its proper use of space. In late July, Piland reported to Killian, in even more emphatic tones, of a need to curtail military involvement in space: "at the present time there is no seriously proposed weapon system, or military operation, which requires the development of a manned satellite." He went on to say that "no reasons have been advanced which indicate that this research and development activity is 'necessary to make effective provision for the defense of the U.S.' other than the '*feeling*' that the military ultimately will require manned satellites or other space vehicles."[65] While the administration shuffled most space projects into ARPA and NASA hands, the Air Force retained authority for its space glider, although the very concepts of space bombardment faced attack and the role of the military in space exploration endured criticism.

July became August, and Eisenhower's advisers continued studying the role and needs of reconnaissance satellites. Richard Leghorn sent David Beckler revisions of a 1955 memo about unauthorized overflight of the USSR. Leghorn had written in July 1957 proposing a reconnaissance satellite, an idea that he had already supported for a decade. Now in July 1958, he reminded the PSAC member that "we should not plan on conducting unauthorized [aircraft] overflights indefinitely." Predicting current unauthorized overflight capability to diminish in about 1959 or 1960, he suggested various "supplementary systems" including reconnaissance satellites, very-high-altitude jets, and ram-jet missiles, but he did not mention a boost-glide vehicle. Like Killian earlier, Leghorn questioned the real utility of aerial inspection to guard against surprise Soviet attack. The ten-hour-long flight time between targets did not completely preclude sudden redeployments and attack, and aerial inspection was an episodic (rather than regularly recurring) method in any case. Fortunately, he expected reconnaissance satellites to provide "fully operational capabilities in 1960."[66]

Accurate intelligence information was a constant and ongoing requirement. "Intelligence" based on guesswork and supposition could lead to cripplingly misdirected spending and even to dangerous strategic decisions. The combination of guesswork, fear, and opportunism had created the myth of the bomber gap in the middle of the decade, and parallel fears could run rampant about missiles as well. The Air Force held no monopoly on alarmism, which was also evident and pronounced in Congress, the media, and elsewhere. The CIA, for example, had estimated in August 1956 that the Soviet Union would possess 470 bombers and 100 ICBMs by the middle of 1958. Two years of reflection, with the benefit of repeated U-2 flights, caused a complete revision of these estimates by the time 1958 actually came, however. In June 1958, the CIA reported its belief that the USSR actually possessed 135 bombers and no operational ICBMs at all.[67]

Analysis in August indicated that reconnaissance satellites would provide more information about Soviet developments *in progress* than about finished achievements already deployed. "Repeated overflights of USSR test ranges with high resolution cameras" would yield valuable information on Soviet missile development but would be less effective identifying weapons that were already built, installed, and camouflaged. Envisioning a "substantial number of satellites" equipped with infrared detectors, "a detection of the launching of ICBMs may be possible by 1963."[68]

Leghorn recognized that improved technical overflight capabilities had to be complemented by political initiative to protect U.S. intelligence gathering. He wanted "to minimize tension with the Soviet Union resulting from the unauthorized overflight program [and] to establish our legal right to be, when unarmed, anywhere in free space," which he personally considered an altitude of about twenty-five miles.[69] Quarles voiced similar ideas to Eisenhower at an NSC meeting on July 31. Recalling that the U.S. satellite project predated the IGY, he explained that from the start it had been "motivated by the desire to have a vantage point from which to view what goes on behind the Iron Curtain. In order to operate reconnaissance satellites for this purpose, however," he continued, "it would be necessary to establish the doctrine of freedom of outer space."[70]

The Sputniks had inadvertently helped set a precedent favorable to Eisenhower's freedom of space policy at present, but its future remained unguaranteed. Opponents such as Stuart Symington cried for ever greater U.S. military capabilities. Having spearheaded congressional angst about the supposed "bomber gap" during Eisenhower's first term, he promoted unrest about an equally specious "missile gap" during the president's second term. Symington leveled caustic statements against the administration, questioned intelligence estimates of Soviet strength, and demanded that Eisenhower act and spend spectacularly more on defense.[71] These noises in Congress became popular among other Democrats who were, like him, interested in proving their mettle as Cold Warriors. Perceived military weakness carried political dangers. The potential of military vulnerability carried existential risk, and it could not easily be discounted.

The power to deny space to the enemy, or the capacity to control it, might prove useful if a general war were to come. Officials considered the technical possibility of denying the use of space. By 1958, researchers anticipated that "a small number of megatons exploded at a suitably high altitude [would] make manned space flight impossible for a period of time." The Argus tests, a series of nuclear explosions triggered in space, aimed to determine the exact impact.[72]

Dyna-Soar represented a far more ambitious alternative for controlling space than a denial technique of in-space detonations outlined in Argus. Joint Chiefs Chairman Gen. Nathan Twining told the president and the National Security Council that U.S. "scientists were not really in sympathy with the military objectives in the exploration and exploitation of outer space." Science dollars thrown into the sky might be gone forever; they would not necessarily land back on earth.

Eisenhower remained unmoved; Killian noted that "fifty percent of our American scientists were now working in one way or another for the military services." Regarding space, Eisenhower emphasized centralization of research under civilian (ARPA or NASA) authority. He noted that "not all of these space projects were going to turn out to have military implications, at least at the outset." Therefore, NASA (not yet officially established) would explore space technologies, proving "the military practicability or feasibility of a given space project or activity," after which DOD would gain authority over it. Defense Secretary McElroy doubted that his subordinates would wait patiently for a civilian space agency to decide whether to cede control of projects to the military. Eisenhower answered that he "couldn't at this time help but look upon these initial activities in outer space as we used to look at wildcatting in oil in a former day."[73]

Eisenhower officials had worked to preserve space for the arrival of reconnaissance satellites, and this meant impeding the aerospace ideas touted by the Air Force during the first months after Sputnik. Eisenhower had become increasingly committed to the idea of centralizing space projects in an office outside the control of the military branches. The military might conceivably be granted authority over technologies *after* their applicability had been confirmed elsewhere. The president's overarching policy was that the military would not develop space technologies, since he expected that space projects run by the armed forces would burn through valuable funds and would be more likely to endanger his policy than to actually contribute to national security. Dyna-Soar slowly, quietly, and in the face of impediments, continued.

8

"Satellites Are Our Last Chance"
Pursuing the Need for Reconnaissance

Ok says P[resident Eisenhower], but still thinks it should
be out of Defense. Pres[ident] wants satellite orbiting and
working before Defense takes it. McElroy defends recon satellite.
Pres[ident]—have we proved we have the satellite we want—lets
[sic] get what we want and then give Def[ense] application.

—MEETING NOTES, PRESIDENT EISENHOWER WITH HIS ADVISERS,
NOVEMBER 28, 1958

The Search for a "Psychological Victory"

The 1958 midterm election unambiguously reflected the public's shaken confidence in the president and his policies.[1] Eisenhower commented in his diary that the period from mid-October through early November saw such "intense political activity . . . that it was impossible to make daily notes." The recession that had started in the latter part of 1957 lingered. Democrats rejected Eisenhower's fiscal restraint, predicting that bold spending initiatives would counteract Soviet strategic challenges and simultaneously reverse the galling effects of recession. Meanwhile, a corruption scandal engulfed White House Chief of Staff Sherman Adams, and Vice President Nixon found that his recent uptick in popularity vanished and that his attempts to assist Republican candidates that fall had been in vain.[2] Ten incumbent Republicans were unseated in the Senate as Lyndon Johnson's narrow majority grew to a formidable 65 of 96 total seats; in the House another Texas Democrat, Sam Rayburn, oversaw the dramatic widening of his party's majority.

President Eisenhower (shown here during the first press conference after the Sputnik launch) developed a space policy nested in his national security priorities during the mid-1950s. Contemporary aerospace doctrine conflicted with this, posing serious challenges for Eisenhower from the time Sputnik altered the national discourse until the close of his presidency. *Library of Congress*

Although prepared to insist on a long-haul approach to waging the Cold War, Eisenhower was reluctant to flout what he regarded as an expression of the national will. The U.S. position in space was not actually weak. On Sputnik I's first anniversary, only a single Soviet satellite remained in orbit, while three U.S. satellites were also there. Nevertheless, as White House personnel recognized, U.S. space accomplishments garnered less attention even if they were more scientifically useful. In this environment, administration officials and their advisers would take steps against military space projects, but only within the parameters set by Eisenhower's reluctance to defy a gathering national demand. PSAC chair Killian's assistant Robert Piland had already counseled against military interests in space earlier in 1958, and he next turned his attention to Dyna-Soar.

On November 3, 1958, Piland scathingly criticized Dyna-Soar in a memo to James Killian. "The question of the need for a satellite vehicle capable of maneuvering and landing upon re-entry appears to be confused with the need for a *glide* missile," he expressed. Condemning the program as ill conceived and claiming that "the Dyna-Soar concept has been studied for five to ten years," Piland challenged the early idea of extending vehicle range by skipping along the earth's upper atmosphere. Satellites potentially nullified the relevance of a vehicle's "range." Piland dismissed Air Force interest in Dyna-Soar's future reconnaissance and bombardment capabilities by pointing out that satellites and "ballistic missiles, including Minuteman," would address these respective roles.[3] Although both a reconnaissance satellite and a functioning Minuteman were seen to be feasible, neither one yet existed. Piland's memo painted the portrait of Dyna-Soar as a misfit. "The conceptual test vehicle," he continued, "is something short of a weapons system but decidedly different from the X-15 research airplane concept."[4] Piland noted that "the NACA/NASA position is that a glide vehicle capable of speeds of 20,000 feet [per second] or thereabouts is a reasonable extension of the research airplane concept . . . valuable in studying and evaluating problems of flight at high altitudes," but the civilian agency would not comment on "the military utility of the vehicles." This was in keeping with the standard practices of NASA and of NACA before it. Piland bluntly added that Dyna-Soar's "desirability as a weapon system has not been clearly established in comparison with reconnaissance satellites and ballistic missiles."[5]

Killian professed an open mind to future possibilities, and he said that R&D regarding weapons development "should [be] predicated on an over-all grand strategy . . . which includes as an inherent concept the recognition that the strategy must be modified as new technological opportunities arise."[6] Officials identified reconnaissance, communications, meteorology, and navigation as "areas which obviously represent potential operational systems that have military application." Payload capacity for boosters expected in calendar years 1958–60 ranged between 22 pounds and 2,700 pounds, although future systems "will eventually allow a satellite payload capability of 6,000 to 8,000 pounds." A NASA working group anticipated launching six 300-pound satellites per year starting in 1959 and three 3,000-pound satellites per year in 1961. The planned growth was ambitious, but it highlights the challenge of launching the more massive Dyna-Soar craft.[7]

While science advisers considered the need for "an adequate successor to the presently operational special reconnaissance aircraft," effort focused on "a new, small and reasonably lightweight *aircraft* [emphasis added] carried aloft to supersonic speed by the B-58 as a mother aircraft."[8] The SR-71 Blackbird, and a variant carrying a piggyback drone, emerged from these plans. Work toward a U.S. satellite continued as well, in parallel with the expectation that the USSR intended to collect intelligence information by constructing a "photo-reconnaissance satellite," which one high-ranking Soviet rocket authority called "a crucial intelligence tool."[9] The U-2 was a stop-gap, and the technological challenge of building a reconnaissance satellite reinforced the expedient of illicit overflight, even so far as to favor construction of a special performance aircraft over a piloted space concept that was still more complex than the Corona program.

Space remained a political football when the Dyna-Soar program first entered the president's line of vision in November 1958. Johnson had started hearings on space a year before, and eight months of testimony had urged weaponization of space. Johnson took advantage of a misunderstanding as another opportunity to lift his stature by addressing the UN General Assembly on November 17. Whereas at the start of the year he had asked White what work would be vital "to control outer space," to the UN in November he sang a very different tune: "Until now, our strivings toward peace have been heavily burdened by legacies of distrust and fear and ignorance and injury." His next sentence, "those legacies do not exist in space," somewhat conveniently glossed over some of his own earlier public statements. The following comment that these attitudes "will not appear there unless we send them on ahead" said more than he perhaps was aware.[10] It was the dour perspective of human nature and the confident expectation of technological triumph that fueled aerospace thought. It was the force behind Dyna-Soar and the desire to "control" the "high ground" that was space.

Eisenhower first discussed Dyna-Soar with his advisers on November 28, 1958. Two "exotic programs" were mentioned during a working vacation in Augusta, Georgia. One was ANP (aircraft, nuclear powered); the other was a very minor program called Dyna-Soar. ANP had been on the Air Force wish list for years, and tests had proceeded from 1955 into 1957 to confirm that a working nuclear reactor could be carried on a very large aircraft and that extensive shielding could protect the crew on board. Progress stalled with the problems of developing a reactor small enough to be carried on a plane yet powerful enough to propel it. Eisenhower asked what Dyna-Soar, was, and he was told that it "follows on [the] X-15" as a "high altitude & high speed [vehicle] moving to an [sic] 4, 5, 6, 100 miles altitude."[11] This description followed NACA's, rather than the Air Force's, interpretation of the program's relevance.

Dyna-Soar did not resurface during the discussion, but Eisenhower's philosophy was reflected in discussions about ANP. Eisenhower asked why NASA and AEC did not get control over nuclear propulsion research, rejecting Quarles' counterarguments and emphasizing that "it should be out of [the] Defense" jurisdiction. Furthermore, Eisenhower "want[ed a] satellite orbiting and working before Defense gets it." McElroy defended the reconnaissance satellite's position, too, but Eisenhower responded, asking whether "we [have] proved we have the satellite we want" and explaining, "lets [sic] get what we want and then give Def[ense] application" of it.[12]

Because Eisenhower aimed to set his own policy but was unwilling to defy what appeared to be the national will, he was sometimes willing to bend to the suggested need for space propaganda stunts. Two December launches did little to assuage aerospace advocates, who claimed to speak on behalf of a nation that they claimed demanded more activity in space. In the late summer of 1958, Eisenhower had reminded the National Security Council that "we are struggling for a psychological victory."[13] A third, heavier Sputnik had been put into space that May. A "psychological victory" had to be in keeping with national policy, aligning with the president's hope to ensure and expand the freedom of space. Organizational models and administration rhetoric meant to point to a "national" space effort. Goodpaster informed press secretary Hagerty that "accomplishments [by the United States in space] should be announced as *U.S.* accomplishments, *not* as accomplishments of particular agencies or organizations such as the Army, the Air Force, etc." Two launches at the end of 1958 sought to provide the psychological victory the president wanted in order to assuage the public.

The first of these launches illustrated the dangers of space. A U.S. launch at 4 a.m. on December 13 was intended to mark the first time that a primate entered space and the first time that a creature would be recovered alive from space. The squirrel monkey, called "Gordo," rode inside the Jupiter rocket's nose cone and entered space, but a parachute failure precluded his survival and the capsule's recovery. Alert to dangers and the potential media fallout of test animal fatalities, Assistant Secretary of State William M. Roundtree had successfully prevented the use of a rhesus monkey, given the "respect, and in some cases, veneration" that these creatures enjoyed in southern Asia. The Operations Coordinating Board suggested that, in the future, efforts should be made to "avoid personalization of the monkeys" and that they should be referred to only by lab names. NASA preferred to "call them Able and Baker . . . lest the press originate fanciful names for the monkeys" to be used subsequently.[14] Problems successfully recovering a small monkey from a short ballistic space trip underscored the technological challenge that deploying a piloted space bomber would entail.

The second of the December launches represented a visible success but little in the way of major scientific achievement. Somewhat ironically, Quarles had reported the very opposite to Eisenhower on August 7, advocating the launching of a "stripped-down ATLAS" to place a nine-thousand-pound satellite with a one-hundred-pound payload "into orbit by late 1958." Quarles noted that such a large object would be visible with the naked eye and "would provide important engineering data of use to the planned 'Man-in-Space' program in terms of guidance and control." Quarles acknowledged the "high risk of failure" of this but "considered [it] worth the risk involved." He also mentioned that the estimated cost for the payload development and booster modification amounted to $600,000, to be provided by ARPA.[15] Quarles identified this project as something capable of fulfilling the "psychological victory" that Eisenhower had that same day told the National Security Council he needed. The action was dubbed Project SCORE.

Piland disparagingly described the project to Killian as the "Atlas 'Carcass' Satellite." Although the warhead was to be replaced by a package not containing any weapon, the vast bulk (nearly 99 percent) of the satellite would be its fuselage rather than its payload, and Piland found this reminiscent of military presence in space. Atlas was, after all, the country's ICBM at that time. "I believe the possible reaction to the U.S. putting a *missile* [emphasis in original] in orbit to pass over other countries might cause adverse reaction," he wrote.[16]

Quarles' perspective won the day, and Project SCORE closed 1958 with a dramatic and visible success. One significant item in its payload was equipment to allow the relay of a message from earth to the satellite and back down to earth. It consisted of fifty-eight words, spoken by the president, proclaiming the scientific marvel and "convey[ing] to . . . all mankind America's wish for peace on earth and good will toward men everywhere."

Project SCORE was an encouraging psychological coup; it was not a knockout blow. Nor did the Eisenhower administration see psychological space achievements as being capable of offering the opportunity for such a blow in the Cold War. An OCB report at the start of 1959 illustrated the administration's perception on this point: "There is less tendency—particularly in the responsible press—to treat each latest triumph as indicating final 'victory' in the space race, or as giving proof of a commanding lead."[17] The administration hoped that it could answer calls for action sufficiently to dispel problematic levels of domestic opposition, permitting the president to wage the "long haul" Cold War campaign against the USSR as he intended to.

In Need of Satellites

Meanwhile, Discoverer tests still failed to lead to a functioning Corona, and ARPA director Roy Johnson deflected media questions about the WS-117L satellite project. While referring to "military missions where you would like to get back the data that you acquired and examine it visually," he refused to say "anything specifically" to a follow-up question suggesting that the government was working on satellite reconnaissance.[18] Johnson did note that Discoverer had been separated from WS-117L, but of course he did not mention that the Corona reconnaissance program secretly piggybacked on the unclassified Discoverer.

C. G. Villard, a member of the Space Science Board of the National Academy of Science's National Research Council, read between the lines and worried about the long-term repercussions of trying to hide strategic reconnaissance in ostensibly scientific projects, which had been mentioned by *Aviation Week* magazine. Villard cautioned that the government must resist the "considerable temptation to pass off what was in fact a military reconnaissance satellite, as just another scientific test. . . . Speaking as a scientist, I would hate to be a party to such a deception, however desirable it might seem at the time." Open societies cannot keep secrets forever, and the leakage of a deception would tarnish what Villard considered "one of the greatest psychological assets of the Western powers, [that] the circumstance that in general their public announcements are truthful."[19]

A simultaneous danger was of falsehood being proclaimed to, and by, the press. In the fall of 1958, Senator Symington accused the administration of incompetence, relying on information illicitly received by a former assistant working at the Convair aircraft company and then passed forward to the senator. Two days before Project SCORE, the director of central intelligence briefed the vocal Symington about what the intelligence apparatus knew of Soviet missile development. Presented with CIA figures on Soviet missile testing, or rather a lull in their missile firing, Symington's assumption that the USSR would have five hundred ICBMs by 1962 began to seem somewhat silly. Relying on his own experience in industry and dealings with corporations, Symington knew that it was difficult to imagine the USSR relying on missiles without testing them, but CIA analysis found no evidence of tests. Soviet missile development tests had lapsed after the Sputniks and were not occurring at the time of the briefing. Indeed, they would not resume in earnest until March 1959.[20]

According to a staff member's notes, "Considering the estimated cutback in bomber production in conjunction with our figures on ICBM firings, it sounded [to Symington] as if Khrushchev was violating Teddy Roosevelt's principle of speaking softly but carrying a big stick." But later during that same briefing, Symington expressed his continuing belief that the Soviets were flying a nuclear-powered aircraft. Administration personnel conceded to being "puzzled" by some of the intelligence information, but Symington believed in rounding any confusion upward so as to support intensified Air Force growth. He refused to seriously believe any of the intelligence information conflicting with his presumptions of massive Soviet military buildup.[21] Eisenhower "called Symington 'neurotic' and a 'demagogue' who 'leaked security information' that 'mislead' the public."[22]

Even the best intelligence estimates about the Soviets were in fact necessarily based on guesswork. Intelligence gathering—and digestion—is almost always a messy task, and an intelligence picture can never be available in a complete, clear, and simultaneously timely form.[23] The CIA anticipated 130 Soviet ballistic missiles operational by 1962, and the Air Force and Senator Symington each insisted that Soviet strengths were growing at geometrically faster rates than believed by the CIA. In fact, Soviet missile buildup during the latter years of Eisenhower's administration was fairly scant. With sporadic U-2 flights overflying the USSR's enormous landmass in small strips, U.S. intelligence personnel could develop a picture of Soviet strength and development, but it remained a picture with gaps. A bare fraction of the USSR could be studied, with only a narrow sector of the country overflown on any single U-2 mission. Herbert York later described the task as "trying to prove a negative on the basis of a very small sample."[24]

In discussion with George Kistiakowsky in January 1959, Eisenhower conceded the scientist's belief that the USSR might possibly possess "an operational long-range missile force." Kistiakowsky was the PSAC member set to succeed Killian when he stepped down as chairman that summer and make his long-postponed return to MIT. As Kistiakowsky had mentioned, long-range ballistic missiles seemed "in fact [to be] a focal point in their whole defense concept." But Eisenhower recognized the vital questions: What were the numbers of these missiles? What was their degree of accuracy? And, crucially, what did the USSR expect to happen afterward if it struck the United States? Even if the USSR had such missiles and used them, the president pointed out that "they [the Soviets] would still be exposed to destruction" from the rump of the U.S. deterrent forces.[25] Eisenhower did not evince anxiety that the nation needed a revolutionary new generation of weapons systems to preserve its deterrent posture.

Despite his personal confidence on security issues, the president felt constrained in space, confiding to close advisers on February 17 that he was "not disposed to challenge the space program conception, or to try to put a stop to a major program," because "the psychological build-up would be untenable." Eisenhower felt "sure that the Congress would break loose under the pressure" and observed "that world psychology on this matter has proven to be tremendously important—even if it is not too well informed." Killian added, "We may have a recurrence of the Sputnik hysteria if the Soviets get a 'man in space' first."[26]

Public opinion gave the president pause in his space programming. Eisenhower expressed unwillingness to dramatically alter space projects that enjoyed substantial public support. Dyna-Soar remained obscure in the waning years of the 1950s. Nonetheless, Eisenhower's confidential statements indicated a concession that he might bend rather than break in the wind of popular insistence in this area. He did not expect Soviet space successes to bring the prospect of an existential threat, however. This was true from the first days following the Sputnik launch and remained so for the rest of his presidency. Eisenhower regretted feeling compelled for domestic political reasons to partially placate the public in its hysteria. He felt able to do so because he did *not* share that hysteria.

OCB's task was to coordinate various policies and report to the NSC, and in March 1959, it predicted a Soviet manned circumlunar mission "with reasonable chance for success" as early as 1961. However, a Soviet human lunar landing seemed unlikely at least in the next six years. In the event, the USSR accomplished neither of these tasks, but OCB's predictions show a sober respect for Soviet technical capability. Trends indicated a prototype U.S. "strategic weapons delivery" system in 1961 (assuming that it won the endorsement of policymakers) and a "maneuverable type" manned satellite in 1962 or 1963. In contrast, OCB considered a Soviet manned capsule possible between 1959 and 1960, "glide-type vehicles" as early as 1960 or 1961, and 25,000-pound satellites between 1961 and 1962.[27]

In keeping with Eisenhower's own thinking, OCB did not assume a Soviet lead in space to constitute an existential security threat to Americans on earth. Soviet use of ICBMs, according to OCB estimates, indicated that "Soviet space exploration programs and military programs are complementary" and that "future Soviet programs probably will be established for fairly specific scientific and/or military purposes in accordance with a planned, step-by-step progression from one achievement to the next." OCB suspected that "scientific purposes" as well as "political and propaganda gain" lay at the root of an organized Soviet space plan: "We believe that the Soviets intend to pursue an active space flight program designed to put men into space for scientific and/or military purposes" and that the Soviets intended to reach the moon, Mars, and Venus with space probes.[28]

Significantly, OCB thought that "immediate or known military considerations may have no bearing on the decision to develop certain types of space vehicles, although the successful development of these vehicles could result in military applications."[29] In this respect, the OCB's portrait of the Soviet space program compares interestingly with the contemporary U.S. space program. Few doubted that the Soviets would leverage space technology for military purposes if it proved useful to do so. Eisenhower and his advisers generally did not expect it to prove useful. However, they were unprepared to invest the political capital needed to defeat (rather than contain) the Air Force's doctrinal and technological interest in space.

NASA's Mercury gathered momentum when, in April 1959, Eisenhower granted it top development status. Three months earlier, NASA and DOD officials had met to coordinate and deconflict their efforts. The central topics of study included Mercury and Discoverer, but they felt that "the requirements of the DYNASOAR Program . . . [had] not yet been thoroughly defined." Furthermore, they "will be reviewed in order not to duplicate the Mercury Program equipment." Space planners would continue to associate Mercury and Dyna-Soar with each other as if they were equivalent first-generation space vehicle projects.[30] This linkage ignored the enormous gap in complexity and in expected capabilities of the two vehicles, it overlooked how a crewed capsule had actually been established as a preliminary step toward the more complex vehicle, and it encouraged assumptions that Dyna-Soar was potentially redundant.

"Satellites are our last chance," warned Richard Leghorn. Consistent with his earlier messages, Leghorn believed that "the problem is not a problem of technology. It is not a problem of vulnerability to Soviet military measures. The problem is one of the political vulnerability of current reconnaissance satellite programs." He had predicted previously that the U-2's capabilities would decline between 1959 and 1960, and by April 1959 the United States found itself already in that window. He warned, "Should reconnaissance satellites be 'politically shot-down,' no scientific or technological opportunity can be foreseen to obtain this security information during the forthcoming years." The solution would be to negotiate the freedom of space, to establish a satellite detection capacity of nuclear tests, and perhaps to offer satellite surveillance to the UN.[31]

Killian had considered complementary issues in March, when he had written to Eisenhower about a possible agreement banning nuclear tests in the atmosphere. Killian had emphasized to the president the importance of establishing a "formal agreement (rather than . . . unilateral action)" that included "some system of monitoring" compliance. The scientific adviser hoped that "such an agreement might include specific provisions for a phased, evolutionary extension of the test ban to include coverage of testing underground and at high altitudes when controls adequate to detect such tests became technically feasible." A formidable hurdle in contemporary discussions with the USSR was the lack of feasibility and dependability of test detection. But in his memo, Killian identified another challenge: "The limit of the 'atmosphere' will probably be difficult to establish."[32]

Defending Sufficiency

The administration flexed its policymaking muscle during the spring. During the National Aeronautics and Space Council meeting of April 27, Eisenhower commented "that any space program, whether it be military or civilian, should come before this Council for recommendation with final determination by the president."[33] This suggested some interest on the president's part in retaining control over space policy, his February remarks about "world psychology" regarding space being "tremendously important" notwithstanding. Director for Defense Research and Engineering Herb York had already impeded the Dyna-Soar program. On April 13, he reemphasized the priority of hypersonic flight research over studies that would support weapons systems variants. He was assisted by the fact that the Air Force had, on Secretary Douglas' advice, revised its Dyna-Soar plan the year before to stress the suborbital aspects of Dyna-Soar I. The Air Force had then intended to prevent York (who was, at that time, chief scientist at

ARPA) from separating Step I from the rest of the program and grabbing it for either ARPA or NASA. Now the price of keeping the program together, the emphasis on suborbital hypersonic flight research, became clear.

In line with Eisenhower's policy and with his policymaking style, Deputy Defense Secretary Quarles delivered a one-two punch to Dyna-Soar early in 1959. First, fresh on the heels of York's directive, Quarles informed the Air Force that the approval he had earlier given to Dyna-Soar's development was only with respect to its research and development as a hypersonic vehicle. Approval of this aspect did not, he declared, imply approval of the weapons system that the Air Force envisioned with Dyna-Soar III.[34] Now, it seemed, Dyna-Soar I and its Step II and III derivatives were effectively separated from one another anyway.

Quarles also emphasized that "space is a place in which old missions may be more effectively carried out with new tools." Therefore, Quarles proposed reprogramming DOD funds, based on defense priorities at large, without preferential treatment of some projects based on "their competitive position on a scale of Defense values rather than on their relationship to the rest of the space program."[35] Dyna-Soar now inhabited the worst of two worlds: approval of the weapons system that the Air Force desired was still not forthcoming, and consideration of the weapons system would be challenged by comparison with other deterrence technologies already in place or under development. Since Piland's excoriating November 1958 memo, Dyna-Soar had increasingly been seen by policymakers as a research vehicle, and as such it was subject to comparison with other research vehicle proposals such as Mercury.

That summer, Eisenhower strove to ensure that the U.S. space program focused on crucial items whose costs could be justified. Eisenhower also described Congress's apparent fickleness, commenting that it was "showing some tendency toward cutting the space budget," although "the Senate Majority Leader [Johnson] had made great speeches when the country was stirred up on this matter, calling for going all-out on crash programs. Now the [Democrats] in the Congress [seem] to be cutting the program back." The president's response kept faith with his general outlook; he "told Dr. Glennan [NASA administrator Keith Glennan] that we must be very sure that we can justify every program. He said he had no idea of cutting any particular item at the present time himself. He felt it was necessary, however, to have a range of projects," from a short list comprising projects "wholly justified for operational or scientific purposes," to a longer list "including some that are planned for psychological reasons."[36] Justifiable, manageable costs remained on the president's mind that afternoon, when he noted that basic research absorbed 40 percent of ARPA's budget, and he told a National Aeronautics and Space Council meeting that basic research "is the area in which unnecessary duplication of effort should be particularly guarded against."[37] Preventing overspending would go hand in hand with safeguarding important projects.

Centralization constituted a cornerstone of that effort. In mid-July, Eisenhower and Killian again voiced agreement "that ARPA should take control of space research throughout the Department of Defense."[38] Eisenhower cautioned NASA officials Dryden and Glennan against "'stunts' such as firing a vehicle into interplanetary space with no real purpose in mind," explaining that he "has always been to favor pushing out beyond what is now known, but only where there is reasonably something to be learned through the effort."[39] Eisenhower explicitly combined fiscal and organizational issues at an NSC meeting in August, noting his interest "in saving money. Some people say that the country can afford anything," but the president knew better. With his New Look in mind, he explained, "Our whole economy was as important as continental defense." However, "The question was, how do you do the job best but cheapest?" When Eisenhower lamented that more work had gone into building missiles than into determining their organization, McElroy praised Eisenhower for his work in preventing Congress from overspending on unreliable and rapidly obsolescing first-generation missiles.[40]

Despite ARPA's oversight of military space spending (projects totaling $500,000 or more were subject to ARPA review), Eisenhower told Killian that "the space program in the Department of Defense may be too ambitious and too costly. I have not seen evidence that it has had the hard-boiled technical review to determine what is realistically possible that has been taking place in the civilian Space program."[41]

Erratic patterns in rhetoric and signals heightened the dangers and uncertainty of the Cold War, and Khrushchev often displayed a testy demeanor in the summer of 1959. With Secretary of State Chris Herter and CIA Deputy Director Bissell, Eisenhower weighed the risks of authorizing each U-2 flight over the USSR. Bissell expressed worry "that the Soviets have a fighter which could probably zoom to the altitude of" the U-2, and Eisenhower responded by remarking "that Khrushchev seems almost to be looking for excuses to be belligerent."[42] Eisenhower met with Bissell to discuss successors to the U-2. Satellites would offer the long-term solution, but still the many bugs in satellite reconnaissance drove decisionmakers to talk about the possibility of a new spy airplane, either based on the B-58 manned bomber or using "a very advanced turbojet."[43] In keeping with OCB predictions about development schedules, the reconnaissance application of Dyna-Soar never came up during the discussion. More complicated than an unmanned satellite, it could hardly be expected to become operational earlier.

In late summer and into the fall, the president further attempted to clarify his views to advisers and to exert more control over space policy. If not for George Kistiakowsky, the departure of James Killian as PSAC might have been a severe handicap. Killian had done able work heading the Technological Capabilities Panel in 1954 and 1955, contributing to his selection as the first PSAC chairman. Killian reflected afterward, "I supported and perhaps had some influence on President Eisenhower's policy of moving slowly into space." Killian had decided to stay on as PSAC chairman ten months longer than he had originally planned, despite his interest in returning to MIT in time for the fall 1958 semester. When he stepped down as PSAC chairman on July 15, 1959, the president wrote a personal letter thanking Killian for "assum[ing] the complex responsibilities of coordinat[ing] . . . the governmental activities" in space, and helping to "develop programs, that, while adapted to requirements, were not dictated or designed in an atmosphere of panic." It had been a time, Eisenhower wrote, "when millions, startled by sputniks, wanted to plunge headfirst and almost blindly into the space age." Yet "no one," the president said, had done more than Killian "to bring reason, fact and logic into our plans for space research and adventure."[44] Killian's advice helped keep Eisenhower's hand as steady as it was. The pressures were certainly enormous and would continue to be so. Fortunately for the president, George Kistiakowsky was also an adviser upon whom he could rely.

Not everyone in the administration saw space policy or space programs in the same light. Differences surfaced at the NSC meeting on July 30, 1959, when Eisenhower told the National Security Council (as John Foster Dulles and Stans had done the previous summer) that the national space program ought not to be based on competition with the USSR "when we did not know exactly what the USSR was doing in outer space." It was not yet clear "whether we were in a 100-yard dash or a mile run," either. The NSC had called for the United States to become a "recognized leader" in space, and Bureau of the Budget officials hoped "to hold to the 'a leader' language." Hugh Dryden responded to the president by agreeing "that it was not necessary for our policy to be stated in terms of what other countries were doing," but Eisenhower answered that the majority proposal had nonetheless done exactly that.[45]

Defense Secretary McElroy suggested that because "we were approaching the point of operational capability for such military programs as surveillance satellites, communications satellites, etc. . . . it was time to effect a clear separation between military space projects and other space projects." Eisenhower answered that an operational capacity would let "the Defense Department . . . stop buying reconnaissance planes and buy reconnaissance satellites instead." The president aimed to avoid duplicating basic research, and he rejected McElroy's philosophy of undertaking parallel military and nonmilitary space research and development programs: "Any purely research activity should be conducted by the Space Agency while applied research looking toward a military capability should be conducted by the military."[46]

Eisenhower interrogated the NSC about the psychological aspects of space programs. Planners agreed on the need "to establish the U.S. as a recognized leader in this field" of space.[47] More contentious was the issue of where U.S. space achievements needed to stand, relative to those of the USSR, to make the nation "a recognized leader." The NSC had split on the issue, and the president registered disapproval with both majority and minority positions. In particular, Eisenhower questioned the NSC whether "comparable" U.S. accomplishments implied "equal" U.S. feats to Soviet ones. The majority had included in paragraph 62 of the Basic National Security Policy document NSC 5906, "In formulating and implementing [space] programs, due consideration should be given to the psychological potential of solid scientific and technical achievement." The State Department and others pressed to interject the phrase "where there are projects of comparable scientific and technical value." If "comparable" implied "equality," Eisenhower asked, then "who would determine which . . . projects had the greater psychological value? For example . . . [did] flying to the moon or visiting Venus [have] the greater psychological value." Eisenhower concluded that "the discussion was turning toward hypothetical situations," and this emphasis was counterproductive. Perhaps the psychological issues "pertain only to NASA, whose whole program was based on psychological values."[48]

Dryden objected to Eisenhower's premise. NASA, he insisted, "rendered very important support to the military." The president would not be swayed, saying that "the furor produced by Sputnik was really the reason for the creation of NASA." He repeated his challenge to explain how programs could be compared or considered "comparable." Dryden answered that "the psychological value of a project really depended on whether or not it was successful. What is done must be successful or it will have no psychological or scientific value." It was decided to pursue "scientific, military, and political purposes," with "the term 'political' includ[ing] consideration of psychological factors." Within this context, "a military space program" would be "designed to extend U.S. military capabilities through application of advancing space technology, *without invading the responsibilities of the National Aeronautics and Space Administration* [emphasis added]."[49] Eisenhower's statements indicate that he saw NASA's budget as a psychological investment rather than a scientific one. They also showed that the president knew that his hand could be forced by what he believed to be an unassailable public pressure.

Essentially, the July 30 notes show Eisenhower struggling to avoid waste, either of extraneous or counterproductive programs or of duplicated efforts. Faithful to the president's ideas, one discussion paper that fall deliberating the merits of a space race advocated "*a vigorous national space program of our own design which emphasizes scientific achievements. . . . More effort should be put into the refinement of smaller payloads and the earliest acquisition of scientific data. . . . Ambitious 'one-of-a-kind' shots, like circumnavigation of the moon, should be planned cautiously, because they open us to unfavorable comparisons with the Soviets.*"[50]

Service rivalry represented another species of waste. Kistiakowsky reported to Goodpaster that military space projects led "to strong competition among the Services to stake out the largest possible claims 'in space.'" At the end of September, Kistiakowsky suggested to Eisenhower that it might "be preferable to aim at fewer weapon systems projects, especially those with such ambitious requirements . . . that operational availability is many years off." Such constraints would allow "more effort [to] be directed to projects whose military usefulness is around the corner" so that "an effective modern force-in-being could be assured for the immediate future."[51] The trajectory of Kistiakowsky's advice clearly boded poorly for futuristic concepts like Dyna-Soar, which could in no way be considered "around the corner." The scientists' advice against the military services could draw wrath from each of the branches. When Army Maj. Gen. John Medaris deemed York as "second rate," however, Eisenhower's principal scientific advisers rallied to the DDR&E's aid.[52]

Technological Factors

The first technical challenge in utilizing space is simply getting there. The development and possession of powerful rocket boosters therefore influenced the direction of space exploration. Representing NASA and DOD, Administrator Glennan and Deputy Defense Secretary Thomas Gates reported to Eisenhower on October 30 of a need for "at least one super booster" and for responsibility to "be vested in one agency," in keeping with Eisenhower's own outlook. Glennan and Gates continued: "There is, at present, no clear military requirement for super boosters, although there is a real possibility that the future will bring military weapons systems requirements. However, there is a definite need for super boosters for civilian space exploration purposes, both manned and unmanned."[53]

From the vantage at the top of NASA and at DOD, a potential need for weaponization of space simply could not be dismissed outright. Several figures, including Glennan and Kistiakowsky, had earlier that month recommended progress on the Nova booster "as insurance against failure of the Saturn." Eisenhower approved the joint memo by Glennan and Gates on November 2.[54]

The Glennan-Gates memo, in the wake of Eisenhower's discussions with advisers, aptly describes the administration's perceptions of the space program. Dyna-Soar's as-yet-nascent stage of development meant that the program got little specific attention in White House circles into 1959. However, the deployment of a Dyna-Soar–type weapons system would have raised policy complications for the satellite reconnaissance that enjoyed the president's attention and priority. Despite all this, U.S. officials could not definitely rule out the possible future need for weaponized military utilization of space. As long as this possibility remained, programs including Dyna-Soar could not be entirely cancelled.

Glennan eagerly used the latter part of 1959 to press for greater NASA control in space issues. In late September, Eisenhower noted to advisers that "other countries did not react to the Russian Sputnik the way the U.S. did" and that "in fact, it was the U.S. hysteria that had most affect [sic] on other countries." Kistiakowsky told the president that DOD had declared "that if Dr. Glennan does not push fast enough in space activities, Defense will do so." Saying that "it would be desirable to delineate, publicly, a program of non-military space research, development and exploration," Glennan urged that the president move to amend the year-old Space Act to "make NASA responsible for the nation's space program including the development of all *new space vehicle systems* whether for use by NASA or the military services."[55] Clearly, any such adjustment would have brought Dyna-Soar under NASA rather than Air Force control. Such a move would certainly have prompted redirection of the program, transforming it into solely a research vehicle, a follow-on to the X-15.

Glennan also urged Eisenhower to follow his upcoming 1960 State of the Union address with a request for more than a billion dollars in space spending. "No development in modern times," Glennan declared, "has so stirred the imagination of the peoples of the world as the beginning of the exploration of space." Regardless of the major mission envisioned, the NASA administrator wrote that "we find that our capabilities are determined in the first instance by the available thrust of the first stage rocket booster." Eisenhower shared part of this view—that "the key to our space program was the Saturn or the big booster"—but Eisenhower did not want to give such a substantial boost to the profile of space activity.[56]

Repeating himself about the need for fiscal stability, the president noted on November 17 that "Sputnik gave a surge to defense spending from which we have not recovered." A scant week later, Eisenhower told the National Security Council that the Soviets' first satellite "induced a Sputnik psychology in this country. One had only to say 'moon' or 'missile' and everyone went berserk. Sputnik coincided with a recession resulting in tremendous pressure to spend additional money. The peak of our anxiety is now past and people are taking things into their stride." Goodpaster's notes showed that, on November 17, Eisenhower had ruminated "that if he has to approve another unbalanced budget he would be obliged to regard his Administration as discredited."[57] Unnecessary programs were expensive. American solvency and security were too precious.

At NASA headquarters, human spaceflight planner George Low noted in mid-November that "to date, the Discoverer project has been plagued with engineering difficulties, and none of the instrumental payloads have yet been recovered." The one really crucial instrumental payload in Discoverer was the Corona satellite reconnaissance project. Attempts from February onward had brought slow progress and extensive frustration. Simultaneously, the final draft of the U.S. Policy on Outer Space addressed, among other issues, the potential military uses of space, and it seemed to open the door to an eventual Dyna-Soar weapons system. The document acknowledged that military applications aimed to "enhance military capabilities" and were "being developed for use as operational systems." The paper identified a list of military applications "that are expected to be available earliest," consisting of meteorology, communications, navigation, mapping, reconnaissance, early warning, and inspection. Afterward, it noted that, additionally, "future military possibilities under study include: passive and active defense systems to detect and to destroy enemy missiles or space vehicles; *space to earth weapons systems to diversify further our strategic deterrent posture* [emphasis added]; electronic countermeasures," and potential lunar military bases.[58] Space-to-earth weapons capabilities were envisioned for Dyna-Soar, but this was nonetheless separated from the upcoming applications such as communications and reconnaissance, and it was restricted to a single brief paragraph about other "future military possibilities." Eisenhower's space security policy stood to fail if Discoverer/ Corona remained technologically stalled while Dyna-Soar foreclosed the geopolitical opportunity to deploy reconnaissance satellites in a space for peace environment.

NSC executive secretary James Lay penned a memo outlining the Dyna-Soar program less than two weeks later. Lay wrote to Eisenhower's assistant Gordon Gray, describing a recent briefing of the Space Council that had discussed the hypersonic glider and its potential role as a weaponized exponent of aerospace power. The Dyna-Soar program "was said to combine the operating capabilities in both air and space" and to lengthen flight duration and range. The Air Force divided the Dyna-Soar program into three phases, described by Lay. The first phase, cited as costing an anticipated $397 million and extending from 1960 through 1964, consisted of drop-tests from airborne bombers, in the same way that the X-15 was being tested. Then flights would be boosted by Titan missiles; unmanned versions were to be boosted in 1962, and the first manned boost-glide flight was scheduled for mid-1963. "The first-step test objectives include an effort to determine the military value of this project."[59]

A second phase, lasting from 1963 to 1966, would cost $241 million and would see the use of Saturn as a booster, with the objective of bringing Dyna-Soar into orbital flight. "The final step which is not yet programmed will be the development of an operational military system." Dyna-Soar's advocates expected the vehicle to "have mission adaptability because additional weight can be added onto the tail of the vehicle." Lay listed six potential future uses: satelloid reconnaissance; reconnaissance; satellite inspector, interceptor; orbital bomber; multi-purpose space station; and logistics maintenance and rescue functions.[60]

Although the list included several items, the orbital bomber function won the primary attention. Dyna-Soar would "provide 'three-dimensional dispersal' of weapons and continuous alert through orbital flight." The memo further explained, "It is contemplated that the vehicle might carry three [nuclear] bombs into orbit. Then, each vehicle and crew could be rotated once a month, while leaving the three bombs in orbit." To preempt potential Soviet objections that orbiting U.S. nuclear weapons intruded on their aerospace sovereignty, "these bomb loads might be carried in an equatorial orbit which would not go over the Soviet Union." Expanding beyond the bombardment role, "it was also suggested that if a system of space stations were established, the Dynasoar vehicle could be used as a taxi between earth and each of these stations."[61]

Remembering the four military space possibilities mentioned in the November 12 policy paper—detection and destruction of enemy satellites, deterrent diversity through space-to-earth weapons, electronic counter-measures satellites, and lunar military bases—the advocates' envisioned role for, and the significance of, Dyna-Soar becomes clear. It would fulfill its primary purpose by extending the U.S. strategic deterrent, and it might provide initial capability as a "taxi" to space bases and perhaps serve as a first-generation satellite interceptor. In so doing, it would answer three of the four potential uses of space identified as subject to study. The contractor selection for Dyna-Soar was completed in early November, on the eve of the policy paper and the Space Council meeting.

For two years, space policy had impacted the fate of the Dyna-Soar program. Lay's memo to Gray clearly shows that a future Dyna-Soar could dramatically affect future U.S. space policy and shape the diversified deterrent upon which U.S. security rested during the Cold War. By the close of the 1950s, Dyna-Soar remained obscure, but space policy won confused attention from a public that recognized space projects and presumed visible activity to indicate coherent policy. The course of space policy and the status of Dyna-Soar were intertwined, and although administration actions gave real reason to doubt the fate of the program, other factors suggested its survival. The Air Force's top officers were more determined to save Dyna-Soar than Eisenhower was personally committed to canceling it. Furthermore, the president was unwilling to confront the full force of what he expected to be a widely angry public if he were seen as totally antithetical to the national will.

Charyk ordered the Phase Alpha study at this juncture, buying time for the administration and for the Corona satellite's development. Dyna-Soar and the aerospace perspective it represented ran counter to the president's preferred freedom of space policy and the prospect of satellites providing the reconnaissance that Open Skies had sought to gain. Phase Alpha would stall the Air Force's space advocates for months, kicking the can down the road into the upcoming election year. But Dyna-Soar, and the problems of aerospace thought, could not be kept submerged forever.

9

"Slipping Out of Control"
The Struggle to Define Security in 1960

The military services feel trapped by the lack of a definitive, approved space program.
—AVIATION WEEK, NOVEMBER 28, 1960

Public opinion pressure frequently reveals a problem but rarely a solution.
—PUBLIC OPINION QUARTERLY, FALL 1960

"Decade of Decision"

"We should frankly admit that we are indeed in a race with the USSR," William H. Pickering told the assembled members of the ARS at their fourteenth annual conference, held in Washington, DC. The New Zealand–born Pickering, serving as head of the California Institute of Technology's Jet Propulsion Lab, was working primarily with NASA when he spoke in early November 1959. He had touched on an issue important to many Americans on the cusp of an election year. The existence of a space race left U.S. policymakers two alternatives: "Either pursue our space developments actively and successfully, or . . . declare ourselves completely out of the space race."[1] The answer, from the podium as in the hall, seemed obvious. The only serious choice was engaged and earnest competition against the Soviets in space. November 1959 saw the selection of Boeing as contractor for Dyna-Soar and of Martin for the Titan A booster rocket.[2] The Air Force pressed ahead with its space agenda, which administration officials sought to contain by ordering Phase Alpha.

Critics of Eisenhower's approach to space argued that the United States lacked "clear national goals, management and funds to support them on a long-term basis and public understanding of the importance of the program and the time and effort required to conduct it." General Boushey, also speaking before the ARS, called "a military space capability . . . a matter of urgency for our national survival." The president had disparaged the idea of "using or misusing military talent to explore the moon," wanting the military instead involved with "only what is their problem and not anything else. The rest stays under civilian control and that is the reason for having this agency [NASA]."[3] The ARS had urged the president after Sputnik to organize a civilian space agency, but in the closing weeks of 1959 it also proved willing to invite the Air Force's director of advanced technology to make the case for vastly greater military involvement in space.[4]

Defense, space, missiles, technology, and research were important issues as party primaries and then the general campaign proceeded in the close-fought campaign of 1960. Contemporaries expected aerospace overall to be profoundly affected by the outcome of the election, and the B-70 manned bomber system won high-profile attention because of the campaign.[5] Space received attention as well. Candidates such as Massachusetts Senator John F. Kennedy used the space issue as part of a palette for painting Eisenhower's leadership as insufficient and for insisting that the United States needed to take more prominent, active steps in this and other Cold War contests.

Senator Kennedy was a young and charismatic man, one of a host of Democrats with presidential aspirations throughout Eisenhower's second term. For his own part, well before the 1960 race, Kennedy had been working to join the two separate threads of defense criticism Eisenhower faced: that aerospace was being starved, and that conventional arms were being ignored. Kennedy embraced both, repeatedly condemning the New Look while insisting both that "we must have more and better missiles . . . and warning systems" and that U.S. security demanded moderniza-tion and growth in its ground forces. To reinforce this drumbeat, Kennedy compiled a book of selected speeches on foreign relations and security issues to showcase his rhetoric and perspective. One speech, for example, first delivered in the months following the Sputnik launch, warned against Americans "deceiv[ing] themselves about Russian intellectual achieve-ments" in the sciences.[6]

Fears abounded of a U.S. lag or gap and of the specter of the Soviet system's superiority, and an Associated Press report in January 1960 suggested that the USSR was "seven to eight years ahead of the United States in developing a Dyna-Soar type manned boost-glide space vehicle." Though rarely front-page stories, such reports reverberated among aviation trade journals and mainstream news outlets such as the *New York*

Times and the *Washington Post*. Ominous words from Khrushchev about "a new space weapon that was 'more incredible' and 'more formidable' than any developed so far" seemed to refer to "a manned orbital bomber of the type being developed in the United States in the Dyna-Soar program."[7] It seemed safer to many people to prepare for the worst possible scenario, guessing that known Soviet feats were just the tip of the USSR's technological iceberg.

The U.S. fear of Soviet prowess fit into a pattern of mistaken beliefs in democratic societies that various nondemocratic systems possessed efficiency through bypassing the democratic process. Some outsiders insisted that the hands-on approach of propaganda-hungry Nikita Khrushchev was more efficient than the U.S. model. Strangely, such observers did not seem to understand that the assertion of Soviet efficiency, and implicitly of Soviet superiority, manifested an antidemocratic assertion—a rejection of a Western tradition laid down by the likes of James Madison or John Stuart Mill that dissenting and conflicting voices can balance each other and even strengthen a society. The fears were also inaccurate (because the supposed communist consensuses and efficiency did not exist), but Soviet leaders were happy to leave outsiders with that impression. Identifiable contradictions between statements of different Soviet officials seemed part of a calculated Soviet disinformation scheme.[1]

Note

1. Scientist Anatoli Blagonravov told the ARS there was no need for a man-in-space program, while aerospace medical czar Andrei Kuznetsov had announced in the summer of 1958 that the USSR had selected four astronauts for a manned space capsule program. "Soviets View Man-in-Space Need," *Aviation Week*, November 23, 1959.

At *Aviation Week*, editor Robert Hotz disdained the frugal approach to the U.S. space effort and the reduction of the B-70 program "to a bare skeleton." *Aviation Week*'s readers included dissenters, such as one man from Los Angeles who argued that "our star chasers" forget that a zooming bomber posed more obvious deterrence and propaganda value than "an abstract picture of the moon's far side." He urged balance in defense spending, "lest we become so sophisticated that we succumb to a row boat invasion."[8] *Aviation Week* rarely printed such letters to the editor, however.

The president and his advisers foresaw the costs of writing a blank check. Dyna-Soar was set to require astronomical funding well before it could explore the heavens or exploit the high ground of space. Dyna-Soar, set to spend $35 million in fiscal year 1960, represented 7.2 percent of the total military space budget and less than 3.3 percent of the nation's total space budget. But projections suggested that by fiscal year 1964, the military space budget was expected to increase by half, and NASA's budget was expected to increase more than 150 percent. And with total U.S. space spending standing at an anticipated $2.1 *billion*, analysts expected Dyna-Soar to absorb more than a quarter of DOD space money and a staggering 9.1 percent of the *total* national space budget.[9] Though a small project at first, at the close of 1959 Dyna-Soar was poised to become the largest single item on the nation's space agenda.

Dyna-Soar was among the most serious possible avenues of space development when the Space Council made its projections in late December 1959. Mercury had begun as an unsolicited gleam in designers' eyes, a preface to Dyna-Soar, adopted by policymakers when they concluded it had political use. Similar conditions existed for Apollo and Dyna-Soar at their outset: they were concepts waiting in the wings, proceeding slowly and awaiting the green light. Early studies enabled policymakers to change their minds and change policy, spending cash to avoid losing time when changing course. This responsiveness to policy change, whether in rocket science or other fields, does not come cheaply. Funds spent on what become dead ends constitute only a small part of the price. Flexibility allows responsiveness to policy changes, but it can also tempt policymakers to defer decisions. Advisers recognized a need to keep studies about the weapons-related potentials of space on a back burner, because while a decision for "space to earth weapons systems to diversify our strategic deterrent posture" might one day be made, it contradicted Eisenhower's priorities, for both military and budgetary reasons.[10]

Answering press questions about an overall transition from planes to missiles, Eisenhower recalled an old aphorism, "Be not the first by which the new is tried, nor the last to lay the old aside," explaining that limited aircraft research (the reduced work on the B-70) complemented work to get "these missiles perfected to the point where we think that the deterrent itself needs nothing more." Air Force generals sought to reinvigorate interest in the B-70 by presenting it as capable of an increasing range of roles.[11] Although LeMay compared the B-70 and Dyna-Soar as competing alternatives to succeed the B-52, other Air Force planners saw the Mach 3 bomber and the boost-glide bomber as complementary, even cooperative, projects. To help safeguard both projects, the Air Force considered multimission capabilities for both craft. Fundamentally, however, both were bombers meant to maintain and diversify the national deterrent. The B-70 could also become a "recoverable first stage booster for satellites and Dyna-Soar-type boost glide vehicles." Parallel with LeMay's estimate, the enormity of the B-70 opened the possibility of its later variants "incorporating nuclear jet engines."[12] The B-70 was presumed to be a relatively safe bet, in terms of technology, with more immediate launch capabilities for Dyna-Soar than possible with contemporary liquid-fueled rocket boosters.

Aerospace advocates greeted the 1960s as the "Decade of Decision," and Hotz asserted that "while the space issue may not be the most important issue of the 1960s," U.S. handling of space issues would model approaches to other "significant problems in the technical-economic-political trinity" as well.[13] For this constituency, space would not only carry inherent importance in terms of technology, defense, prestige, and propaganda, it also would be emblematic of the era as a whole. *Aviation Week* and *Rockets and Missiles* magazines each had subscriptions numbering in the tens of thousands, and other science publications also captured attention and niches in the print publication market.

The price tag for the as-yet-unflown Mercury project soared, amid rumors of Soviet Dyna-Soar–type craft and advertisements that implied the U.S. boost-glide project's inevitability. NASA had needed to double the number of Mercury capsules to contract from McDonnell Aircraft, and building a global tracking network added further costs. To meet these Mercury expenses, funds were drawn from elsewhere, prompting a "darken[ed] scene in Seattle," where Boeing's KC-135 tanker plane and its "Dyna-Soar rocket orbital bomber" faced rumored cuts.[14]

Eisenhower maintained work on the vital Corona program while continuing to fight a political delaying action against critics stirring popular disapproval. Kistiakowsky confided to the president that although Corona had not yet worked, "each successive launch has resulted in some progress, one difficulty after another being eliminated." Perhaps a launch the following month would meet with complete success. Eisenhower began a mid-January meeting with his space officials "by saying he felt we are going to be impelled to put an additional $100 million into [the] Saturn [rocket], . . . the most important project before the Space Agency at the present time," because "if we do not take the initiative on this, the funds will be pushed on him, and this is an area in which he does not wish to be on the defensive."[15]

The president remained on the defensive, as Air Force brass publicly condemned cuts made to the B-70 program and the aviation trade press slammed the administration for moving funds and the Saturn project from Army control to NASA to merely give the illusion of a redoubled space program.[16] *Aviation Week* also informed readers that Dyna-Soar faced the major Phase Alpha review, that North American Aviation had proposed building a modified X-15 in place of the Boeing Dyna-Soar, and that Philip Farley, the secretary of state's special assistant for atomic and outer space issues, had endorsed work to ban space weapons that made *sustained* travel in space. This would preserve the role of ICBMs, but it undercut Dyna-Soar II and III. *Aviation Week* insisted that any arms control agreement that extended beyond what technology could confirm and verify was "patently ridiculous."[17]

The Question of Obsolescence

Eisenhower did not ignore space as his detractors insisted, and he concurred with Kistiakowsky's suggestion to quietly "create a capability to incapacitate satellites" while emphasizing that "we should simply state we are investigating outer space." Eisenhower's vulnerability to defense critics grew. Senators Symington and Johnson were joined by other powerful Democrats such as House Speaker Sam Rayburn and House Armed Services Committee chair Carl Vinson and by influential scientists such as William Pickering, ARS president Howard S. Seifert, and MIT professor H. Guyford Stever.[18] Five Republican notables openly opposed Eisenhower's defense position in the spring. These included former Gaither Committee director Robert C. Sprague, former Defense Secretary Robert Lovett, and IBM president Thomas Watson.[19]

PSAC deemed Dyna-Soar "very costly" and lacking a "potential military use." Citing the project's small but growing price tag, PSAC believed that "orderly follow-ons of the Mercury and X-15 projects" would be cheaper and just as beneficial. It "recommend[ed] that the Dyna-Soar project either be cancelled or limited to paper studies at this time."[20] But although Eisenhower trusted PSAC, the Dyna-Soar program was not ended.[21] Pulling the trigger on Dyna-Soar cancellation, even with the support of his most trusted scientific advisers, seemed impossible to a president sustaining sharp criticism in an environment where the national culture appeared to support the aerospace perspective.

Cultural evidence suggested that space was on the national mind. Rod Serling's classic television series *The Twilight Zone* premiered on CBS on Friday, October 2, 1959. Its first episode, "Where Is Everybody?" explored the mental agonies of a man in a deserted town, and the closing scene revealed that the previous half hour had been the man's nightmare as he was tested in an Air Force isolation booth as an astronaut candidate. Altogether, seven of the *Twilight Zone*'s thirty-six shows in its first season that year dealt with space environments. In general, however, space was the setting for stories with messages that were very much directed toward Earth. Two episodes in January 1960 and another in March 1960 focused on the human condition; outer space and other planets were, for Serling's program, essentially a backdrop.[22]

While the new *Twilight Zone* showed that viewers would accept space as one of the environments for a philosophical science fiction series, another contemporary CBS program suggested a niche of viewers interested in space itself. *Men into Space* debuted on September 30, 1959, just two days before the first *Twilight Zone*. *Men into Space* was a product of Ziv television, which from 1955 to 1957 had made nearly eighty episodes of a half-hour program called *Science Fiction Theater*. Although lacking the philosophical drama of Serling's programs, *Men into Space* reflected a far more conscious effort to focus on space issues and worked to get space science as "right" as possible.

The first episode of *Men into Space* borrowed actual high-altitude flight suits from the Navy, and it also credited the Air Force's research and development command, its surgeon general, and its school of aviation medicine for special support. Subsequent episodes would adopt prop spacesuits, but Air Force advice and support earned recurrent thanks during the show's thirty-seven aired episodes. Whereas *The Twilight Zone* confined narration to the introductory and closing moments, *Men into Space* used narration liberally as a means of explaining topical space science issues, including the need for artificial gravity, the dynamics of floating outside a spaceship in weightlessness, and the constraints of carrying the necessary air supply. As the Phase Alpha review got going, one *Men into Space* episode explained the atmospheric skipping flight pattern that had been a feature of early boost-glide research.[23]

Men into Space repaid the Air Force for its support during the show's year-long run by presenting a highly positive image of Air Force astronauts as family men, ingenious adventurers, and knowledgeable scientists. Characters echoed Air Force attitudes that "you [have] to make [space flight] routine to play . . . this game," and that "space flight is only a natural, inevitable step in evolution." In an episode from mid-November 1959, the show's protagonist voiced the Air Force belief that "a robot wouldn't know what to do in an emergency, and on a mission like this you have to expect a few emergencies." This logic stood at the core of the Air Force's insistence that it needed *piloted* aerospace vehicles. A space-hungry youth in an episode is advised, "You're young, healthy, and obviously very bright. I should think that the quickest way to get into space would be to join the United States Air Force."[24] Again, these words were a tailored presentation matching the Air Force's view of its role as the explorer and defender of the heavens. The nature of the television program's corporate sponsorship, Lucky Strike cigarettes and Gulf gasoline, suggests that the show aimed to include a grown-up (voting) demographic. The program's quest for realism and adventure plots both reflected and reinforced the presence of space in the national consciousness.

The media helped produce an image of social consensus favoring much more energetic space activity. This exerted a counterpressure against the president's preferred space posture. Eisenhower had been described as "the captive of those who controlled the presentation of scientific advice within his administration." Until the formation of PSAC, the Atomic Energy Committee's test-ban opponents had Eisenhower's ear, overcoming his proclivities toward a nuclear test arrangement. PSAC freed Eisenhower's hand by showing the president that scientific advice was divided on controversial issues.[25] The superpowers had already stumbled into voluntary test moratoria during Eisenhower's second term, but frictions surrounding the U-2 shootdown in May 1960 halted his gesture to expand and formalize a ban. Eisenhower had viewed nuclear weapons tests as a source of anxiety rather than crisis, and domestic politics required that he obscure his feelings from the public early in his tenure. He became more overt in his activity to curb the arms race and arms testing when public concern appeared about strontium radiation contaminating foods and causing cancer.[26]

The existence of competing scientific advice, coupled with public division, permitted Eisenhower to abandon the test-ban opponents of the AEC. The same was not the case for persuading Eisenhower to deal a death blow to Dyna-Soar. Administration officials, from Douglas to York to Quarles to Charyk, had taken steps as early as December 1957 to impede Dyna-Soar and thereby uphold Eisenhower's freedom of space policy. They understood that Dyna-Soar III challenged the freedom of space and the safety of Corona. PSAC advised Eisenhower to pursue a test ban and to cancel Dyna-Soar. Eisenhower's lack of personal action to cancel Dyna-Soar shows that although the PSAC mattered, it was not enough on its own. Eisenhower was not the puppet of his science advisers, but he was corralled by his perception of national will, which seemed to favor a ban on nuclear arms testing and to support greater endeavors in space. Although he privately questioned how well informed the public might be, Eisenhower believed that enough of the nation felt this way that the chief executive could not prudently defy it. The hidden-hand leadership style that he had embraced cemented this conclusion further. Eisenhower's style of leadership and the public voices urging aerospace activity saved Dyna-Soar from destruction, despite Eisenhower's feelings and the recommendations of the PSAC.

Thus, when Phase Alpha concluded in April 1960, Boeing began moving ahead again on the program. The *New York Times* reported these events to an incompletely informed public referring (incorrectly)

The lifting body M2-F3 is on display at the National Air and Space Museum in Washington, DC. *Author's collection*

to a "Dyna-Soar missile" on April 23 but (accurately) to the "Dynasoar space glider program" five days later.[27] As Phase Alpha concluded, one episode of *Men into Space* described a futuristic spacecraft's skipping flight pattern, and in another episode a prop model resembled the booster rocket configuration later adopted for the Titan IIIC in 1961.[28] *Aviation Week* recognized the importance of surviving Phase Alpha. Contracts for major subsystems could not be awarded until decisions favoring Dyna-Soar's medium lift-to-drag concept were confirmed over semi-ballistic, high-lift, or high-drag alternatives. The trade journal noted that "while there is no military requirement for a Dyna-Soar vehicle at the moment, USAF feels that this glider will meet the general military needs for a hypersonic vehicle as they are now conceived."[29]

Spared destruction, Dyna-Soar had been stalled, but ARPA had relinquished its oversight prerogatives during the Phase Alpha study. Furthermore, when Air Force planners convened at Maxwell Air Force Base in mid-May to determine the service's long-term structure, DDR&E approval had been secured, and Boeing had just negotiated a contract for building the Step I craft. *Aviation Week* predicted that by about 1966, over $600 million would be invested in the Dyna-Soar's research and development and another $300 million in procurement if the program met success.[30]

Dyna-Soar would need a powerful boost to get into space—not only from policymakers but from a large rocket as well. The Saturn rocket came under NASA's control in March, and the following month, Air Force Secretary Douglas wrote Eisenhower that the Saturn "has potential military application and the Department of Defense will continue to follow the project with considerable interest." Just prior to Phase Alpha, NSC executive secretary Lay informed Eisenhower's assistant that the Air Force considered using the Saturn as a booster for Dyna-Soar.[31] With Dyna-Soar emerging from Phase Alpha by late April, perhaps it could employ the big booster under development. Glennan "clear[ed] the air" in a meeting days later with the Space Science Panel, where there was "complete agreement" of a "lack of a payload for Saturn" and speculation that "the decision to go ahead with Saturn [was] a politically motivated decision in attempting to compete with the USSR."[32] It seemed logical (to the Air Force) that the Saturn be given a purpose and Dyna-Soar be given a booster.

Eisenhower first discovered NASA's lunar intentions in May, and NASA did not unveil its Apollo project until July. Even then, NASA's announcement of a project did not constitute a national objective, and it did not guarantee long-term funding. What the early work of organizations accomplishes is to provide details about what may be possible, so that poli-cymakers can then decide from such options that contribute to preferable directions in policy. Upon learning of the NASA aspiration, the president directed his science adviser to examine "the goals, the missions and the costs" that NASA hopes might entail. Donald Hornig, a veteran of the Manhattan Project, was to direct a committee of six to discover answers for the president. They would report their findings to Eisenhower weeks after the fall election.[33]

May started badly, when Soviet missiles vectored against an intruder succeeded in downing a spy plane (and, inadvertently, also a Soviet fighter scrambled to intercept it). The American pilot, Francis Gary Powers, had survived, and enough of the broken plane had survived to feed Soviet suspicions. Twenty-four similar surveillance missions had been flown, rarely but deliberately, in the previous four years. Soviet defenses had detected these flights but had not proven able to intercept during that time. The administration's thin cover stories unraveled; *Aviation Week* was incensed to have been duped by the initial fabrication that while flying for NASA, the pilot lost consciousness and drifted into Soviet airspace. The journal cried for "a congressional or some other 'watchdog' operation over [the] CIA." Eisenhower eventually acknowledged authorizing the flights.

Eisenhower's official statement, which noted that "we do not use our Army, Navy or Air Force for this purpose," publicly reemphasized his rejection of the Air Force conducting strategic reconnaissance over Soviet territory.[34] The press speculated that U-2 flights would continue "despite Russian retaliation threats until an effective surveillance satellite system becomes operational and/or an arms inspection agreement is reached."[35] Critics condemned the president's positions but disagreed about which course to adopt. Aerospace advocates such as Hotz urged combining civilian and military space projects to expedite satellite development. Democratic Senator William Fulbright of Arkansas, chair of the Foreign Affairs Committee, criticized the use of NASA as a "cover" for the flights as "a rather dangerous undertaking."[36]

Aviation Week proclaimed that the U-2 had reached "diplomatic obsolescence," declaring that satellites presented the future path and pointing to the infrared detection system MIDAS and the photo and electronic reconnaissance SAMOS, which the Air Force had not concealed.[37] The president demanded to know how SAMOS had "slip[ped] out of control." As defense secretary, Neil McElroy had been more sympathetic to aerospace ideas than the president, and he had dispersed operational use of satellites rather than keeping them as centralized as Eisenhower wanted. DDR&E York, Goodpaster told the president, could oversee and corral projects during their development stage, but "the sky is the limit so far as the operational use is concerned."[38]

Nonetheless, the pattern was one of centralization and oversight outside of military control. The president told Defense Secretary Gates that "SAMOS and related projects . . . should be brought before the National Security Council for careful consideration and review."[39] Centralizing nascent space projects allowed the administration to stall the potentially problematic ones in the "developmental" stage for extended periods. Fortunately, from the administration's perspective, the technological problems of hypersonic flight alone were sufficient to prevent a vehicle from being built quickly. When officials familiar with Corona still wanted to slow the Dyna-Soar program down, they could point to the technological feasibility questions, as Charyk did when he initiated Phase Alpha.

Aviation Week celebrated SAC's capabilities and technologies and the Air Force's space program throughout the summer of 1960.[40] As with the rumored Soviet Dyna-Soar in January, general readership newspapers occasionally pushed growing stories into a wider limelight. Nuclear-armed bombardment satellites sparked debate in the aerospace community, as proponents claimed that orbital travel reduced their vulnerability to interception and skeptics questioned the targeting accuracy of space-based systems and pointed to possible anti-satellite weapons, anticipated to appear by 1970.[41] Dyna-Soar advocates, meanwhile, asserted that piloted satellites could steer varying, unpredictable orbits to avoid interception.

News reports and advertisements in the *New York Times* and *Washington Post* mentioned Dyna-Soar and reflected the program's apparent outlook. The second half of 1958, when Eisenhower felt compelled to assuage an anxious public by launching Project SCORE, had seen an uptick in advertisements displaying Dyna-Soar. Ad emphasis tapered off toward the summer of 1959, as Quarles and York redirected the program into something more technically oriented and less "sexy." Ads and reporting remained at a fairly low level for the next year.

Dyna-Soar reporting, and especially advertising, ramped up in national news outlets in the summer of 1960. Businesses used it as an exotic example of a company's technical achievement and to advertise jobs. "Dyna-Soar" was a name worth ad space in aviation trade journals and national newspapers. June 1960 saw the beginning of another bump in advertising and news reporting that continued through the remainder of Eisenhower's term and beyond the first year of Kennedy's presidency. Eisenhower noticed this trend of reporting and advertising paying attention to weaponry and space topics. His comments on it would be among his last remarks as president, and his awareness of this media activity influenced his ideas about national opinion and his handling of the Dyna-Soar program.

Wary of the risks of war, startled by Soviet outrage after the U-2 shootdown, and hoping that Corona would soon pay off, Eisenhower abandoned further aircraft surveillance over the USSR.[42] News on August 18 that the Discoverer XIII reentry capsule had successfully been recovered after its reentry from orbit was certainly welcome. As the news arrived, Discoverer XIV was launched. This satellite orbited for twenty-six hours and was snagged by an Air Force C-119 as it parachuted earthward after reentry.[43] The public was informed of the Discoverer feat but not of the Corona secret on board. News stories also mentioned the Mercury capsule and the Dyna-Soar vehicle, and they distinguished between the ballistic and gliding reentry techniques. ARDC commander Gen. Bernard Schriever said he doubted recent Soviet claims to have recovered animals from space, and he insisted to reporters that "although the successful recovery of smaller animals would aid in the eventual man-in-space plans, there was 'more valuable military application' in the Dynasoar program than might be expected from the Mercury program."[44]

High Technology on the Campaign Trail

Meanwhile, presidential campaigns incorporated defense and space issues in bids to garner enthusiasm and support, but *Aviation Week*'s editor noted that both Republican Vice President Richard Nixon and Democratic Senator John F. Kennedy of Massachusetts lacked the "practical experience with the realities of the Soviet threat" possessed by their running mates, UN Ambassador Henry Cabot Lodge and Senator Lyndon Johnson. To compensate, Nixon soon called for stronger defense efforts, and Kennedy began accumulating defense advisers such as Roswell Gilpatric from the nonprofit Aerospace Corporation and Democratic policy veteran Paul Nitze. Symington's primary hopes had faded; a personal friend of Kennedy but antagonistic toward Johnson, he agreed to help the Democratic candidate review defense policy.[45]

Apollo was a project without a mandate when NASA christened it on July 28, 1960. Mercury designers had tinkered with a Mercury Phase II craft as a transition project, and by August 1960 *Aviation Week* noted that "progress of Mercury and the NASA-USAF Dyna-Soar program will determine whether a Mercury follow-on precedes the more ambitious Apollo."[46] Mercury II seemed likely to become overtaken by advancing technologies of Dyna-Soar and possibly Apollo.

Aerospace's louder advocates occasionally conceded that the U.S. space posture was actually moving "faster and on a broader front than anybody might have hoped in the gloomy months after Sputnik." They still urged bolder steps forward.[47] Meanwhile, dream vehicles such as Dyna-Soar and the B-70 took on additional conceptual roles to insulate them from administration resistance. Air Force planners identified six categories of space activity: targeting; surveillance of Earth; surveillance, inspection, and neutralization of other space vehicles; interception of enemy missiles and space vehicles; "an ability to strike surface targets from vehicles deployed into space for protection against surprise attack"; and space-based weapons to provide "improved enemy warning and reaction time." Space-located "alternative command posts" were considered important, given that "earth based survival techniques and assured earth-to-space communications" were expected to degrade during combat. Lower launching costs, space-to-space and space-to-earth weapons, and routine mission capabilities fit the Air Force outlook.[48]

The Air Force took heart that a Kennedy presidency would promote this vision. Major Abbott C. Greenleaf of the Long Range Objective Group reported favorably to Chief of Staff White on the senator's "views on key defenses policy issue[s]." Greenleaf conceded that Kennedy "has not outlined explicitly the strategic concept he endorses," but Symington's committee advised Kennedy, and "Senator Symington's views and the public announcements outlining the proposals his committee is studying appear to be very close—if not identical in most respects—to the USAF position" on defense issues.[49] The Air Force and Kennedy agreed "that the U.S. must surpass Russians." Regarding the military's role, Kennedy thought "freedom of space must be assured, preferably by the UN. The U.S. must have pre-eminence in security," centralizing missile and space programs and accelerating Saturn development. Significantly, the memo identified the Saturn rocket as a booster for Dyna-Soar. Kennedy's lukewarm support for the B-70, silence regarding counterforce deterrence or "pre-emptive strategic strikes," and enthusiasm for limited war capabilities diverged from Air Force preferences.[50] Overall, however, Kennedy's stated and implicit views seemed to mesh with the Air Force's own position.

White addressed the Air Force Association in San Francisco in late September, warning against the assumption of indefinite stalemate "as a warranty of peace." A stalemate, he said, "is not a long-continuing probability." White anticipated revolutions in military technology. His deputy for development, Lt. Gen. Roscoe Wilson, told a news conference at the same convention that the B-70 "admirably meets our needs for high-performance strategic aircraft during this decade and beyond." Then "he unveiled a model of the Dyna-Soar, a manned boost-glide vehicle destined eventually for orbital flight."[51]

George W. Rathjens, the staff assistant to the Missile Panel, hoped to stop that "destiny." Hypersonic manned technology would ordinarily be a NASA responsibility, but since Dyna-Soar already resided under Air Force auspices, the panel decided that it should be left there. Despite the lack of sufficient boosters, hypersonic research should continue, perhaps with an unmanned, lighter, expedited version of Dyna-Soar. The panel recommended, however, that the program be reoriented away from an operational capacity, a shift that DDR&E York had tried to enforce the previous year.[52] Rathjens wrote to Eisenhower's science adviser Kistiakowsky two days after making the report, expressing his sharper opposition to Dyna-Soar. He explained to Kistiakowsky that the panel's report was "essentially a compromise between my own views and what I believe is the view of the Panel." Although seeing "a good case . . . for a research program on the properties of materials under conditions that might be encountered by DINOSAUR [sic], . . . I see no reason for the present emphasis on manned flight nor for the Panel's insistence that this be part of the present program."[53]

Four days later, Rathjens wrote to the Strategic Systems Panel, discounting Dyna-Soar's "operational capability, politico-psychological advantages, and biomedical experience" on the assertion that the Mercury and possible Apollo programs were already set to fulfill these tasks. Man-rating space vehicles—the process of upgrading the reliability of rocket technologies so that humans could more safely ride atop them—slowed their development and dramatically increased their cost. Rathjens worried that the Air Force might attempt to manipulate the Missile Panel's views into a seeming endorsement of the program. Having compared notes with the president's science adviser, he also wrote that Kistiakowsky "is concerned that it may develop into another gigantic program with emphasis on a poorly-defined or nonsensical strategic operation requirement, when we are already confronted with what might be an unreasonably large proliferation of strategic systems."[54]

Despite this criticism, the Air Force had a definite role in mind for the ultimate manifestation of Dyna-Soar: an orbital space bomber, with potential to do other yeoman work as a space pioneer, logistical, or reconnaissance vehicle. While skeptics could call this a "nonsensical mission," and it definitely conflicted with the core of Eisenhower's preferred national policy, it was hardly the vague vision as suggested by Rathjens in 1960 (and by Defense Secretary Robert McNamara in the years ahead). Its multimission profile stemmed from its advocates seeking to demonstrate the utility of a craft that already had a vital primary mission in mind. The Air Force saw a need for an orbital bomber.

In September Eisenhower and his advisers pondered the best way to word the president's idea that "all launchings of space craft should be verified in advance by the UN." In a bow to the joint chiefs, who were expected to reject foreign inspection of SAMOS, "it was concluded that SAMOS activities apply to exploration of our own earth and therefore are not subject to verification and disapproval." Eisenhower "decided that any public statements should be as low key as possible but that it must be acknowledged that there was involved photographic and related equipment for research and development purposes."[55] Keeping Corona's secret meant little if a separate program (SAMOS) appeared to the world to be what Corona actually was. Congressional testimony by Secretary Sharp, Under Secretary Charyk and Maj. Gen. Osmond Ritland, commander of the Air Force Ballistic Missile Division, "all indicat[ed] that there would be photographic equipment in the SAMOS," and published records described equipment, orbit, and image resolution.

Dyna-Soar continued apace. *Aviation Week* outlined still more ambitious plans for it in September, saying that the "Air Force is pinning its hopes for development of a true space weapon system" on Dyna-Soar and that "long-range thinking on the Dyna-Soar contemplates flight to the moon and re-entry into the earth's atmosphere." Air Force headquarters approved studies for Dyna-Soar Steps II and III the following month.[56]

Commenting on the subject of education, *Public Opinion Quarterly* observed that "Sputnik fever" had declined from its initial "endemic proportions" but that the public pressure on education was largely "unrelated to Sputnik fever." The journal aptly noted that "public opinion pressure frequently reveals a problem but rarely a solution."[57] Regarding space, the novelty and inferred hostile potency of Soviet space successes prompted disquiet that Eisenhower had sought to address with reminders that the strategic situation had not altered. In military terms, the United States retained its strong deterrent posture, but many had lost faith that this was true.

The *New York Times* described the warm, even exuberant reception awaiting Kennedy upstate, where "the enthusiasm and noise were greater than at any Democratic rallies since the days of Franklin D. Roosevelt" and a woman broke through police lines to kiss the candidate on the cheek. The senator criticized Nixon on a variety of issues, ranging from communist advances in Cuba and Congo to layoffs in the steel industry and the elderly's need for medical assistance. Kennedy saw defense spending and contract distribution as among the methods for alleviating unemployment.[58]

Nixon also suffered unintentional blows from Eisenhower, whose growing determination to assert his active charge of the country undermined Nixon's campaign claims of experience and leadership from his years as vice president. On August 10, Eisenhower had testily told a reporter, "As long as any question is put up before me involving what I believe is the good of the country, *I'm* going to decide it according to *my* judgment [emphasis added]."[59] On August 24, just two weeks after insisting to the press that he and not Nixon was the decider in the executive branch, Eisenhower again sought to revoke the implicit inactivity that was part of the image he had built. A reporter asked Eisenhower if he could describe how Nixon really fit into the decisionmaking process. Eisenhower answered that Nixon, like all the president's advisers, had participated in discussions. Pressed to identify a policy idea that had originated with Nixon, Eisenhower interrupted the questioner to say, "If you give me a week, I might think of one. I don't remember."[60] The news conference ended there. Eisenhower had damaged Nixon's campaign without convincing the country that he himself had a steady hand as commander-in-chief.

Neither candidate impressed Hotz with its eleventh-hour antics regarding aerospace and national defense issues: "Now in the waning weeks of September and the early days of October this same Administration has been thawing out these frozen funds and quietly trickling them into the very programs that the critics have been howling about since last January as requiring more funding and higher priority. . . . Polaris, Minuteman, SAMOS, MIDAS, B-70, MATS [Military Air Transport Service] modernization, modernized Army equipment, anti-submarine warfare, Dyna-Soar, Skybolt and other key research areas."[61] Nixon quickly presented specifics, ranging from new attention on civilian aviation to work on anti-ballistic missile systems and an increased capability to wage limited war. Kennedy, his rhetoric pointed at the purported missile gap, urged acceleration of "ultimate weapons" such as Polaris and Minuteman missiles (although military technologies experts shied away from believing any weapon to be truly "ultimate"). Fulbright and Kennedy accused Eisenhower of eroding international respect in the country. In the final days before the election, Nixon rejoined that U.S. prestige had never been higher and accused the penny-pinching Truman administration of starving the missile research budget and "ignoring the implications of the long-range rocket."[62] The vice president insisted that "America will be second to none in the long stride into space" and that "we will launch in the period from 1966 to 1968, manned circumlunar flights. By the early 1970s we will launch manned space ships to land and return from the moon."[63] Eisenhower's reluctance to step past the Mercury demonstration of human spaceflight had been abandoned by both candidates.

As voters moved to register their say on the nation's future, *Public Opinion Quarterly* pondered impending reaction to upcoming military developments in space. Yale political scientist Gabriel Almond discussed reconnaissance satellites and bomb delivery satellites. (Unbeknownst to Almond, satellite reconnaissance had already been conducted.) The article identified reconnaissance as "so near to reality, without unusual public response, as to suggest that we will move into the age of satellite reconnaissance without a notable opinion reaction."[64] Indeed, the low-key element of reaction about reconnaissance satellites was hardly accidental; Eisenhower's team had been struggling to establish a favorable environment for them.

Almond, like many, believed the weaponization of space to be almost guaranteed in the future. "It is less easy to appraise the consequences of the complication of the deterrent picture by the development of bomb-carrying satellites," he wrote mildly. Some had already suggested that either power's stationing deliverable nuclear warheads in orbit could "represent a step which neither side could tolerate, or which might raise anxiety to panic levels and trigger military action." In fact, the two issues of the surveillance satellite and the weaponized space system intertwined in that deployment of the latter would preclude any implicit or explicit "space for peace" agreement. At a stroke, weaponization would establish an Air Force–style definition of aerospace as an operational continuum. Corona would then have no more diplomatic safety than that possessed by the U-2 in the wake of Powers' intercepted flight. In a Cold War environment, opening the door a crack to a space arms race could easily lead to it being thrown wide open.

Eisenhower took this possibility seriously and would register horror at this thought. The Dyna-Soar vision represented a particular danger to the legacy of peace, security, and solvency Eisenhower had striven for eight years to establish.

Almond suggested that space weaponization might fit, albeit uncomfortably, into the larger Cold War context of ominous weapons deployment and extended armed tension: "The question here is whether the development of a satellite-bombing capability would differ in any striking way from the development of other delivery systems such as the ICBM or the Polaris." An endless anxiety worse than that prompted by the existence of nuclear-armed submarines might result from "the constant passage . . . of a fleet of bomb-carrying satellites" overhead. Similar psychological strain had been predicted regarding the development of hydrogen weapons and ICBMs. Although destructive yields and weapons' ranges had grown by stunning proportions, the truly "ultimate" weapon had as yet not been developed. Therefore, Almond said, "it might be argued that the development of bomb-carrying satellites would be greeted at first with anxiety and attention, and then gradually would be assimilated into the monstrous family of threats and deterrents among which it is the fate of modern man to live."

Almond's was a reluctant strain of the technological determinism that characterized Air Force doctrine. He concluded with hopes that although space weaponization would present another onerous psychological burden, it would not prove to be the fatal step catalyzing oblivion and that future generations would learn to cope with the danger. Almond's article left little room for the possibility that space might not become weaponized. By November 1960, the Air Force was busily considering a more powerful Titan II booster for Dyna-Soar's suborbital and orbital missions. By December, in the wake of Kennedy's victory, the Dyna-Soar office had completed a stand-by plan to leverage the new booster to accelerate the program development. This promised to overcome the major obstacles put in place by Eisenhower officials since Sputnik.[65]

In the environment of global armed tension, Almond seemed to expect that surveillance satellites "so near to reality" would enter operational status more or less imminently and that more sophisticated space weapons systems would take longer in arriving but would eventually be realized as well. Policymakers, by definition, have a responsibility not to be inert passengers of fate. Citizens in a representative system of government have a complementary responsibility. Facing the last weeks of his presidency,

Dwight Eisenhower conceded the finite span of his remaining authority. He also recognized the ongoing responsibility of citizenship. If it was too late to cancel or derail any ill-conceived projects, it was not too late to finish preparations on a message to remind and advise an anxious nation.

Secretary of Defense Gates added his perspective to the mix on October 25 while addressing the Postmasters of America. In the count of satellites launched, remaining in orbit, and functioning, the United States consistently led the Soviet Union.[66] Gates assured his audience that the administration was far from complacent, characterizing the era as a "period of scientific and military transition." He noted that "it takes courage to change, to cancel an expensive program that has been overtaken by events. . . . Sometimes it takes more courage to do this than to cross a new threshold." He described national defense as "the burden of everyone—every citizen, every member of the press, every member of Congress," though pointing out that policymakers "who carry the burden of the responsibility must answer 'yes' or 'no'" to proposed projects and policies. He also reminded listeners that "defense [policy] is not made in isolation" but is instead a "part of national policy."[67]

Gates urged vigilance, tempered by composure. "The struggle with Communism will be long and costly," he reminded the postmasters. "Our national strategy and policy gathers our resources for a prolonged test of endurance. We must run our defense on a balanced, sensible, and timely basis for the long pull."[68] Eisenhower was hard at work on a farewell speech addressing many similar points.

10

"A Thousand Drawing Boards"
Eisenhower's Farewell Warnings

General White expressed the view that the Dynasoar program was vital in order to keep the U.S. in the technological race. The President said that his comments on Dynasoar had been based on his view of the national security rather than the technological race.

—NATIONAL SECURITY COUNCIL MEETING NOTES,
DECEMBER 8, 1960

Repeatedly, I saw Ike angered by the excesses, both in text and advertising, of the aerospace-electronics press, which advocated ever bigger and better weapons to meet an ever bigger and better Soviet threat that they had conjured up.

—JAMES KILLIAN

Aerospace Initiatives Continue Onward and Upward

In his last statements as president, Dwight Eisenhower warned of a number of profoundly important problems, the best remembered of which is the "military-industrial complex." Many opinion leaders resented or discounted these "Parthian shots"[1] as the broodings of a chief executive whom they dismissed as inadequately engaged with events. Eisenhower's behind-the-scenes method of governing guaranteed this public image in the wake of supposed technological crisis. In the waning days of the Eisenhower administration, with the presidential election imminent, the military seemed incorrigible in its claims to space. Public support for national defense and affinity for technology made development of the B-70, its nuclear-powered derivative, and Dyna-Soar appear probable.[2] The Air Force was even in the process of considering an even more ambitious project than Dyna-Soar. On October 31, 1960, *Aviation Week* reported on Air Force plans for a "radical new space plane." The Space Plane would weigh an extraordinary half-million pounds (about thirty-five times more than Dyna-Soar weight estimates). An alternative method of airbreathing launch meant it would require "no large rocket booster such as Mercury, Apollo and Dyna-Soar do."[3] The news story's use of the present tense for the three space systems is interesting; at this point, Mercury had carried two macaques in separate suborbital flights, and Apollo and Dyna-Soar remained essentially drawings.

The Space Plane's jet engines would lift it like an aircraft to an altitude of 285,000 to 380,000 feet; meanwhile, scoops would collect atmospheric oxygen that, combined with on-board liquid hydrogen, would constitute chemical rocket fuel for space travel. *Aviation Week* predicted that the "Space Plane probably would be the last chemically powered spacecraft because it is believed that nonchemical systems in the nuclear and nuclear-electric fields will be sufficiently developed in the mid-1970s to power a follow-on craft." Although advocates conceded upcoming adjustments during "the study phase and the research and development phases," they insisted that "the system is based on studies already completed" that demonstrated its feasibility.[4] In contrast to Dyna-Soar, which seemed pedestrian by comparison, the purpose of the Space Plane was not altogether evident. Aerospace visionaries were not fettered by reality. Despite the atmosphere of technological determinism, the Air Force's Scientific Advisory Board would become nervous about *Aviation Week*'s repeated publications on the conceptual aerospace plane, because "too much emphasis may be placed on the more glamorous aspects of the Aerospace Plane resulting in the neglect of what appear to be more conventional problems."[5]

Within the administration, the NSC Planning Board briefed officials on DOD space activities on November 3. Meeting notes revealed that "Gordon Gray was concerned with the impression given that prior to Sputnik I we were doing nothing in space, whereas immediately thereafter we have projects across the board." A technical assistant on the Ad Hoc Panel on Man-in-Space told Kistiakowsky that the briefing did little to explain "where there is duplication in the DOD and NASA programs or in which projects DOD is invading NASA territory." DOD representatives at the meeting "began to get into trouble," complaining of hazy boundaries that separated NASA and DOD responsibilities, but "refus[ing] to accept orders and stay out of the space exploration business." They noted instead that military systems needed independent design and extensive study programs that were separate from NASA efforts. Planners for NASA's man-in-space efforts worried "that these studies are getting out of hand and are monopolizing an inordinate amount of technical manpower." One DOD presentation of its space activities had "listed such clearly NASA responsibilities as Man-in-Space."[6]

Aviation Week presumed that civilian control had been a mistake rather than a deliberate (let alone defensible) policy decision. The November 7 editorial deemed it "now apparent that the original concept of NASA as handling the total U.S. space effort was a rather shortsighted outlook, although understandably so in 1958." NASA's "abrasive approach to its military and industrial cohorts" met criticism. Soviet obfuscations concealed failures and preserved U.S. technologists' fears of Soviet advantage. On October 24, Soviet missile boss Marshal Mitrofan Nedelin and one hundred other rocket scientists and technicians were killed by a launchpad mishap. Nedelin himself was vaporized. Yet when the Soviet news agency TASS issued the cover story of an ordinary air crash killing Nedelin, *Aviation Week* accepted the contrivance at face value. Simultaneously, the trade journal reported on the possibility of "a 'revolutionary' third-generation nuclear weapon" able to contain a nuclear war (and thereby make nuclearized conflict more attractive to strategists). AEC commissioner Thomas Murray predicted that international moral condemnation of such weapons would be less than that of second-generation (hydrogen) weapons and that the new warheads would become "a new symbol of strength and the massive weapons a symbol of weakness." *Aviation Week* speculated that "the Soviet Union probably has actively developed nuclear technology along the lines of the new weapon."[7]

John F. Kennedy won the November 8 presidential election. His campaign had linked defense spending with economic recovery from the lingering effects of recession, and Kennedy had furthermore joined the chorus of voices interpreting space issues within the larger realm of defense concerns. U.S. policymaking in space, and thus the Air Force's bid to harness aerospace, would be in his hands. For the ten weeks prior to Kennedy's inauguration, Eisenhower would oversee a country without realistically being able to determine its upcoming policy. Despite the severely constrained authority involved, it was not a responsibility to be shirked.

Eisenhower's meetings on election day reflected his efforts to prevent strange, expensive, and unnecessary new projects in space. That morning he met with National Security Advisor Gordon Gray. At the close of their meeting, the president said he "felt that the only place we ought to be even in a clandestine way contesting with the Soviet Union is the development of the big engine," while in contrast he repeated "that little would be accomplished by putting a man into space." Budget director Stans noted recent reporting on a "space plane" and asked to know specifically what relationship this had to Dyna-Soar and why DOD (rather than NASA) controlled the project. John Rubel, York's deputy, answered that the Space Plane had not received approval as a project. It represented a vocalized Air Force wish rather than anything officially authorized. Gray told the group that DDR&E York and NASA administrator Glennan were on top of the task of keeping bogus projects corked, and Eisenhower "said that he was delighted with the mechanism that had been created."[8]

Aviation Week urged greater military involvement in space and recognized that Glennan's impending departure from NASA meant that Kennedy's selection of a new administrator would "to a large degree" determine "the results of this NASA-Air Force battle" over U.S. activity in space. In late November, the journal condemned "the present official attitude that there is no demonstrable military need for man in space" and identified the space plane as the most recent victim of Eisenhower's bungling on space. "Defense has been unable to convince President Eisenhower on its space role," and aerospace advocates believed that the military space role was "restricted primarily not by finances, but by a national philosophy which virtually excludes identification of military with space."[9]

The quest for economic vitality demanded sufficiency in place of extravagance with respect to security spending. December 5 saw more White House discussions about streamlining the proposed fiscal year 1962 budget. Eliminating or reducing several projects could reduce the coming defense budget by $2.5 billion. Titan could be capped at 10 squadrons to yield a $500 million saving. Minuteman could be kept to four hundred missiles until it was a proven system, saving $400 million. Confining SAC airborne alert to its current level, reducing B-58 strength, and trimming overhead at arsenals could save further money. Keeping the Hound Dog standoff missile in the arsenal obviated a need to develop the Skybolt. The F4H jet could operate in place of a new short take-off and landing plane. Conventional long-range interceptor aircraft could preclude development of the exotic F-108. Ending the B-70 program was worth $400 million, and eliminating one of the two nuclear engine programs could save between $20 million and $50 million as well. Canceling Atlas would yield nearly $200 million.[10]

Canceling Dyna-Soar would save $146 million, a significant—but by no means gargantuan—sum when compared to the average contemporary annual defense budget of about $40 billion.[11] The report included the typed notation, "Eliminate—no military value." In part, civilian space projects served as an instrument by which to help short-circuit popular insistence on support for the military's ambitious space projects. Both NASA and the Air Force wanted a piloted orbital craft, and NASA supported research on hypersonic flight. However, NASA was reported to feel unable to "support Dyna-Soar as a true space program until the Air Force gives the vehicle a mission and the mission gets the approval of the Air Force, Defense Department and White House," and, in keeping with York's strictures, DOD had approved only a suborbital research craft.[12]

White House approval was presumed to occur following Kennedy's accession. Journalist Edward Kolcum informed the public that the Air Force "conceivably could try to skip the suborbital vehicle . . . and go directly to an orbital craft," although it would likely encounter objections from NASA since its Saturn would be the only booster capable of lifting a heavy Dyna-Soar to orbit. Although vehicles and boosters required "a high degree of reliability," NASA saw them as "still research tools, or extensions of research programs." National security depended, in part, on NASA working to "*push* and *strain* [emphasis in original] the state-of-the-art with its research programs."[13]

Journalists repeated the "publicity offensive" waged by the Air Force. John W. Finney of the *Times* noted the Air Force's enthusiastic appraisals "that President-elect John F. Kennedy had indicated a realization that military supremacy in space as [sic] essential to our security as military supremacy at altitudes near earth." The Air Force found, and the *New York Times* quoted, Senator Kennedy saying, "Control of space will be decided in the next decade. If the Soviets control space, they can control the earth." It was little more than an echo of the assertions of others, prominently Vice President–elect Lyndon Johnson, who in turn had been referring to Chief of Staff White. Finney noted that the Air Force had prepared its publicity campaign for some time and that "in some ways the Air Force offensive is a direct challenge to the law of 1958" that had empowered the new civilian space agency. *Aviation Week* declared that "an expanded military space program is almost certain to result from the combination of legislative and executive department changes expected during the early days of the Kennedy administration."[14]

Key technology questions awaited the president-elect, while Eisenhower stood by the policies he had formulated. The NSC meeting on December 8 included a heated debate between President Eisenhower and

Air Force Chief of Staff White about the Dyna-Soar program. In the course of discussing defense projects, Eisenhower had inquired about the final objective of Dyna-Soar. DDR&E York replied, describing it as a military man-in-space program and a follow-on to the X-15 and the Mercury. The president, minimally committed to Mercury, saw investment in a successor as a waste. Eisenhower said that "Dynasoar would be a desirable project to play around with if unlimited funds were available."[15] But Eisenhower was a budget-conscious chief executive, and in 1960 the effects of recession lingered. York, whose opinions Eisenhower trusted, said that Dyna-Soar had less promise than did MIDAS or SAMOS. Reconnaissance, not combat, was the field of promise for Eisenhower and his chief advisers.

When White insisted that the Dyna-Soar was a crucial program, Eisenhower flatly disagreed and replied that "insufficient discrimination had been used in establishing priorities" for research. White tried to present Dyna-Soar as one of several important "building blocks" for future national defense, but the president answered that monetary health, national defense, and economic strength were interrelated issues. Military safety was vital, but in a Cold War framework the president doubted that an "absolute assurance" could be guaranteed. The "principal objective must [therefore] be to convince the Soviets that they cannot attack us with impunity."[16] Eisenhower had sought to explain exactly this perspective to the nation in November 1957, just after the first Soviet space successes. The president maintained the priorities of the long haul. But this debate with his senior Air Force officer at the close of his second term shows how little headway he had made with the armed aerospace advocates.

Aviation Week commented on December 19 that the "Space Plane has become a controversy before it has become a project," and it reported former PSAC chairman Killian's caution "against excessive stress on [the] man-in-space program." Reiterating his position consistent with speeches from years past, he supported space exploration and contended that "we must never be content to be second best, but I do not believe that this requires us to engage in a prestige race with the Soviets." Killian reminded the public of ideas that advisers had discussed with Eisenhower two years earlier: "We should pursue our own objectives in space science and exploration and not let the Soviets choose them for us by our copying what they do." He noted that Soviet accomplishments had focused on "spectacular accomplishments in space technology as an index of national strength, and too often the press and the public at large" bought the image that the Soviets sold. Kennedy, even prior to the Democratic National Convention, had decided to bring change at PSAC. He selected MIT electrical engineering professor Jerome Wiesner to succeed Kistiakowsky as chairman. By year's end, the media reported Wiesner criticizing PSAC for constraining defense science projects.[17]

Senior Air Force commanders and other advocates debated the relation of offensive and defensive roles regarding the service's most exotic projects. In the Senate, Symington tried to rescue the troubled B-70 program by suggesting an interceptor variant of the pending bomber; this would sacrifice the exotic F-108 interceptor in order to secure the exotic multirole B-70. North American Aerospace Defense Command head Gen. Lawrence Kuter even asserted the F-108's greater importance relative to a Mach 3 bomber. White rejected both ideas. Kuter also said, less controversially within aerospace circles, that "defense leads to space, and in space the offense and the defense merge." He fretted about the possibility that "the Air Force would be left with no space mission."[18] Generals retained hope that Dyna-Soar as a pioneer space vehicle would be even more versatile than the exotic aircraft that they were loading with so many potential missions.

Eisenhower was disinclined to drive deeper into the manned space field, regardless of whether the pilot wore a uniform or the vehicle was intended to carry a weapon. In late December, Eisenhower and his advisers reached consensus, as shown by notes from NSC meetings of December 20 and 21. The president approved findings that "there appeared at this time to be no psychological or scientific reason for carrying on the 'Man in Space' program beyond the completion of Project MERCURY, and that ways might be sought publicly to disclose this administration's views on this subject." The State Department, the National Science Foundation, NASA, and the Joint Chiefs of Staff (with the likely exception of White) were believed generally to agree.[19] Apollo, Mercury II, and Dyna-Soar each contradicted President Eisenhower's outlook. President-elect Kennedy's tone had indicated far more strident support for such projects across the board. While Eisenhower did not intend to set last-minute formal policies that would simply be undone and create confusion, he meant to leave a lasting mark as chief executive by delivering a potent message to his countrymen from the bully pulpit.

Crafting the Farewell Address

As early as May 1959, seventeen months before the end of his tenure, Dwight Eisenhower had begun thinking about his own concluding remarks as president. He told his speechwriter Malcolm Moos that he hoped "that the Congress might invite him to make a 10 minute farewell address to the Congress and the American people." Moos deemed his boss's idea "brilliant . . . if it can be carried off with a minimum of fanfare and emotionalism." Work began by collecting ideas from which to hone speech drafts. Special Assistant Frederic Fox, in April 1960, suggested that George Washington's farewell address might "perhaps serve as a guide for [Eisenhower's] final statement in January 1961."[20]

The problem of the "military-industrial complex" was one of the key themes of the speech from its earliest conception. The founding fathers had sought "to make sure that no military group arose to challenge the civil authority, and that no segment be allowed to develop which was permanently and exclusively concerned with building the weapons of war." Formation of "a large and permanent military establishment" was a regrettable and potentially dangerous necessity, imposed by the tense state of world affairs and the possibility of a sudden global nuclear war. The speech initially declared that "we shall need all the organizing genius we possess to mesh the huge machinery of our defenses with our peace-oriented economy so that liberty and security are both well served." However, "we must be especially careful to avoid measures which would enable any segment of this vast military-industrial complex to sharpen the focus on its own power at the expense of the sound balance which now prevails."[21] This original draft spanned a scant three pages, but it was, in Eisenhower's words, a "start."

Eisenhower "plan[ned] to go from here to the Scientific Revolution and the twin dangers of government dominating scientific research through purse power, and of the generalists becoming captives of the technical specialists." In most respects, the outline's similarity with the speech's final form is striking, although until mid-December, the president also intended to discuss the international security issues related to the decolonization of the so-called third world.[22] A stream of new drafts passed across the president's desk throughout December and early January. Seemingly cosmetic adjustments and semantic choices actually were important, because Eisenhower needed his message to be heard and understood to be effective. Edits reflected the care taken to show Eisenhower's concluding remarks as the reasoned conclusions of a circumspect mind and not simply be dismissed as Luddite prattle.

Regarding research, the president regretted that "formula has replaced empiricism."[23] Concurrently, he disliked that "the major impetus to research now comes not from private individuals in pursuit of knowledge for its own sake, but from public agencies in pursuit of specific, predetermined results." To assure his audience that he did not dismiss technology, he also noted that "research is the pathfinder of progress. Where it leads, all else must follow. Yet we must also be alert to the opposite danger that public policy may itself become the captive of technological opinions and pressures."[24] This was in essence what was occurring with the national space projects. The drafting process saw Eisenhower and his writers replace an allusion to "the certain terrors of nuclear war" with the phrase "the certain agony of the battlefield." The less vividly visceral image of the substitution also reflected Eisenhower's own belief (despite what some advocates of limited war might claim) that if a war erupted, it would engulf the world and involve all kinds of weapons. Revisions reinforced the emphasis on the military-industrial complex and the scientific-technological elite. These issues dominated a considerable part of Eisenhower's attention in the present and his concern for the future.

Much of the effort during the next two weeks aimed to express one last time why economic factors were a part of, not a distraction from, national security calculations. "Government and citizen alike" were cautioned against "the impulse to live off of today for tomorrow—to plunder the precious resources of our children and grandchildren." Prosperity was important, but so was vigilance. Eisenhower, the father of the interstate highway system, considered telling his countrymen that "the role of government is not that of a giant turnpike, well stocked with comforts, but rather a role that maximizes liberty and opportunity—the twin goals of all free societies."[25] Strength was a tool because "our moral, material, and military strength should give us confidence as we go to the conference table."[26] The president's most trusted brother, Milton, contributed his input in the last days of the year, and at this point, "technological opinions and pressures" became the phrase ultimately delivered: "technological elite." An ungainly paragraph about technological revolution disappeared, and the closing statement, "Not goodbye—but onward" was cut.[27] It rang hollow: Eisenhower was going home, Nixon had been defeated, and Nixon had also edged away from aspects of Eisenhower's policy. The military-industrial complex and the scientific-technological elite that Eisenhower identified had predated his administration, but they had grown throughout its duration.

While honing the speech, Eisenhower reiterated his earlier thinking about space exploitation in a discussion with Gordon Gray. On the morning of December 29, while the two men discussed the NSC's recent decisions,

Gray "told the President that [he] was having difficulty with the paragraph relating to the man in space program[,] as it might involve projects beyond the completion of MERCURY which had been discussed in the NSC meeting of December 20." Gray noted that, within the OCB and among department heads, voices had called for further consideration of future space projects. Eisenhower suggested language that Gray noted "would be objected to by the Department of Defense because it would probably rule out further development of DINOSAUR [sic]."[28]

The president "said that he questioned whether we should proceed with DINOSAUR [sic] in any event." Still, Gray remembered Eisenhower commenting that the fiscal year 1962 budget "was not the vehicle to use for stopping that program if indeed he wanted to."[29] Corona reconnaissance satellite images confirmed what Eisenhower policymakers had suspected: the "missile-space gap" that the political opponents had screamed about and that Kennedy had used to help secure his slim victory over Vice President Nixon was a complete untruth.

Although the president disdained Dyna-Soar, NASA administrator Keith Glennan told the president that NASA supported the Dyna-Soar project, deeming it an aeronautical program rather than a military space vehicle. Glennan's December 28, 1960, report to Eisenhower on the "accomplishments of the National Aeronautics and Space Administration" proudly identified the speed and altitude records set by the X-15 thus far and noted that this had been accomplished with an interim engine to be replaced by a more powerful X-15 power plant. A separate section entitled "Manned Space Flight" reported on Mercury but not on considerations about Apollo within NASA or Dyna-Soar or SAINT in DOD.[30] None of these latter vehicles had yet flown, and even Mercury had not yet launched a human into space. The X-15 thus represented the only actual U.S. accomplishments in manned aerospace flight, so the trajectory of this project spoke powerfully to the potential future of space flight.

Reminiscent of its advisory predecessor organization, NASA declared that "as was the case with earlier experimental airplanes, the basic purpose for which the X-15 was built is the extension of our knowledge of the conditions to be encountered in the control of an aircraft in flight at increasingly high speeds and altitudes. Much higher speeds and altitudes are in prospect during the coming months. Research in aerodynamics ranged widely from vertical take-off and landing aircraft to Dynasoar, a joint project with the Department of Defense for a man-maneuverable vehicle to explore hypersonic flight up to orbital speeds."[31]

Glennan's report to Eisenhower did not dwell on the weaponized features the Air Force envisioned for Dyna-Soar III. NASA saw Dyna-Soar for the technological issues relevant to that agency, and Air Force leaders saw Dyna-Soar for its Phase III objective and its successors beyond that. The transition from bedeviled research vehicle to operational system would be greatly assisted by a green light from the chief executive's office in Washington.

The State of the Union address, encompassing a broad range of the nation's challenges and opportunities, the general situation, and the president's own exhortations and proposals, provided the outgoing president with a forum in addition to his farewell. This time, with Eisenhower's tenure coming to a close, there would be "not proposals, but a review of the record." Almost all of the speech's envisioned fourteen sections would conclude with a "statement of principles."[32] The first days of 1961 saw Eisenhower's writers turn their attention to this work. The speech celebrated the administration's commitment to national security, pointing to the Atlas ICBM and the Air Force's Thor and the Navy's Jupiter IRBMs, and alluding to aircraft-borne weapons as well. One draft promised that "more and better ICBMs are on the horizon." The draft boasted of a 700 percent increase in spending toward "the exploration of space for peaceful purposes" since 1958; "Our goal always has been to add to the strength of our great nation." Though the sentence was later removed, drafts declared that "fictions like 'bomber gap' and 'missile gap' have been exploded."[33] The draft signaled Eisenhower's success at protecting national security and investing properly to establish a secure strategic footing.

The Dyna-Soar program, though not a topic within the State of the Union, was affected by the policies that the speech summarized. Revisions made to the speech on January 3 reinforced Eisenhower's theme of contemporary strength, tempered by reminders to avoid the pitfall of waste. The assertion that "every dollar uselessly spent on military mechanisms decreases our total strength and, therefore, our security" echoed a theme that spanned Eisenhower's presidency, evident in the "cross of iron" speech he had given just three months after taking office. "The nation can ill-afford to reverse a national policy which so adequately provides for a constant, steady level of effort for the long pull," since the Cold War challenge promised to be an ongoing one. Americans were entreated to feel confident that U.S. policy was working: "Cracks were opened in the Iron Curtain through a comprehensive agreement" with the Soviets "for cultural, technological and educational exchange." Separately, "despite the absence of permanent agreement, West Berlin has remained free."[34] Abandoning the balanced approach to national security was unwise.

Eisenhower did not forget Dyna-Soar, and he did not get a chance to try. On the afternoon of January 3, Gray spoke with Eisenhower again on the subject of future manned space flight. This topic demanded the president's repeated attention during the preparation of the final budget address. The manned space issue had earlier been discussed by the NSC, which had moved toward closing the door on manned programs extending beyond Mercury. But language in the upcoming budget speech touched on this too. Gray showed the president an excerpt from the contemporary draft: "Pending further testing and experimentation, there appears to be no valid scientific or security reason at this time for extending manned space flight beyond the Mercury program."[35] Gray told Eisenhower "that in any event the word 'space' should be inserted and that the phrase 'or security' probably impinged upon Defense programs such as DINOSAUR [sic] and the Space Plane."[36] Including the phrase "or security" would be tantamount to publicly proclaiming what Eisenhower and the NSC had concluded by December 5: that Dyna-Soar was of no military value. The governmental entities most involved in space security issues—DOD, the State Department, NASA, and PSAC—agreed that language that "might raise issues rather than put them to rest" did not belong in a speech being delivered less than one hundred hours before the president's retirement. Eisenhower conceded to removing the words "or security," but he continued to assert that no self-evident *scientific* reason existed to reach past Mercury.[37] On the question of security rationales, he would maintain a restrained silence. The theme of sufficiency in defense spending gathered steam during the several rounds of revision on January 5 and 6.[38]

The farewell address, meanwhile, incorporated one flourish of unmatched audacity. The president wanted to explain that government-directed research teams seeking specific outcomes were displacing lone inventors focused on scientific knowledge. At first, the draft asserted that "for every blackboard there are now a thousand drawing boards." It was a highly improbable claim, but as a rhetorical tool it served its function. Over time, it became "thousands of drawing boards." Now, on about January 6, the phrase reached its ultimate form: "for every blackboard there are now hundreds of electronic computers." This rhetoric made its point, but it was patently, even farcically, incorrect: computers of that era tended to be enormous machines that consumed large amounts of electrical power and required entire teams of specialists to operate. The CIA had conducted a study on the development of computing power in the United States and the Soviet bloc, and it had estimated the number of electronic computers in the United States to be three thousand. The remark that hundreds of computers had replaced each blackboard suggested that there were no more than thirty schoolhouse blackboards in all of the country's now fifty states. One administration reader caught this exaggeration, writing in the margin, "Probably not. These are million-dollar pieces of hardware."[39] But the new wording stood, an uncharacteristic example of the speech embracing rather than rejecting passages that lent a seeming intemperance to the president's parting message.

The United States, both in relative and in real terms, was indeed prosperous in the years following the Second World War. In fact, Eisenhower and his confidants had struggled to find the best way to explain the country's postwar wealth and power (outcomes due in part to the war itself) and also the nation's revulsion for future war. Industrial power and technological skill had been hallmarks of the U.S. contribution to victory. Technology in the home had increased the comforts of daily life, and medical advances such as the Salk vaccine had increased the possibility of living life to its fullest. Technology had accomplished marvels. Eisenhower had specifically acknowledged some of these accomplishments when he spoke to the nation in the wake of Sputnik I and II.

However, he noted, "In meeting the many foreign and domestic crises of our time, there is a recurring temptation to feel that some forward action could become the miraculous solution to all current difficulties." Biased parties sought to sell incomplete solutions, seasoned by false assurances, to a worried populace: "A massive increase in newer elements of our defense; a spectacular attack on our deficiencies in education; a rapid expansion in basic and applied research—these and many other possibilities, each desirable in itself, may be suggested as the open sesame to the road we wish to travel."[40] Each proposal had, however, to be judged "in the light of a broader consideration."[41] Science was a tool, not a deity, and even its welcome accomplishments did not guarantee that they would be worth their price.

Although sufficiency had been a cornerstone of Eisenhower's defense philosophy,[42] its rhetorical power was doubtful, and over the following days, the word was avoided in sections of the speech about deterrence. Earlier drafts cautioning that "further revolutionary changes in the structure [of DOD] at this time would be a mistake" now bowed to the seeming inevitability of upcoming adjustments, and new words appeared instead: "gradual improvements in its structure and procedures are to be expected."[43] Indeed, Kennedy had gotten himself elected in part by promising that big changes would happen.

Eisenhower's final considerations about space topics reflected this. U.S. space accomplishments had been scientifically significant. Moreover, as of August 1960, U.S. satellite technology was strategically significant. But the former activities, such as measuring the Van Allen belt in space or determining the exact contours of the Earth (slightly pear-shaped) did not make glaring headlines or produce evocative images that could compete with reports of ponderously large and menacing Soviet satellites. In short, much of the American public would not agree with Eisenhower that U.S. space feats had been "dazzling." The speechwriters may have recognized this when on January 9 and 10, they decided to refer to national space projects as "new" rather than "dazzling." Reference to unmanned projects bound for Mars and Venus remained.[44] In effect, the existence of some space projects would be reiterated, but little real effort would be made to "sell" space. It would have been anathema to the president's outlook to promote an expensive race for only the most conjectural of purposes.

Language in Eisenhower's fiscal year 1962 budget delivered to Congress on January 16 complemented the messages from his final State of the Union delivered four days earlier and was consistent with the farewell address to be delivered the following day. He still believed that the Mercury program should mark the endpoint of U.S. human spaceflight, but he bowed to apparent national will: "Further testing and experimentation will be necessary to establish whether there are any valid scientific reasons for extending manned space flight beyond the Mercury program."[45] A potentially unpopular sentiment was molded into a minimally unpalatable form.

The budget message also explored the promise offered by unmanned space science, including earth-orbiting satellites, the upcoming Ranger, Surveyor, and Prospector lunar satellite vehicles, and Mariner's program goal of journeying to Mars and Venus in 1962. "We have just cause to be proud of the accomplishments of our space programs to date and can look forward with confidence to future achievements which will succeed in extending ever further the horizons of our knowledge."[46] The president aimed to encourage satisfaction with recent progress and its current pace, support the legacy of his policymaking, and hopefully dampen the expectant demands for "dazzling" feats in the future.

Just before delivering his farewell on January 17, Eisenhower decided on the final wording of a key message in the speech. Eisenhower asserted that "only an alert and knowledgeable citizenry can compel the proper meshing of the huge industrial and military machinery of defense with our peaceful methods and goals, so that security and liberty may prosper together." It was not simply a harangue about the military-industrial complex. The scientific-technological issues were the pressures put on the administration and the public by publicity-hungry weapons and vehicles advocates; this, in fact, had been a key theme of the speech from its first drafting. "In the councils of government, we must guard against the acquisition of unwarranted influence" by intertwined special interests. To help make that warning heard, Eisenhower added the phrase "whether sought or unsought" regarding the military-industrial complex's power. This was intended to excuse him from a need to prove corporate or military malevolence, instead focusing attention on the structural problem he identified.[47]

Reaction and Transition

Listings in the *New York Times* and *Washington Post* indicated that viewers could watch the farewell address on three New York and four Washington stations or listen to it on six Washington radio stations. But in New York City, two stations skipped the speech. The use of television to project the farewell "seem[ed] fitting" to *Los Angeles Times* journalist Cecil Smith, who explained that "the span of the Eisenhower administration is virtually the span of television as a potent force," with the 1952 nomination coverage standing as a crucial marker. Despite the expected jabs by the late-night comedians, networks demonstrated respect for the transition process. The National Broadcasting Company cut an eight-minute comedy sketch, to be performed by Art Carney and Lee Remick, about John and Jacqueline Kennedy entering the White House.[48]

The *Washington Post* jabbed the president just prior to the address, saying he "firmly rejected any need for a recession cure and left President-elect Kennedy with a forecast of a thinly balanced budget. . . . Mr. Eisenhower said pointedly that his budget 'reflects my confidence in the strength of our economy now and in the years to come.' Outside his official family, however, virtually all authorities agree that business is currently slumping." The *New York Times* had already voiced a similar conclusion in a story titled, "1960, a Year of Great Promise, Developed 'Hidden Recession.'"[49] During the campaign, Kennedy had capitalized on the lingering effects of recession that had started late in 1957, coincident with the space race. It was another area where Eisenhower seemed inactive, and

Democrats had gained political traction by arguing that increased defense spending could *lift* the economy while tackling security issues like the supposed missile-space gap. This outlook prima facie rejected Eisenhower's perspective on defense and on national security at large. The *Los Angeles Times* more favorably noted a parallel between Dwight Eisenhower and George Washington, praising "the striking similarity in the way these two General-Presidents regarded their high office." The paper anticipated, however, that "there are some smart ones, to be sure, who compose cruel parodies of Eisenhower syntax and smirk at savage caricatures of 'Old Bubblehead,' who quote archly all the clever columnists and churlish wisecracks by the current late club bores."[50]

The farewell address evoked some diametrically opposed responses within the media. The sympathetic *Wall Street Journal* advised readers that "the American people, who have so often listened to this man, might do well to listen again." Rather than "raising any specter of a power-hungry military," Eisenhower noted "that as the enormous defense program begins to work its way into the warp of our whole economic life . . . we risk having defense come insidiously to dominate our thinking in matters actually remote from military needs. America as a permanent armed camp could become a very different America." The *Journal* conceded that Eisenhower's message sounded "so simple—we are almost tempted to say obvious—that to self-styled sophisticates it may sound like just another homily out of a discarded copybook."[51] This, in the *Journal*'s view, stood at the core of its veracity.

At *Aviation Week*, editor Robert Hotz was outraged. He declared that Eisenhower had "never really understood" the concerns held by people in the defense sector. The expanded ties between the military, industry, and research lab had begun before Eisenhower's presidency, but they had strengthened during it and had even been advocated by then-General Eisenhower just after the Second World War ended. *Aviation Week*'s editorial repudiated Eisenhower's message wholesale: "It will probably come as a considerable surprise to the millions of people in the military-industry complex of the defense effort and the smaller group of scientists and engineers who have worked so hard in the face of so many official obstacles to bring this country into the forefront of the new technologies to learn from President Eisenhower that he regards them as serious threats to the future liberty of this nation." Hotz declared, "We doubt if any significant percentage of the American people will share the outgoing President's fears and we certainly don't."[52]

Eisenhower's warnings were in some sense disorienting, since they ran contrary to his presidential employment of the military-industrial complex. The origins of this partnership preceded him, but it grew steadily under his watch. By 1958, the nation's leading defense contractors accounted for thirty of the country's top fifty corporations. And Eisenhower in the interwar era had written as a student at the Army War College and the Army Industrial College about the *need* for closer cooperation during peacetime between these two groups.[53]

Alexander de Seversky, a Russian expatriate, aircraft manufacturer, and longtime advocate of American airpower, published a polemic in 1961 entitled, *America: Too Young to Die!* It included de Seversky's account of a conversation with Eisenhower in 1945: "Ever since [that time] . . . I had looked upon Eisenhower more with incredulity than any other reaction." De Seversky deemed the general's estimation of the airpower contribution to victory to be inadequate. A decade and a half later, de Seversky proclaimed, "Here, I finally realized, was a man [Eisenhower] who had just won the greatest war in the history of mankind, and he had no idea of how he had done it!"[54] Eisenhower's leadership also meant that such attacks would not be as damaging under ordinary circumstances. They certainly would have made little headway in 1945, when Eisenhower was Supreme Allied Commander and de Seversky's conversation with him allegedly took place. By 1961, however, Eisenhower's minimal self-aggrandizement, combined with his deliberate work to seem unflustered and the supposed security crises of the bomber gap and missile-space gap, opened the door to wild criticisms.

Eisenhower conducted his last presidential press conference on January 18, the morning after his farewell address. Lillian Levy of the *Science Service* inquired about the "scientific technological elite" portion of the speech. Since Eisenhower saw this as a problem, what was the answer? Eisenhower replied, "I know nothing here that is possible, or useful, except the performance of . . . responsible citizenship." This alone, he answered, could prevent the dominance of such groups. Taking a shot at the trade journals, although not mentioning them by name, the president illustrated his point: "When you see almost every one of your magazines, no matter what they are advertising, has a picture of the Titan missile or the Atlas or solid fuel or other things, there is becoming a great influence, almost an insidious penetration of our own minds that the only thing this country is engaged in is weaponry and missiles. And, I'll tell you we just can't afford to do that."[55] He concluded by explaining that "the reason we have them here is to protect the great values in which we believe, and they are far deeper even than our own lives and our own property, as I see it."[56] It was one more unscripted apologia for sufficiency and the long haul. It was through these explanations that Eisenhower came the closest to tipping his hand—to showing the world what was behind the grinning face and the below-the-radar approach to statecraft. When Hotz described the exchange to *Aviation Week* readers, he guessed the outgoing president's frustration at having failed to bring about arms control had prompted Eisenhower "to lash out again at his final press conference."[57] Rather than being confounded by the prospect of new technology, Eisenhower resisted the counterproductive trends he anticipated emerging from several categories of technologies.

Hotz reduced the issue to a simple "age versus youth" dichotomy, simplistically and inaccurately dismissing the legitimacy of debate. Other journalists also noted contrasts between the perspectives of Eisenhower and Kennedy. Comparing Eisenhower's final speeches in office and Kennedy's first, Chalmers M. Roberts of the *Washington Post* declared, "Listeners and readers will be pardoned if they have trouble realizing the two presidents, only a few days apart, were talking about the same country." The two presidents did publicly agree that the United States now led the Soviet Union in space, although congressional Republicans were annoyed by Kennedy's pointing out the Soviets' lead in lifting large vehicles to orbit. The *Los Angeles Times* contrasted Eisenhower's tone that "he was leaving the country in pretty round shape" with Kennedy's statement that "the American economy is in trouble and steps are needed immediately to bolster it." The report also emphasized a contrast on missiles. "Mr Eisenhower said the 'missile gap' shows every sign of being a myth. Mr Kennedy decreed a speed up of 'our entire missile program.'"[58] The high-technology advocates had reason for enthusiasm.

Two leading lights in journalism pointed to continuities and gave less emphasis to the purported contrasts noted by so many. Taking a long view, Arthur Krock noted "the constancy of the basic policies and purposes of the American people" and "of the fundamental aspirations of the United States."[59] For his part, Walter Lippmann noticed that Eisenhower had "dwelt on a question, never before discussed publicly by any responsible official, which is of profound importance to the Nation's future." He saw the outgoing president identifying "the danger of unwarranted military influence—or to use an old phrase for it, the danger of militarism."[60]

Eisenhower's decades of military service had provided a useful vantage point from which to consider the issue of U.S. military preparedness and its implications, but they also served as a touchstone for supporters and detractors alike. Lippmann saw an old *soldier*, whereas Hotz—who dismissed Eisenhower's hesitance to embrace bold new defense technologies as evidence of an old man with an antiquated outlook, "vexed by energetic young men whose discoveries spawned perplexing new problems for the chief executive's administration"—saw an *old* soldier.

In the context of the Cold War, Lippmann wrote that military strength could not safely be diminished. Despite his concern about the "military-industrial complex" and the painful budget cuts the Army endured (to save money while favoring the Air Force's ability to project nuclear delivery capabilities), Eisenhower had not ignored U.S. security needs. They had, instead, been central to much of his thought and attention as president, and national security and solvency were among the core reasons that Eisenhower decided to run for office. President-elect Kennedy was determined to increase the military across the board. "Only by making civilian influence greater, not by reducing military power," Lippmann asserted, could the nation maintain national security and insulate the national character from militarism. "The true solution of the problem that President Eisenhower warned the country against is to be found in civilian appointees who are confident and willing to command. When such civilians are in office, it will be possible for the administration to wean the Congress and portions of the press from their undue reliance upon the military establishment as the true source of the true American policy."[61] Civilian appointees thus represented a key element in staving off militarism, and the character and strength of these selected individuals would have a tremendous impact on the country's future.

Lippmann respected Eisenhower's parting message, and he found reason for encouragement in the incoming Kennedy team: "There is, I am convinced, solid ground for confidence in the administration which Mr. Kennedy has organized." Among his reasons were that "it is not an administration led by corporate executives," a distinct contrast from the Eisenhower cabinet, sometimes dismissed as nine "businessmen and a plumber." Furthermore, "it is not an administration composed primarily of professional politicians, although the head of it is among his other aptitudes a professional politician of the first order." Lippmann wrote also that "it is not an administration made up of professors drawn out of an academic life of scholarship and research. It is, for the first time in our history, an administration manned primarily by professional public servants—by men whose primary careers have long been the public service."[62] Ironically, Lippmann overlooked the prewar Harvard teaching career of incoming Defense Secretary Robert McNamara, in addition to the fact that he was a rising executive in (and ultimately president of) Ford Motors when he was tapped to head DOD.[63] By taking a firm hold over the defense establishment, McNamara seemed able to achieve the kind of civilian authority that Lippmann felt imperative in preventing the militarism Eisenhower considered both possible and dangerous.

Lippmann said the new officials "will need public support. They will need a lot of luck. But I do not know of any administration in our times in which the level of competence has been so high."[64] Kennedy's incoming administration had campaigned on accomplishing far more in many realms, including prominently defense- and space-related fields. Now, with its entry into the White House, the Kennedy team would have the opportunity to try to shape policies guaranteeing national safety and interests while also appealing to the public's culturally informed perceptions about themselves, their country, and the potential and promise held by technology.

11

"Equal Attention to Both"
John F. Kennedy's Activity in Space

*So far the Air Force is precluded from the man-in-space field.
But with each Soviet feat there is a growing impetus on Capitol
Hill and in the Pentagon to accelerate the Dyna-Soar project.*
—JOHN W. FINNEY, *NEW YORK TIMES*, AUGUST 7, 1961

*Disaster will follow if we are as negligent [with the Dyna-Soar
program] as we were with ICBMs 10 years ago.*
—FREDERICK PILCHER, LETTER IN *AVIATION WEEK*,
NOVEMBER 27, 1961

Kennedy's First Steps Toward a Space Policy

For John F. Kennedy, leadership meant *movement*. Dynamic, young, with the polished charm and charisma of a born politician, Kennedy was an ardent Cold Warrior and a fervent believer in American potential. The presidential role is never an easy one, and the Cold War imposed extraordinary and peculiar burdens on it. As with any leadership role, the rigors of the presidency have the potential to lay bare and even spotlight any shortcoming. Although he focused on movement, Kennedy allowed *direction* to take a back burner. Because of how he interpreted the country's great potential strength and because of the ambiguities and perils of the Cold War, he was loath to make specific crucial decisions in space security policy and stick to them.

Candidate Kennedy had promised action, and with his inauguration in January 1961, President Kennedy's supporters enthusiastically anticipated just that. Aerospace advocates, their dreams frustrated and curtailed in the Eisenhower years, looked forward to dramatic endeavors. For a moment, the advocates of both civilian space programs and military exploitation of space would think their aspirations enunciated and their plans on course toward fulfillment. That moment was 1961.

Visible changes arrived quickly. After the all-too-public launch failures of Vanguard, U.S. space sorties had been made with little pre-launch fanfare. Prior announcement would drum up expectations as well as raise the profile of space activities, and both results flew in the face of Eisenhower's intentions. With Kennedy's inauguration two weeks away, the incoming team moved to end the quiet approach, calling instead for NASA launches to be announced before they occurred rather than afterward. Space advocates in Congress urged Kennedy to increase NASA funding that would support commercial satellites, and trade journals lauded the "new vigor for [the] space program."[1]

The relationship between civilian and military space projects was crucial but unsettled. Louisiana Democratic Congressman Overton Brooks, who supported a $400 million expansion in NASA's budget (an increase of about one-third), was the first to insist that increased military space activity not come at the expense of civilian efforts; Kennedy's response was that "legitimate missions in space" existed for the military.[2] Both NASA and DOD requested more money for space, with NASA asking for $120 million to support communication satellites and as much again for Mercury. Notable DOD requests included $146 million for Dyna-Soar and $20 million to study the space plane concept.

The year 1961 marked the point at which increased spending on Dyna-Soar truly began. During Eisenhower's tenure, the program had annually cost single-digit millions until 1960, when the budget reached $58 million. Eisenhower had been sorely tempted to cut the Dyna-Soar program entirely, but with the end of his own presidency imminent and the president-elect's strident criticism of perceived technological gaps, eleventh-hour cancellation of Dyna-Soar had apparently seemed pointless. Instead, Eisenhower's fiscal year 1962 budget conceded $70 million to a project that fit snugly into the category of programs he had warned against in his farewell address. For a project of "no military value," $70 million was a waste, but the Air Force wanted twice that much. Seattle's economy hinged on Boeing contracts, and Air Force approval "to work on the Dyna-Soar space glider project" was welcome news there.[3] Building Dyna-Soar in turn would spark production of rocket boosters.

Although contemporary opinion leaders Arthur Krock and Walter Lippmann emphasized in positive tones a continuity between the admittedly different outgoing and incoming administrations, crucial distinctions existed. Both Eisenhower and Kennedy were Cold Warriors, but their attitudes and assumptions affected the different policies they set. In his inaugural message, Kennedy had spoken boldly "to pay any price, bear any burden, meet any hardship, support any friend, oppose any foe, in order to assure the survival and the success of liberty."[4] No one could yet know that this thesis would be tested in the diverse terrains of South Vietnam. The logic drew on Kennedy's image of a virile nation capable of undertaking the global commitments that had been implicitly accepted by Truman in NSC-68. Eisenhower never believed that any country could actually absorb the economic and psychological strain NSC-68 imposed in a Cold War environment. No country could do all things at all times for an indefinite period. As early as his inaugural address, Kennedy showed his zeal for American success through competition; "Let both sides seek to invoke the wonders of science instead of its terrors," he proclaimed, testing for a positive Soviet reaction.[5]

In New York, the *World-Telegram and Sun* described "a new administration, a new man with a freshness in look, in some ways a freshness in ideas, a new president of eloquence . . . attuned to our times." In Baltimore, the *Sun* praised Kennedy's being "somber without despair, firm without bellicosity, bold without arrogance." The *Detroit Free Press* deemed Kennedy's inaugural address "stirring [and] thoughtful," and the *Salt Lake City Tribune* called it "a fine start." The *New York Times* also repeated laudatory remarks from other newspapers, projecting them so that they reached an even wider audience.[6]

National prestige ranked first in Kennedy's justification for space activity. "Second," his new PSAC chairman Jerome Wiesner reported, "we believe that some space developments, in addition to missiles, can contribute much to our national security." Scientific exploration came third, followed by "important practical non-military applications of space technology" such as communications, navigation, meteorology, and mapping. These applications could be used for both military and nonmilitary purposes, and application in one realm could foster eventual adoption by the other, as military satellite navigation would ultimately do at the end of the century. Last among Wiesner's rationales was that "space activities, particularly in the fields of communications and in the exploration of our solar system, offer exciting possibilities for international cooperation with all the nations of the world."[7] Indeed, "space projects" could involve progress in any of these various directions, and often in several of those directions. But it would be difficult to kill all birds with the same stone. Some situations, such as NASA's interest in the Air Force's Atlas-G rockets as space boosters, suggested that synergistic relationships could be built. However, conflicting policy implications could erase the promised dividends of technology transfer.

Kennedy's first State of the Union address hinted that he was not cognizant of this tension. Kennedy noted that U.S. space technology outpaced the Soviets but that the Soviets led in booster technology. Declaring that "both nations would help themselves as well as other nations by removing these endeavors from the bitter and wasteful competition of the Cold War," he also hastened to announce that he had "directed

President John Kennedy inspects *Friendship 7,* the Mercury space capsule in which John Glenn undertook the first American human orbit in 1962. Kennedy's aim for the space program was national prestige, whether through competition against or cooperation with the Soviet Union. *John F. Kennedy Library/photo by Cecil W. Stoughton*

prompt action to accelerate our entire missile program." Referring to the presidential seal, he noted that the emblazoned eagle clutched olive branches as well as a brace of arrows. Kennedy vowed that his administration would give "equal attention to both." Congressional Democrats were reported to "generally hail" the speech "as a sober, courageous challenge to the Nation," whereas the media characterized Republican reaction as somewhat resentful. But the message of movement triggered enthusiasm on the stock market, particularly in aviation and other defense-related sectors such as steel, electronics, and oil.[8]

With prestige, and presumably national security, at stake, equipment reliability was vital. Each Saturn launch was expected to cost $20 million, and reliability analysis costing $150,000 promised to save money in the long run.[9] However, Saturn's payload remained to be determined: the Mercury capsule would never require such an enormous booster. The Dyna-Soar glider and the Apollo capsule were, respectively, military and civilian vehicles eligible for coupling with NASA's super booster. Ballistic Missile Division commander Maj. Gen. Ritland and NASA associate administrator Robert Seamans arranged to discuss "Saturn use in NASA and Dynasoar programs" and other topics in February.[10] The Aeronautics and Astronautics Coordinating Board meanwhile sought to bring coherence to various space efforts. NASA deemed the board's "most significant achievement . . . the formulation of a National Launch Vehicle Program Summary containing the currently approved vehicles of both DOD and NASA and also the follow-on coordination procedure whereby NASA and DOD shall obtain the approval of each other for new space launch vehicles prior to giving departmental approval or allocating funds for hardware development."[11]

Aerospace outlets and politicians leveraged domestic politics by insisting that space issues stood above partisan concerns. Wiesner prepared to pour cold water on some of the more extreme space ambitions, and aerospace advocates suspected that Wiesner was concerned that "the new Democratic administration would get the blame if this first attempt at manned space flight fails." *Aviation Week* editor Robert Hotz expressed discomfort with Kennedy's choice of business professional and political angler James Webb as the new head of NASA, although Webb proved to be a forceful and energetic administrator whose efforts would quickly and chronically grate on the new president.[12]

Meanwhile, aerospace advocates in the media scorned Eisenhower as an antiquated Luddite whose reticence and "gimmicks" had made the country vulnerable. Only Eisenhower and a handful of presidential advisers had been privy to satellite photographs showing that the myriad missile bases the USSR was thought to have did not actually exist. The false image of Russian missile strength was in the late 1950s and early 1960s largely a Potemkin village, built by Soviet leaders in order to mislead. The "missile gap" had also been a key pillar of Democratic opposition to Eisenhower. Now president, Kennedy directed Defense Secretary Robert McNamara to investigate critical defense issues, and by early February a DOD study showed that Eisenhower had been correct and the Democratic campaign season charges had been false, just as CIA briefers had told Kennedy the previous summer. The *New York Times* noticed the report, but Kennedy officials promptly denied that the missile gap concept (and their campaign allegations) had been undercut.[13]

Despite having access to a clearer intelligence picture, Kennedy in the remainder of 1961 would mostly follow the current on the missile-related field of space. Air Force officers and political figures (including Kennedy himself) had utilized the culturally induced beliefs in technology in order to decry the supposed inactivity of Eisenhower and to promote themselves and the alternatives they touted. The president was now caught in a current he had helped encourage as a candidate. As president, however, he had a special position from which to direct the contour of that current.

In February, the Air Force selected Titan II as the Dyna-Soar's booster, hoping to prepare the program for an anticipated acceleration in its development schedule. *Aviation Week* described Dyna-Soar as on the cutting edge in a series of aerospace pioneer craft, beginning with the X-15 rocket research plane that had first flown in June 1959 and including the ballistic capsule Mercury, which "fits roughly in the middle."[14] *Aviation Week* continued: "Beyond the Mercury goal is the Dyna-Soar program, which although currently planned and funded through suborbital flight, is aimed at eventually providing manned vehicles that can perform useful missions in orbital flight." Mercury data would "provide the foundation on which Dyna-Soar, Apollo and other manned scientific and military space vehicles can build." NASA's Director of Advanced Research Programs Ira Abbott, as well as boost-glide design veteran Eugen Sänger, chimed in.[15] Meanwhile, space advocate and Peenemünde alumnus Krafft Ehricke emphasized to congressional representatives in the Joint Committee for Atomic Energy that it was "almost axiomatic, that techno-scientific (in distinction from a purely scientific) progress in a specific area is accomplished faster and more economically if the work is tied to a major mission requirement." The Air Force continued to assert mission requirements for aerospace programs.

Air Force planners envisioned establishing a manned global surveillance system "by 1966 or earlier," but this sounded excessively optimistic to industry observers who knew that only the X-15 had flown and that beyond it, "there is only the Dyna-Soar boost glider, and it will be essentially a research vehicle through its initial stages."[16]

Representative Brooks voiced his "concern over the 'quasi-public fashion' in which military use of space was being promoted." Leaks about the B-70 program annoyed McNamara, and he initiated a crackdown following an Air Force general's announcement (in approved remarks) stating that Dyna-Soar would fly within three years. McNamara "was reported to feel that too much weapons information was being cleared for release in generals' speeches." Brooks, meanwhile, feared "that the executive branch was contemplating 'a radical change in our national space policy' that would accentuate military space 'at the expense of civilian and peaceful uses.'"[17] Policy remained in flux, and Kennedy's upcoming speeches showed that he had not yet determined whether the military or the civilian agency would take the leading position in a partnership.

During the transition period, civilian planners pondered the best steps to move the country forward—although Kennedy had not yet explicitly decided which direction that would be. In late November 1960, Robert Seamans and outgoing deputy administrator Hugh Dryden met incoming administrator James Webb and discussed industry proposals for a possible Apollo space capsule. Two stood out in particular: "Martin led on technical approach, North American led on technical competence." Although the deliberations would be second-guessed in the wake of the fatal Apollo I mishap six years hence, both contractor alternatives seemed able, and by March 1961 Webb had chosen North American to build the successor capsule. A lunar project did not yet have Kennedy's endorsement, and new budget director David Bell noted that the extent of future space exploration would alter the practicality of different propulsion research programs. Nuclear rockets, for example, might be useful for interplanetary ventures but were unnecessary if the country's space efforts focused on lunar or Earth orbit arenas closer to home. With the direction not yet set, Bell advised steering a middle course. He reminded the president that "the basic question is whether the tangible and intangible benefits of manned space flight to the moon and beyond are worth the cost and effort."[18]

Information and advice conflicted—a normal challenge for policymakers —and discovery that the missile gap was a falsehood likely accentuated this confusion. Speaking to Congress in late March about defense spending,

Kennedy reaccelerated the retirement of B-47 medium-range bombers from SAC service, and he abruptly announced the development of a B-70 Mach-3 bomber to be "unnecessary and economically unjustifiable" and curtailed ANP and moved it to the non-military AEC "where it belongs."[19]

But space was another story, and Kennedy said as much. Dyna-Soar was one of four space systems Kennedy named as "related to our strategic and continental air defense forces which I find require additional support." March brought disquieting news of the Soviet orbit and recovery of a dog and several other creatures, providing the Soviets with an opportunity to predict a manned spaceflight in the future. Although Kennedy's rhetoric supported the aerospace advocates, he split the difference between the $70 million that Eisenhower had reluctantly earmarked for Dyna-Soar, the $130-million that Congress had supported, and the $146-million that the Air Force had hoped for. The administration meant to provide $100 million for Dyna-Soar, which, along with the other space projects, shared a modest paragraph at the end of the deterrent forces section of Kennedy's March 28 speech to Congress.[20] Kennedy also described base closures, which were necessary for funding investment in other aspects of national defense and made possible because the missile gap had been a fiction. Although willing to spend larger sums to keep space defense options open, Kennedy had not yet committed to an actually defined space policy.

The notion of "aerospace," introduced into the popular lexicon by Gen. Thomas White in 1958, gathered increasing cachet and momentum. The New York Times reported in March 1961 that the Keystone Low-Priced Common Stock Fund S-4 had established a new category designated "aerospace—to list portfolio holdings in the space fields," and Aviation Week noted that "Kennedy budget shifts" could "accelerate [a] slight rise in aerospace sales."[21] A month later, the New York Times commented on the "new world, [and] new words," explaining that "the Air Force, in its space program, has definitely outsoared the former limits of our language. . . . They haven't hesitated to coin whatever words they needed," among them "aerospace," which meant "an upper region in which the atmosphere ceases to function as a factor in flight." The report defined the Dyna-Soar as "a projected manned orbital glider or bomber now under study."[22] Space-minded publications told readers that Boeing was at work on "space bomber concepts" including a "Dyna-MOWS" (manned orbital weapons system) that would be a refinement on Dyna-Soar III.[23]

Shooting for the Moon

In terms of evocative imagery, winged piloted space vehicles overshadowed orbital capsules. On April 12, the *New York Times* published a report identifying hopes to lift a manned Mercury capsule into space by year's end. Nonetheless, the X-15 was described as "only slightly less ambitious than Mercury." Moreover, Dyna-Soar was identified as "a long step beyond Mercury . . . much more agile" than the capsule, and "under energetic development by the Air Force." The report also mentioned a NASA initiative for a three-man capsule for two-week-long Earth orbit missions or for "a week-long trip around the moon."[24] A bigger surprise still lay in store that day, however: Yuri Gagarin, a 27-year-old Soviet air force major, orbited the earth in a five-ton spacecraft. The president had already repeatedly conceded "months" of Soviet superiority in booster development, and journals such as *Aviation Week* had estimated the Soviets to have a four-year lead there. With Gagarin's flight, the Soviets tallied yet another spectacular first.[25]

Technological determinism and Soviet secrecy combined to prompt fear that any Soviet accomplishment was just the tip of a technological iceberg, and this seriously complicated space policymaking and program setting. For years, the Soviets hid the fact that Gagarin had parachuted from the Vostok craft separately after reentry and before landing, despite agreed IGY provisions defining a first human spaceflight. If an ejection system had not been devised to jettison the cosmonaut to parachute separately from the capsule at an altitude of two miles, a cosmonaut surviving space travel would die upon the impact of landing. With the Voskhod series of vehicles in 1964, the Soviets finally developed a "soft landing" system that permitted cosmonauts to occupy spacecraft for the duration of a mission.[26] The Soviets concealed the rudimentary nature of the first-generation Vostok capsule, and the secretiveness sparked speculation in the United States and Britain about Gagarin having perhaps ridden a winged craft far more advanced than the Dyna-Soar.

Military space advocates redoubled their efforts, and the president pondered his options. Hotz compared Gagarin's orbit to the Wright brothers' first flight and regretted that the accomplishment came from the Soviet Union. "Determination" by the nation's "top leadership" could allow the United States to catch up, *Aviation Week* promised, although it told readers that "proponents of a bigger military space effort expect the Soviet success in recovering a man from orbit to help them promote an expanded program." General White declared "Soviet aerospace strength to be perhaps the greatest threat in the history of the country," and one MIT professor ascribed Soviet space success to superior management arrangements. Kennedy assigned Vice President Johnson leadership of the space council, calling the body's authorization "a key step toward moving the U.S. into its proper place in the space race."[27]

Race imagery was potent but deceiving, as Kennedy now knew. U.S. satellites outworked and outnumbered Soviet counterparts, while the Soviets seemed consistently to steal the show with spectacular "firsts." *Aviation Week* suggested that "as usual—Russian success makes a U.S. space push suddenly more popular," and it accurately reported Kennedy fretting "whether there is any program now, regardless of cost, which offers us hopes of being pioneers in a project."[28] Kennedy publicly conceded that "it is possible to spend billions of dollars in these projects in space to the detriment of other programs and still not be successful." The Air Force Systems Command's assistant for bioastronautics warned that a Soviet lead in life sciences could lead to a Soviet capability to deny U.S. vehicles' access to space in the future.[29]

Seeking bold initiatives in space, Kennedy gave a five-point memo to Johnson on April 20. "Do we have a chance of beating the Soviets" at tasks such as launching a space lab or venturing to the moon? "Is there any other space program which promises dramatic results in which we could win?" Prestige demanded boldness, but how bold would be necessary— and possible? Earlier Soviet space accomplishments had also led Robert McNamara to wonder whether a lunar venture lay too close to existing Soviet technological capabilities. The objective needed to be bold enough to stand well beyond both U.S. and Soviet capacity; otherwise, there could be no prestige-garnering race. McNamara therefore pressed officials to instead consider "whether we shouldn't embark on a planetary program."[30]

Wernher von Braun, rarely bashful about promoting his aspired projects, promptly informed the vice president that "we have a sporting chance of sending a 3-man crew *around the moon* ahead of the Soviets" and that "we have an excellent chance of beating the Soviets to the *first landing of a crew on the moon*" with a safe return. Presidential historian Michael Beschloss suspects that Kennedy's thoughts in space owed more to the defeat of anticommunist Cuban exiles at the Bay of Pigs than to Kennedy's sincere and coherent concern about Soviet space advances. Beschloss persuasively suggests that Kennedy hoped for "a quick, theatrical reversal of his new administration's flagging position" on the eve of its first summit with Soviet premier Khrushchev that summer.[31] The president likely had a number of contributing motives.

Dyna-Soar's developers at Boeing introduced their Project Streamline on May 4, suggesting alternate accelerated schedules to help Dyna-Soar recover from the time lost by the Phase Alpha delay the previous winter. Streamline looked beyond the Titan II boosters, whose power would still be insufficient to take the heavy Dyna-Soar into orbit. From a policymaker's standpoint, an expedited space project might offer an opportunity to seize a Cold War initiative at precisely the time that the president wanted to do so.

The boldest of the three Streamline alternatives (called 3A) proposed using a Saturn rocket to lift Dyna-Soar as early as the first quarter of 1963 and introducing a piloted version of this configuration by the end of that year. Greater maneuverability would be offered by provisioning the Dyna-Soar vehicle with a maneuverable engine at the end of 1965.[32] Conspicuously, schedule 3A completely bypassed the suborbital stage, which Eisenhower officials had so carefully emphasized to imprison the troublesome program. Boeing also outlined an intermediate schedule (3B) and a comparatively conservative acceleration schedule (4), which retained the suborbital test phase and anticipated using NASA's massive (and as-yet-unlaunched) Saturn booster in the mid-1960s.[33]

The vital importance of a big booster grew as Dyna-Soar's weight increased apace. The glider itself remained fairly steady at about five tons (the anticipated weight was 10,608 pounds). However, the provisions needed for multiorbital flight that would be stored in the transition section, which connected the boxy glider to its round booster rocket, almost doubled that section's original weight of 4,663 pounds. Thus, Dyna-Soar I and its transition section could be expected to weigh between 17,600 and 22,000 pounds. Astronaut Alan Shepard made the first U.S. suborbital flight in a Mercury vehicle one day after Boeing submitted Streamline. Due to its detaching safety system, the Mercury capsule weighed less than 3,900 pounds at launch and under 2,300 pounds upon splashdown.[34] Dyna-Soar's booster would have to be big, powerful, and reliable.

The Air Force developed an astronaut pilot training course to reverse NASA's domination of U.S. man-in-space efforts. The *New York Times* referred to the Air Force contract awarded to Martin for studying a lunar landing in military terms, noting the service's interest in "mounting a manned assault on the moon." The Air Force anticipated that a human lunar landing might cost $40 billion. In the Eisenhower era, NASA had estimated that a human lunar landing might be undertaken sometime "beyond 1970."

Kennedy officials reportedly were hinting that it might be accomplished by 1969.[35]

On May 25, the president gave his second State of the Union address in barely four months, explaining that the country faced "extraordinary times." A discussion of domestic economic and social topics gave way to foreign policy and eventually to military needs for alliances, intelligence, and civil defense. A passage addressing disarmament provided the transition to the speech's bold challenge to conduct a human lunar mission by the end of the 1960s. Kennedy unambiguously asserted that space accomplishments "impact . . . the minds of men everywhere, who are attempting to make a determination of which road they should take" for their political and economic future during the Cold War. He advocated a lunar goal because "no single space project in this period will be more impressive to mankind, or more important for the long-range exploration of space." Although indicating that it would entail unparalleled costs, the president provided no explicit figures in his speech beyond the price tag for the coming fiscal year.[36] Kennedy did, however, offer precise figures on some supporting space projects whose costs were dramatically less consequential. Work on the Rover nuclear rocket would cost $23 million, communication satellite work would cost $50 million, and $75 million would be required for weather observation. Unless the nation was intent on *completing* the lunar undertaking, Kennedy advised in his address, "in my judgment it would be better not to go at all."[37]

Even in this iconic speech remembered for its boldness, Kennedy hugged political cover. Speechwriter Ted Sorensen remembered the president deviating from the text of a formal address only once during their many years of work together; this speech was that occasion. Immediately after warning that the lunar quest ought to be either fully embraced or completely spurned, Kennedy extemporaneously disposed of his ownership of the initiative: "This is a choice which this country must make. . . . It is a most important decision that we make as a nation. But all of you have lived through the last four years and have seen the significance of space and the advantages in space, and no one can predict with certainty what the ultimate meaning will be of the mastery of space." Still deviating from the text, he noted, "I believe we should go to the moon. But I think every citizen of this country as well as the Members of Congress should consider the matter carefully. . . . If we are not, we should decide today and this year."[38] This address avoided explaining whether he preferred a civilian, military, or fused lunar effort. The historical record shows that Kennedy was actually much less decisive about space policymaking than he was urging citizens and congressional representatives to be. Despite favorable reception of the speech, Kennedy launched an adventure to which he was not personally committed.[39]

Kennedy's fence-riding as a space policymaker (unlike his decisive talk as a candidate and orator) led him to split the difference on the nuclear rocket effort. Ambitious space initiatives appeared to merit a nuclear rocket, but Kennedy proposed a funding level for the Rover project that lay between the Bureau of the Budget's recommendation for a modest increase and the bolder alternative of a $50 million investment. Advisers such as Jerome Wiesner argued against wildly ambitious manned projects that would suck the air out of other worthy projects in space and earth science.[40] Kennedy sought to buy time to decide on his space policy.

Space advocates were themselves divided about the impact of a human lunar landing. Like Wiesner, NASA director Webb correctly suspected that dominant emphasis in funding and press coverage would undercut NASA's other space science efforts. In the Air Force, however, Gen. Bernard Schriever had been supporting a lunar landing as a means of focusing effort on a high-profile mission that would require innumerable subsidiary challenges to be mastered. "If we had this sort of an objective," he asserted prior to Kennedy's May 25 address, "there would be so many other things that would be required that you couldn't avoid having a major space program." Although Kennedy had not specifically asserted that NASA (rather than the Air Force) should run the lunar program, Schriever recalled during oral histories in the late 1960s that "that never came up" because "there was no argument about who was going to run the program." Simultaneously, Boeing announced that it would spend $40 million on major subcontracts related to Dyna-Soar.[41] The Air Force and its major contractors felt confident in the service's having a strong role in space.

DDR&E York, who had two years earlier confined the Dyna-Soar program to suborbital hypersonic research, told the House Appropriations Committee in 1961 that the Soviets were believed to be interested in an equivalent to Dyna-Soar. Soviet defector Grigori Tokaev's account of Joseph Stalin's interest in "aircraft of the Sänger type" had been available for years, but suspicions by the DDR&E in the United States and by British Interplanetary Society secretary Leonard Carter attracted new attention from the *Washington Post*. York specifically cited Soviet superiority in boosters as enabling more advanced work in such heavy vehicles. York, who had confined Dyna-Soar in 1959, labeled "Dyna-Soar a 'serious program'—

but 'not a high priority program.'" In contrast, Deputy Chief of Staff Lt. Gen. Roscoe Wilson called it "the most important program we have in the Air Force because we will never be able to talk about space flight until we are able to take off at the pilot's option, control the vehicle and return it at the pilot's option." Ballistic capsules, whose launch schedules were at the mercy of weather patterns and whose landings required Navy search and recovery actions, were unsuitable for military space operations.[42] Deputy Defense Secretary Roswell Gilpatric reiterated that "it is unclear at this time" whether Dyna-Soar would become a weapons system following its development.[43]

Among *Aviation Week*'s 50,000 subscribers, opinion was divided. One letter to the editor in June promoted the peaceful use of space as a means by which to curtail "this overdone sword sharpening contest" and compared the arms race proponents to children "want[ing] to do something dangerous or injurious" and in need of "something of more interest to divert his attention." Another reader wrote in to criticize this perspective by insisting that "space is for peaceful purposes, and aerospace power is for the preservation of peace." The journal also ran a letter to the editor from General White, praising the journal for showing that "aerospace is truly our new frontier." Congress, convinced along similar lines, voted added funds for SAC long-range bombers and for Dyna-Soar.[44]

The Air Force seemed preeminent in the nation's space journey. In August the *New York Times* acknowledged that "the actual training of the Dyna-Soar astronauts has not begun, partly because of the uncertainty and controversy still surrounding the project." However, astronaut selection did take place, and by 1962 the Air Force had a cadre of intended Dyna-Soar pilots, including future lunar pioneer Neil Armstrong.[45] When White, who had pioneered the term "aerospace" and had worked to advance Dyna-Soar, retired at the end of June, the National Geographic Society honored him with a bronze trophy of a male figure heaving a Saturn rocket into space. The award would be given annually to uniformed and civilian Air Force personnel for outstanding aerospace contributions. National Geographic Society president Thomas W. McKnew "told President Kennedy that its purpose was to encourage and inspire further conquests of space after the example set by General White."[46]

Aviation Week reported that NASA's deputy administrator Dryden believed that the emphasis on the lunar goal, to the exclusion of concurrent scientific and technological gains from space exploration, was "getting [NASA] in trouble." He told the journal he wanted the focus regarding the

Neil Armstrong, posing at the nose of an X-15, was slated to pilot Dyna-Soar for the Air Force but by 1962 had been claimed by NASA for the Gemini and subsequent Apollo programs. In July 1969, Armstrong would become the first human to set foot on the moon. *NASA/photo by USAF*

lunar mission to resemble that of the X-15, in which "speed advances are secondary to the flight knowledge gained." Meanwhile, Trevor Gardner, a veteran of the Manhattan Project during the Second World War, warned the ARS against timidity regarding technology: "Unfortunately, our society has a consistent record of having under-imagined the possibilities of the technological future, and of having under-reacted to the ominous significance of Soviet technological and military progress." Gardner challenged contemporary estimates of the cost of a lunar landing, remarking that "it seems more probable that the lunar landing and return mission can be accomplished for approximately $10 billion, rather than $40 billion."[47]

Tracking the Space Race

For Boeing and its major subcontractors, Dyna-Soar was a valuable image for attracting engineering talent. An illustration of the glider covered half of Honeywell Aerospace's full-page advertisement, which announced that "new and exciting programs, such as project Dyna-Soar, have created these challenging career openings for engineers and scientists." Boeing ran a full-page advertisement in the same issue of *Aviation Week*, announcing "immediate . . . openings for electronic and electrical engineers" to work for Boeing on Minuteman and Dyna-Soar. Boeing chose to divide its page vertically, using the right half as a form for job applicants. Corporations working on Dyna-Soar recognized its eye-catching potency in attracting new talent.[48]

The sense of lagging in a space race meant that general readership media identified military projects even when reporting on NASA accomplishments. In July, the *New York Times* identified Shepard's May trip and Capt. Virgil "Gus" Grissom's spaceflight (the second U.S. suborbital journey) as "steps in the right direction." Ominously, however, Soviets had already bested these flights in the spring, and "the experts consider the Soviet program as a one-way street with military domination of space the ultimate goal," while U.S. projects were in the works but "several years from fruition." Dyna-Soar, described as "a rocket-launched, manned orbiting bomber," headed this list of projects.[49] A Soviet rhetorical assault on the MIDAS III early warning and TIROS III weather observation satellites charged that these were spy vehicles similar to the U-2 plane. Such accusations implicitly challenged the safety of satellites, lending plausibility to arguments for vehicles able to dominate aerospace. As if anticipating such a development, *Aviation Week* announced that "on September 25, *Aviation Week and Space*

Technology will publish one of the most important issues in its history . . . 'Forging military spacepower.'"[50]

Soviet secrecy, understandable U.S. suspicion, and hysterical press and political claims combined to prompt fear of hostile Soviet space potency. John Finney of the *New York Times* elaborated on this dynamic in the wake of cosmonaut Gherman Titov's seventeen-orbit trip on Vostok II in early August. Finney noted that "the Soviet flight had no immediate military application," and the reporter considered Kennedy's May initiative to counteract potential discouragement about the new Soviet success. But Titov's flight nonetheless "pointed to future capabilities in orbital reconnaissance or bombing and missile-launching vehicles." Military and congressional calls for expanded space efforts therefore seemed inevitable: "So far the Air Force is precluded from the man-in-space field. But with each Soviet feat there is a growing impetus on Capitol Hill and in the Pentagon to accelerate the Dynasoar project," which was described as "a rocket-boosted space glider that would be capable of reconnaissance or firing bombs or missiles on earth targets."[51] Kennedy's May speech provided a temporary inoculation against the worst criticism regarding imminent Soviet space spectacles, but the nation's future course in space was far from settled, even in the wake of Kennedy's lunar quest speech.

Pentagon officials considered the intersecting aerospace ventures even as the executive and legislative branches differed over funding. While Air Force Secretary Eugene Zuckert advised McNamara that the service wanted "to participate in the lunar program" so as to "benefit from the technological fall-out of the lunar program," Gilpatric argued that much of the support work expected of the military would "contribute nothing to our current military strength" and risked undercutting its "primary military mission especially in the field of R&D." To promote some kind of space activity, Congress continued to push $85 million more for Dyna-Soar than the $100 million Kennedy wanted.[52]

During late summer, *Aviation Week* raised alarms of growing Soviet space power. One editorial explored the potential of a Soviet orbital bomber; the USSR had recently boasted of a one-hundred-megaton thermonuclear warhead, and conjecture suggested that such a weapon delivered by a weaponized satellite could be detonated at extreme altitude and spray lethal effects over vast swaths of the planet: "The real value of a weapon of terror such as the Soviets now are constructing is not in its actual use but in its psychological effect on the people over whom it is dangled like a Damoclean sword." Hotz described the Soviet Union as a "beast that is leaping for our jugular."[53]

Aerospace technology advocates preached a gospel of action. *Aviation Week* reported in September that "Soviet space gains threaten U.S. security," and the journal contrasted the activism of Khrushchev with the allegedly slothful "ultra-conservatism" of Eisenhower. Although Khrushchev had somewhat preposterously awarded himself a medal for Titov's space accomplishment, *Aviation Week's* editor believed that "the cost of a medal to Nikita Khrushchev is a small price to pay for the support he has given [Soviet] space technologists." Jerome Wiesner drew fire for leading "the 'kill the Mercury' wing" of advisers to both Eisenhower and Kennedy. The journal's editorial page compared the Vostok's suspected 10-day life support capacity with that of Mercury, in which "a 24-hour orbital flight [would be] a risky business," and the editor speculated that the Soviets might possess capacity "for a quick [manned] round trip to the moon."[54]

Vostok II seemed to warn "that it is impossible for the U.S. lunar program recently established by President Kennedy to beat the Russians to a manned landing on the moon." Air Force plans contrasted sharply with the lagging and lurching character of national space development: *Aviation Week* readers learned that the Air Force hoped to save more than two years of development by bypassing the suborbital phase of Dyna-Soar, and the Air Force had identified Mayaguez in Puerto Rico, Santa Lucia Island in the Caribbean, and Recife in Brazil as potential landing sites for Dyna-Soar I. Garrett Corporation ratified its confidence in the program by using a full-page ad to display its work on the glider's cooling system.[55] Press reports were also capable of undercutting Dyna-Soar, however. Multiple *New York Times* stories that fall referred to Dyna-Soar as a "manned space glider" without mentioning the armed objective, and one article disparagingly noted that the vehicle's name "calls up immediate images of prehistoric bogs and giant beasts."[56]

As advertised, the September 25 issue of *Aviation Week and Space Technology* celebrated military space efforts. "The Military Space Role" was a three-part editorial that preceded and followed the September 25 edition in order to further emphasize that issue of the journal. Hotz's series of essays argued strenuously for the military being granted a more energetic role in space. The first essay, in the September 11 issue, noted that diverse figures such as Bell Aerospace vice president Walter Dornberger, but also Mississippi Democratic Senator John Stennis and NASA's Seamans, "have voiced public concern over the current neglect of military aspects of the U.S. space program." Hotz asserted that two crucial mistakes—a pattern of overly pessimistic calculations of the time required for space research, and the assumption that "hanging the 'peace' label on our space program" would "score a great moral propaganda victory"—had restrained military space development. Like Air Force aerospace advocates, Hotz dismissed the idea that the military could adopt and employ civilian space technology: the requirements were dissimilar and nontransferable.[57] Military vehicles needed to be more resilient and cheaper than was necessary for science projects.

The following week's editorial argued for greater military representation in the decision-making space council and complained that "the USAF Dyna-Soar program . . . is currently stalled in the top Pentagon decision-making bayous." A week later, the trade journal sought to assure readers, many of whom had "apparently misinterpret[ted] our comments as a plea for complete military domination of the space effort." Still, it blasted Eisenhower's approach to space research and development as that of "conservatives" intent on "wait[ing] until all the basic research and development in space technology has been done by NASA and then decide at their leisure whether there are indeed any military applications of this technology." Buildup to the September 25 issue on military space referred to an aerospace plane that would step technologically past even Dyna-Soar, and it included a full reprint of a speech by Walter Dornberger. Not surprisingly, the boost-glide advocate and former Peenemünde boss who had helped "launch" Air Force interest leading to Dyna-Soar advised "establishing a space bombardment system, putting reconnaissance and communication systems as well as logistic, inspection, and maintenance systems in space."[58]

According to *Aviation Week*, Dyna-Soar "bridge[d] a gap between the conventional air-breathing craft of today and the deep-space craft of the future." One article went into substantial detail, identifying the goal of a Step III bomber after noting that Dyna-Soar I "can serve as a laboratory to advance technology, to prove concepts, to test specific systems." *Aviation Week* provided a brief history of the boost-glide regime since the Second World War. Air Force Space Command's "future spacecraft needs" were reported to include the aerospace plane as well as a strategic satellite strike system "for retaliation from space," a recoverable booster, a supersonic transport, and a military test space station.[59]

Despite Kennedy's vow to give "equal attention to both" military and civilian endeavors, Dyna-Soar and Apollo seemed on a collision course. Space-supporting facilities were finite, and experts anticipated "a pressing schedule problem" when Apollo and Dyna-Soar each got on track and competed for launch resources at Cape Canaveral. Overton Brooks' death on September 16 was expected to "result in a radical change of course" in the House Science and Astronautics Committee which he had overseen, since the new chairman, California Democrat George Paul Miller, "firmly believes the U.S. space effort should be a step-by-step program rather than a headlong race to the moon."[60]

The Saturn booster's payload and its niche within the U.S. booster array remained undetermined when the first tests occurred in October 1961. Debate continued whether a lunar venture should be undertaken directly from Earth or by an assembly of components lifted individually into Earth orbit first. The former plan demanded enormous booster strength, while the latter required mastery of new skills such as rendezvous of vehicles in space. AEC chairman Glenn Seaborg told a conference on nuclear aerospace propulsion that development of a nuclear rocket could be accelerated through greater federal funding.

National prestige came first; military applications seemed set to attain prominence as well, as observers inside and outside of government saw science take a back seat. The *New York Times* explained that "in some ways, Saturn has not kept pace with the growing weight of manned space capsules, and in other ways scientific thinking has not kept pace with the weightlifting capabilities of the rocket. . . . Aside from these orbital flights for the Apollo and possibly for the Air Force's Dyna-Soar space craft, no other specific missions have been established for the Saturn rocket. No scientific payloads are being developed for the Saturn."[61] While Trevor Gardner complained about the lack of "large booster rockets designed" to support operational military space vehicles, physicist James Van Allen (discoverer of the radiation belts surrounding the earth) noted that "our national ambitions have greatly outrun our national competence." Van Allen noted "a striking paucity of solid, fundamental literature" about the subject, which accentuated "what I regard as 'the great chasm' in our space exploration by the youngsters of the nation and the general public."[62]

The ARS held a conference at which advocates of the Earth-orbit concept for a lunar mission, such as von Braun and NASA chief Webb, asserted that Earth-orbit rendezvous and assembly should shave two years off the timeline. Nicholas Golovan, a Russian émigré educated in the United

States and working as associate director of NASA, chaired the Large Launch Vehicles Group that sought to deliver recommendations on standardized boosters for the growing space program. *Aviation Week*, noticing "Defense Secretary McNamara's coolness toward the Dyna-Soar program," predicted that the glider's first launches would have to be made with more modest Air Force boosters rather than with NASA's Saturn.[63]

Gen. Bernard Schriever, head of the new Air Force Systems Command, championed Dyna-Soar, announcing at a press conference that a plan for accelerating it would be presented to the Pentagon within thirty days. The *New York Times* wrote that Schriever and other aerospace advocates objected to "the frequent refusal of the Pentagon over the last few years to support ambitious research on manned space flight until a mission requiring such capabilities could be defined." Schriever laid most of the blame on the Eisenhower administration and "went on to say that there [has] 'definitely been a change' with the Kennedy administration," since both President Kennedy and Vice President Johnson supported space research. Still, however, Schriever was far from satisfied. Reporters asked whether projects such as Dyna-Soar were inconsistent with a speech Kennedy had made to the United Nations the previous month, which had called for space to be reserved wholly for peaceful uses. The general denied any inconsistency, arguing that "it is desirable to reserve any medium for peaceful purposes" but that "in a practical world . . . if certain things [for protecting national security] can be done best in space, then certainly we must do them. I don't think the president infers that we won't."[64]

A Space Glider, but Not a Space Bomber

The day after Schriever's speech, the Pentagon approved an Air Force proposal to use a new booster, called Titan III. It was a Titan II modified by the addition of a pair of solid propellant boosters on its sides to provide more thrust. The *New York Times* reported in early November of Golovan's search for rockets "to serve the over-all national space program" to be "used interchangeably in civilian or military programs." But Vice President Johnson believed that it was "not useful to pretend that arbitrary distinctions can or should be made between the two," because "the same engineering, the same scientific knowledge, and the same values of man in space are applicable to defense and non-defense objectives."[65]

That approach would eliminate wasted resources on siloed but parallel civilian and military booster programs, but at the cost of blurring the line between civilian and military endeavors—a result that Kennedy had urged and Eisenhower had consciously resisted. Americans had earlier condemned the Soviets for using identical rockets to lift space probes and to wield nuclear warheads. Expediency had prompted the United States to follow suit, carrying the Mercury capsule aboard modified Redstone and Atlas rockets (both being variants of ballistic missiles). Efforts to consolidate big booster programs went another step further, opening the door to military space vehicles too large to be put into orbit by the relatively modest power of ICBM boosters.

Pushback, ironically at first glance, came from the Air Force. Some of its aerospace advocates sought the development of specialized booster technology tailored to operational needs, and the deployment of mission-oriented boosters seemed preferable to the expedited but expedient use of NASA's big booster. For its part, NASA officials wrestled with the technological challenges of a lunar mission, and they commented that "there will never be an optimized booster" and that "to seek one, would just cause deliberation to string out indefinitely with little, if any, progress being made. *The Dyna-Soar case is a good example of this* [emphasis added]."[66]

Because Kennedy used space to pursue multiple goals, retaining a coherent focus proved difficult. Space provided possible opportunities for cooperation with the Soviets to help lessen international tension while concurrently being a realm in which to compete against the Soviet Union for prestige at home and globally. Kennedy could not quite decide which track he preferred, but his speechwriter Ted Sorensen believed that Kennedy had three overarching goals in space: demilitarization, preserving U.S. access to space, and U.S. scientific prestige.[67] Arguably, either a cooperative or a competitive approach to the moon could achieve these goals. Kennedy viewed government spending, including that on space, as one of the tools for reversing the nation's frustrating signs of recession lingering from 1958, and *Aviation Week* noticed Kennedy's efforts "to harness space and military spending to its economic recovery effort" and "to control [the] space dollar impact." The objective was "to invoke a far larger proportion of the nation in the space program" because of a "fear that enthusiasm for [space spending] will slacken once the true costs are realized."[68]

A San Diego electronics engineer writing to *Aviation Week* suggested that "space competition can take the place of war, as long as space enthusiasm holds up."[69] Such views of the space race as an *alternative* to war presupposed that it would not be a weaponized field of conflict. This brand of space enthusiasm ran counter to the Air Force's beliefs about future uses of space, which lay at the core of its thinking about the Dyna-Soar glider and about aerospace. On November 20, *Aviation Week* announced that Dyna-Soar "is no longer being pushed as a potential offensive weapon system." Instead, "emphasis [was] being placed on surveillance, maintenance and rescue tasks."[70] This was an operational focus, unlike the hypersonic flight research focus insisted upon by the Eisenhower administration, but the new direction of the program reflected the new president's approach to space: it was visibly ambitious but subtly indeterminate.

Technological limitations contributed to the concession. The vehicle's design reportedly housed a 75-cubic-foot space in the "belly . . . for installation of about 1,000 lb of payload," but most U.S. nuclear weapons exceeded this weight, and since a bomb could not simply be "released" into orbit, additional guidance systems would be required to put a bomb through the atmosphere and onto a target. Nonnuclear weapons harnessing the potential of light (lasers) or of microwave energy (masers) remained in the hypothetical realm.[71] Dyna-Soar would need a much more extensive payload capacity before it could ever function as a weapons platform. Meanwhile, McNamara resisted congressional efforts to push more money into Dyna-Soar development.[72]

In spite of the hardships, Dyna-Soar still inspired support from interested members of the public who were frustrated by the seemingly lackadaisical pursuit of the boost-glide pioneer project. A graduate student in physics at the University of Kansas expressed disappointment that the country was "still making only a hesitant and half-hearted effort on the Dyna-Soar program when this project deserves, in fact, the highest possible priority" because of its usefulness as a reconnaissance vehicle, nuclear bomb carrier, and interceptor of equivalent Soviet craft. He predicted "a panic of unprecedented proportions . . . sweep[ing] the country . . . if the Russians get the Dyna-Soar first and start flying nuclear weapons over our skies. . . . Dyna-Soar is the next big advance after the ICBM."[73]

The ongoing possibility of an Earth-orbit rendezvous-and-assembly method for reaching the moon confirmed the need for an interim vehicle between Mercury and Apollo. The twin-seater Mercury Mark II was soon renamed Gemini, possibly so that the exclusive rights that *Life* magazine had bought for the first Mercury astronauts' stories could not extend further. Gemini crews, *Aviation Week* reported, could "come from the X-15 USAF and NASA pilot pool, or from the pilots being trained for the USAF Dyna-Soar program."[74] The most famous of these potential Dyna-Soar pilots

Dyna-Soar got a boost from the publisher Doubleday. Donald Wollheim had written science fiction since 1933 and had reputedly pioneered the paperback science fiction genre. By 1961, the prospect of manned space travel provided a terrific environment for writers seeking to attract youth readers. Wollheim tried his hand at a space-themed science fiction series featuring the fictional Air Force astronaut Mike Mars. A figure worthy of youths' emulation, Mike Mars drank milk instead of coffee (or liquor), displayed modesty and determination, and had minded his parents' advice during his boyhood.[1] Of the eight books ultimately in the Mike Mars series, Doubleday printed the first four in 1961: *Mike Mars Astronaut, Mike Mars Flies the X-15, Mike Mars at Cape Canaveral,* and *Mike Mars in Orbit.*

Next in the series would be *Mike Mars Flies the Dyna-Soar,* published in 1962. This time the Air Force hero would fly Dyna-Soar into orbit (implausibly) using a Titan I booster in order to rescue his colleague Johnny Bluehawk, whose "Quicksilver" (Mercury) capsule was unable to land because of a hurricane, so Mike Mars used Dyna-Soar to conduct an in-space rendezvous and rescue. Published in 1962, the work on this volume of the Mike Mars series would have been conducted while the first four volumes were printed and as the U.S. space program began to truly solidify under Kennedy. Significantly, Dyna-Soar stands as the rescuing *salvation* for the trapped "Quicksilver" astronaut. The hurricane in the story challenged the practicality of oceanic capsule recovery and underscored the advantages that Dyna-Soar's advocates boasted their glider concept would possess. Dyna-Soar's role in the book followed the new vision of the project's role rather than the Air Force's original goal. Early books in Wollheim's series had identified space exploration as a race but had shied away from overt mention of the Soviets.[2] Wollheim's popular children's science fiction works not only played a very indirect role in shaping public opinion about space programs, they also reflect elements of that opinion. The public, Doubleday presumably believed, was ready for science fiction books on space, including works pointing to the shortcomings of capsules and the advantages of a more versatile alternative.

Notes

1. "Donald A. Wollheim, Publisher, Dies at 76," *New York Times,* November 3, 1990; Justine Flood, "The Mike Mars Series Sourcebook," www.thethunder child.com, accessed November 13, 2011, http://thethunderchild.com/ Books/MikeMars/MikeMars.html; "The Mike Mars Children's Series,"

boggsspace.com, accessed November 13, 2011, http://boggsspace.com/ mike_mars.asp.

2. Flood, "The Mike Mars Series Sourcebook."

who in 1962 would be taken into NASA for work in Gemini (and ultimately Apollo) capsules was Neil Armstrong.

Air Force space advocates pressed ahead, though not in alignment with the priorities of the Kennedy administration. *Aviation Week* noted an Air Force concept for a three-man lunar rover "capable of performing . . . logistic functions" to support "a military base on the moon" by 1966. An Air Force study was also under way to consider a prefabricated lunar shelter. In December, representatives of the United States and the Soviet Union "reached agreement . . . on two critical areas—disarmament and the peaceful uses of space" within the context of UN dialogue.[75] A formal agreement against the deployment of weapons in space or in celestial bodies such as the moon was not finalized until 1967, but the Air Force's aerospace outlook flouted any work in this area.

Aerospace advocates welcomed encouraging news at the end of 1961: the Dyna-Soar program appeared to be back on track. The $85 million in additional funds that Congress had voted for Dyna-Soar remained unused, but civilian authorities approved the accelerated development program on December 11, and the *New York Times* informed readers on Christmas Eve that the Air Force's 1.5-million-pound-thrust Titan III had been approved as well. Titan III development was interpreted as "an indication that the administration had acceded to long-held Air Force desires to push the Dyna-Soar project into the orbital phase." The *Times* noted that "this would represent a change in the past Defense Department opposition to military man-in-space missions."[76]

The front page of the *New York Times* soon brought word of a "major revision" in the Dyna-Soar program. On December 26, the Air Force was authorized to skip the suborbital stage of flight and move directly toward an orbital test using the powerful new Titan III intended for the glider. The days of DOD limiting the program "to relatively low-priority research objectives aimed at exploring the technical problems of maneuverable space vehicles" because of "the absence of clearly established military require-ments for manned space flight" appeared to be over.[77] Careful readers could notice, however, the absence of any reference to the armed Phase III system. Dyna-Soar was "a manned space glider" and "an orbital spacecraft," built "to develop and demonstrate the capability of manned space flight in which pilots can return from orbit and land in conventional manner at air bases."[78] The *New York Times* believed that diplomatic challenges in finding overseas landing sites had helped prompt the decision to eliminate the suborbital development stage. Aerospace supporters optimistically noticed the time that eliminating suborbital efforts could save.

Concurrently with these decisions, a new trend seemed to be emerging in the defense sector. *New York Times* reporter Jack Raymond noted that "high officials of the Pentagon are troubled by the number of organizations, some technical, some with business interests and some purely lobbyist, that have sprouted in the defense industries." It was the number of the organizations more than their role that had begun to give pause to some Defense officials.[79] Strangely, the article did not mention whether this was an example of the military-industrial or the scientific-technological dangers that the former president had warned of just eleven months before. Approval of Dyna-Soar's accelerated development and of its more powerful Air Force booster put "the Air Force squarely into the manned space flight business on about equal terms with the civilian space agency."[80] And so it had seemed that the United States would now pursue both arms of space development.

Aviation Week greeted 1962 by noticing "a brighter outlook for manned military spacecraft," citing DOD's "acceptance of a partly accelerated Dyna-Soar project" and NASA's interest in fuller cooperation with the military regarding "manned space flight and rendezvous." Reporter Larry Booda called Dyna-Soar "unique in one respect. It has received a go-ahead at the Defense Department level without the Air Force having to prove a military requirement for it. About a year ago, USAF was forced to try to justify the project by stating a military mission in order to keep the project alive."[81] Dyna-Soar, which had survived the budget-minded Eisenhower administration intent on using space for nonweaponized reconnaissance, had now survived the first months of the new Kennedy administration. Although its mission had been gutted, Dyna-Soar's path to development seemed assured.

Epilogue and Conclusion

Since the general public does not govern, presidential influence is shielded from the vagaries of shifting sentiment.
—RICHARD NEUSTADT, 1959

The Space Legacy

In his farewell address, Dwight D. Eisenhower had declared that the "task of statesmanship is to mold, to balance and to integrate" the forces "within the principles of our democratic system, ever aiming toward the supreme goals of our free society." In the months after John F. Kennedy's inauguration, Eisenhower commented that "as emeritus, I must be silent."[1] While not remaining entirely silent, the former president did restrain impulses to critique the new president in public, even as his successor struggled during a challenging first year. In part, Eisenhower's reticence was required by the fact that much of the population, including his presidential successor and his own heir apparent, viewed his security policy as politically untenable.

If a president was "shielded from the vagaries of shifting sentiment," as political scientist (and Kennedy adviser) Richard Neustadt asserted at the time, the same was certainly not the case regarding concerted initiatives that were attuned to cultural tenets. In space, Eisenhower and his administration officials struggled to retain control of policymaking, while the Air Force pursued an aerospace vision that was steeped in a technological determinism that presumed that policymakers possessed no credible alternatives, save for the ardent pursuit of newer technology. The competition to shape space policy highlighted the shortcomings of Eisenhower's style of presidential leadership, and the secret nature of his space initiatives further disadvantaged him as space advocates, both in the military and in politics, jockeyed for public attention and more ambitious and prominent space activity.

The manipulation of space and missile issues for political advantage was an open secret, and Kennedy's presidential campaign eagerly grasped the "missile-space gap" as a political cudgel. As president, Kennedy found that this tool, so useful during the campaign season, was now glued to his hand. He would be called to act upon and fulfill the demands and expectations of the aerospace advocates, whom he had encouraged and echoed.

In his study of the development of the B-52 bomber (the mainstay of SAC units from the mid-1950s until the command's disbandment in 1992 and still in Air Force service), Mark Mandeles noted that "many weapons or tools have waited years for the emergence of an appropriate social structure to make their employment possible."[2] As impatient as they might have been, aerospace advocates had reasons not to accept "no" as a final answer, because they were convinced that aerospace power held the key to national security. The brief history of airpower was already punctuated by fiscal challenges that had been overcome. The B-17 Flying Fortress had been cancelled on one occasion before the Second World War. Technical problems had dogged the B-29 Superfortress. The B-36 Peacemaker program had been challenged by the Navy and Congress following the war. In the fiscally lean years between the end of the Second World War and the adoption of the policy of containment, development of the B-52 Stratofortress had nearly been ended, too. Yet each survived to personify American airpower, and, following this precedent, aerospace advocates felt some justifiable self-assurance that their vision of future national security would be realized. This was so much more the case since Kennedy had endorsed national defense spending across the board—especially in missiles and space—because the national culture embraced the certainty that technology improved life.

Although initially promoting Dyna-Soar, during 1961 Kennedy essentially began to draw national attention away from military projects to the bolstered civilian space undertakings by NASA. A newsreel celebrating astronaut John Glenn's Earth orbital mission pointed out that "the Russian orbits were in a thick fog of secrecy. The United States stands or falls in the white-hot glare of world publicity."[3] This publicity was a force to which Eisenhower had bent; in an NSC meeting in July 1959, Eisenhower had informed NASA's deputy director that NASA had been established to assuage public outcry for space activity.

Kennedy had actively courted and encouraged that publicity. The space issue was not his creation, and it was not under his control, but it was one of the topics he sought to harness as a candidate and as president. Historian John Logsdon described Kennedy as "a true space pioneer" who was committed to space accomplishments because they could advance the national image and national security.[4] Logsdon observed that there were several times throughout Kennedy's tenure that he was open to a human lunar landing being undertaken as a cooperative venture between the superpowers.

William Duggan later called Kennedy's Apollo quest an example of "strategic intuition," the creative assembly of existing events and entities into a new transformative theme. "Neither the goal to land on the moon nor the plan to get there was Kennedy's idea," Duggan reflected. "Kennedy did not dream the impossible. He brought forward elements from the past that showed the way to a reachable goal." The management lesson Duggan drew from case studies including Kennedy's is that "you look to specific opportunities, not ambitious dreams, to find the way ahead."[5] This was also true of the space program during the Eisenhower era. However, both Eisenhower's policy objectives and his approach to policymaking led him to see and seek specific opportunities in different ways from Kennedy. After Kennedy had exploited discontent with the Eisenhower era to become president himself, he had to cope with the challenges of office while also dealing with the constituencies that he had led to expect his leadership.

NASA director James Webb urged the president to maintain a broad science-based space program in which a human lunar landing was a crucial but not singular goal. In November 1962, Kennedy bluntly told him, "Jim, I think [the moon landing] is the top priority. I think we ought to have that very clear." While other projects could be allowed to lag, Apollo was "important for political reasons."[6] In Kennedy's eyes, international politics drove the imperative of the moon landing. "Otherwise," he told Webb, "we shouldn't be spending this kind of money, *because I'm not that interested in space* [emphasis added]." That was Kennedy's private side, but alert contemporaries could sense the attitude. *Time* magazine noted that "Kennedy seemed to know less about [space], [to] be less interested in it" than even Eisenhower. Kennedy's goal in fact was to emphasize and showcase American power. "We've been telling everybody we're preeminent in space," Kennedy told Webb. "This is the way to prove your preeminence."[7]

President Kennedy chats with astronaut John Glenn in 1962. Kennedy confided only to his innermost circle that he was actually "not that interested in space." *John F. Kennedy Library/photo by Robert L. Knudson*

Two months before lecturing Webb, Kennedy had addressed a crowd at Rice University in Houston, Texas, declaring "the exploration of space . . . is one of the great adventures of all time."[8] Kennedy's public pronouncements reflected his commitment to garnering national political support, but he was not the space pioneer of popular imagination. He was a Cold Warrior bent on achieving national and personal prestige and security. When he became president, he did not have a clear vision about how precisely to accomplish this, advocating a more-of-everything approach to defense. Kennedy first urged greater funding for Dyna-Soar, then proposed a lunar landing, and then his administration expedited (and reoriented) the Dyna-Soar program—all in his first year. This activity reflected the new administration's reluctance to entirely close the door on military activity in space while Soviet actions and intentions remained indeterminate.

Several factors—Soviet inscrutability, mounting space cost estimates, and the undefined strategic utility of space—prompted Kennedy to lurch back and forth between a competitive and a potentially cooperative mindset about the lunar venture. "The point of the matter," he said in 1963, "always has been not only of our excitement or interest in being on the moon, but the capacity to dominate space, which would be demonstrated by a moon landing." Lunar urgency slipped late in his tenure, and John Finney of the *New York Times* reported in October 1963 that "the United States will not abandon the lunar expedition, but it will be pursued with less competitive zeal and a more leisurely pace."[9] Perhaps this might have happened if Kennedy had not been assassinated; he may have permitted the fulfillment of his rhetorical goals (such as space and civil rights) to simmer during the remainder of his presidency, but his martyrdom reignited the renewal of both. In 1961, the Kennedy administration was in the process of successfully, but expensively, realigning national attention and the space effort away from the dominance of the Air Force's aerospace advocates. This alignment was effected less because Kennedy objected to the Air Force's vision than because Kennedy was determined, as the chief executive, to be the one setting policy. The meager specificity of his policy agenda is an irony of his presidency.

The ballistic flight pattern of the Mercury, Gemini, and ultimately Apollo programs progressed through the rest of the 1960s, while the Air Force's vision and its cherished vehicle ran into trouble. Contemporary commentators described the two-man Gemini program as "a bridge" between the single-seat Earth-orbiter Mercury and the three-seat lunar-bound Apollo. Gemini aimed to build toward a lunar landing as quickly and cheaply as possible. In December 1961, NASA was already defining a "project philosophy" where "all systems will be modularized" and "maximum use of available hardware" would be made. Gemini allowed the pioneering of true rendezvous and spacewalking techniques in the mid-1960s so that an expedited lunar vehicle could be assembled in space, conceivably bypassing the development of another generation of superboosters. These were deemed crucial by planners who doubted that crew, spacecraft, and lunar lander could all be sent directly from the earth to the moon on a single rocket booster. Refinement of the Saturn series of boosters yielded the model V, which provided a direct ascent capability, voiding part of Gemini's reason for existence. Kennedy grew increasingly concerned about the mounting costs of the space program, but little evidence suggests that he was set on radically changing its trajectory.[10]

After Kennedy's assassination, determination redoubled, as did NASA spending. Its budget grew to a peak of $5.9 billion during 1966, nearly 4.5 percent of the total federal budget, although the following years all saw a decline in NASA's funding. Apollo conducted a human lunar orbit at the end of 1968 and a human lunar landing seven months later. But

Kennedy had framed the adventure as a race against time (and implicitly against the Soviets as well). The finish line was successfully reached, yet no sufficiently compelling vision had been put in place. The country faced fiscal distress, cultural rift, and political discord in the wake of the war in Vietnam. Little impetus existed to push human exploration deeper into space in the foreseeable future, and even space enthusiasts such as John Logsdon would have to admit, looking back on events, that "the Apollo experience [had] little to teach us beyond its status as a lasting example of a great American achievement."[11]

The Air Force's path in space also brought surprises. Unweaponized—and automated—military satellites proliferated during the coming years of the Cold War; their utility for military position tracking, targeting, and communications would become obvious by the end of the Cold War. But the piloted weapons system that the Air Force desired was doomed to a slow death under Kennedy and Johnson. Dyna-Soar research continued in 1962, but the weaponized objective had been derailed the previous year. To emphasize the experimental research orientation of the project, Victor Charyk rechristened the glider as the X-20 in mid-1962.[12]

Defense Secretary McNamara remained unconvinced of the need for Air Force astronauts or the flexible landing characteristics of a winged space glider, and in mid-March 1963 he directed studies comparing the applicability of Gemini and Dyna-Soar to explore Air Force space needs. The single-use Gemini, which required sea landing and naval rescue of the crew, did not fit Air Force priorities for routine space missions. Meanwhile, Dyna-Soar continued to face severe technological challenges and a high price tag relative to the modest cargo it was anticipated to be able to bring into space. Its cost seemed particularly unsustainable since its weaponized missions had been rejected by consecutive policymakers. President Johnson directed McNamara to shave excess expenditures from the fiscal year 1964 military budget, and within two weeks of Kennedy's assassination, McNamara announced the cancellation of the Dyna-Soar program.[13]

Dyna-Soar had not been helped by recurrent characterizations in policymaking circles of the Mercury and Dyna-Soar as being of a compatible generation of spacecraft. Despite its limitations, Gemini partially accomplished an effect that the Air Force had wanted: space missions were gradually becoming relatively routine. By mid-decade, NASA's public affairs office conceded that "we are in a new phase of our program," that "each flight is not going to be spectacular," and that NASA should "discourage . . . activity, such as ticker-tape parades" for future astronauts.

Worse, for Dyna-Soar, were the studies to equip Gemini with a paraglider contraption to provide a modicum of steering during reentry. "From the beginning of the Gemini Program, one of the objectives was to develop reentry flight-path and landing control," including an offset center of gravity to allow limited forces of lift during reentry that had been lacking in Mercury. These factors helped enable policymakers who wanted to see Gemini and Dyna-Soar as parallel (and Dyna-Soar as therefore redundant) to do so.[14]

Despite canceling Dyna-Soar, McNamara was not entirely ready to close the door on Air Force astronautics, and he announced the start of a manned orbital laboratory (MOL), to be serviced by Air Force Gemini craft, simultaneous with Dyna-Soar's cancellation. Although the application for the MOL remains murky even today, spending for it outpaced the $410 million Dyna-Soar investment by a factor a three. MOL's cancellation was announced by President Richard Nixon's defense secretary a month before would-be Dyna-Soar pilot Neil Armstrong set foot on the moon.[15]

NASA's ballistic capsules won out over Dyna-Soar because NASA's lunar program answered the policy need for activity in space without necessarily opening the door to a space arms race, as Dyna-Soar seemed likely to do. Ted Sorensen explained that "even though the President would stress from time to time that the idea of a race or competition was not our sole motivation, there was no doubt that that's what made it more interesting to the Congress and to the general public."[16] That motivation had been nurtured in part by Kennedy himself. NASA had been created by Eisenhower to placate angry voices in the country after Sputnik, and NASA's Apollo program had won Kennedy's endorsement as the new president grasped for ways to express his own and the county's power and prestige. It did benefit from events such as the 1967 Interplanetary Space Treaty that banned weapons and territorial claims on planetary bodies. It strove unsuccessfully to promote true space science by leveraging the prestigious status it had attained in the Kennedy and Johnson era as a cat's paw.

NASA also worked to weave national policy tenets into its justifications for projects, as when it prepared to frame an earth resource satellite in the context of Eisenhower's "open skies" proposal of twenty years earlier. In selecting the name of the Apollo 11 lander (Eagle) that would touch the moon, NASA officials recognized that "the name of the vehicle should be dignified and hopefully convey the sense of the 'beginning' rather than 'culmination' of man's exploration of other worlds." But the note of finality in the Apollo 11 mission report captured the dynamic more accurately:

"With the completion of Apollo 11, the national objective of landing men on the moon and returning them safely to earth before the end of the decade had been accomplished."[17] Despite Logsdon's suggestion that there is nothing to be learned from the Apollo experience, it in fact helps shed light on how national leaders worked to shape and answer public opinion while charting U.S. policy. Dyna-Soar's development, particularly the underemphasized early years from the first Air Force interest in hypersonic flight in 1954 to Dyna-Soar's project consolidation in 1957 and its redirection in 1961, also speaks to the struggle by leaders to determine and steer policy to ensure national security.

As with the "missile gap," the "space gap" was largely a figment of the national imagination that Soviet leaders were happy to encourage. As late as 1962, Americans envied the emergency escape system of the Vostok spacecraft. It did not occur to them that the Soviets were still years from perfecting a means of recovering a live cosmonaut within his capsule. The three-man Voskhod was developed primarily by inserting two more seats and a larger parachute while removing safety equipment that had been part of Vostok. The world's first "space rendezvous," by Vostok 3 and 4 in mid-1962, was rigged: the craft were launched into similar orbits and floated near each other in orbit, at times a mere four miles apart; however, they lacked a means of maneuvering themselves any closer than that. The first spacewalk nearly ended in disaster when the air inside the cosmonaut's suit temporarily prevented his return through the airlock into the capsule. Soviet science was called upon to produce a new and more spectacular space "first" whenever "the memory of our sensational space feats was fading." No Soviet counterpart to Dyna-Soar seems to have been undertaken until after Dyna-Soar's cancellation, and even this was a craft focused primarily on studying the landing characteristics of "lifting bodies."[18]

Some of the U.S. researchers on Dyna-Soar worked later on the lifting body vehicles such as the M2-F3. The M2-F3, like the Round One (X-1) and Round Two (X-15), is on display at the Smithsonian Air and Space Museum in Washington, DC; the terminated Dyna-Soar (Round Three) coincides with its relative obscurity, despite its significant place in events. The NASA space shuttle, the surviving component of a larger NASA outlook for the 1970s and beyond, was a winged reentry vehicle. Many Dyna-Soar supporters claimed that the shuttle's development would have been easier if Dyna-Soar (which had undergone extensive wind tunnel and other dynamics studies at NASA labs, and whose prototype was half completed at the time of its cancellation) had been completed.[19] Walter

McDougall's study of the U.S. venture into space noted that "Dyna-Soar was much loved and lamented" by its advocates.[20] It was also continually reimagined. Dyna-Soar was typically either forgotten or dismissed as an odd waste of money, while a few writers insisted that its cancellation was a tragic mistake.

The Air Force has continued to eye space since Gen. Thomas White's proclamations about the continuous aerospace realm. The physics problems of accurately "dropping" orbital weapons notwithstanding, the cachet of imagining space as a strategic "high ground" endures. Air Force Chief of Staff Gen. Charles Gabriel referred to space as "the ultimate high ground" in 1982. While Air Force doctrine had been republished in December 1959 to officially incorporate the word "aerospace," the doctrine itself remained unchanged—not surprising, since White had promoted aerospace while explaining that "the basics have been constant—greater speed, longer range, and higher altitude." The Air Force's first manual on space doctrine (AFM 1-6), released in October 1982, emphatically endorsed the incorporation of space across the spectrum of Air Force missions. A RAND Corporation history of Air Force space planning noted that "the Air Force's reversion to the 'aerospace' formula as its organizing theme for air and space beginning in late 1997 turned the clock back to the 1950s." Shortly thereafter, a later chief of staff, Gen. John P. Jumper, counteracted this building swell of aerospace rhetoric by emphasizing the separate "physics that apply to orbital dynamics" as those within the air realm.[21]

Meanwhile, in an ironic reversal of Dyna-Soar's fate, the Air Force's series of unmanned orbital hypersonic gliders flies today. NASA and the Air Force had partnered on a project in 1998 to serve as an emergency vehicle for the International Space Station. In 2003, NASA began winding down its involvement, anticipating a new lunar initiative by President George W. Bush and concluding that the orbital glider did not fit with the rest of NASA's plans. By 2004, the X-37B had been inherited by the Air Force and the Defense Advanced Research Projects Agency, the follow-on agency to Eisenhower's ARPA. The unmanned X-37 is twenty-nine feet in length and fourteen feet across, about two-thirds the size Dyna-Soar would have been. Despite rumors to the contrary, X-37B is probably not meant to become a weapons platform, but space media noted that the Air Force had released much less information to the press from 2004 onward than NASA had been willing to release between 1998 and 2003.[22]

Work has continued under the administration of President Barack Obama. As recently as April 2014, the Air Force would not comment on the vehicle's missions, earlier described as being to "demonstrate a reliable, reusable, unmanned space test platform for the United States Air Force." The vehicle has already demonstrated a capability to conduct missions lasting nearly two years at a time, and Boeing has openly discussed studies for an expanded variant (X-37C) that could carry passengers.[23] The Air Force's interest in space, and space gliders, has not ended.

Space Security and the Presidency

Histories of U.S. space policy of the 1950s and 1960s often describe the overall continuity between Eisenhower, Kennedy, and their immediate successors. For example, in 2012, historian Sean Kalic noted that "Eisenhower, Kennedy, and Johnson [used] the U.S. space program to advance a pacific image of the United States while highlighting the Soviet Union's perceived interest in dominating space." Furthermore, "with the exception of the moon mission, the majority of the space programs advocated by Kennedy had already been under development during the Eisenhower administration." An Air Force paper published in 2005 puts the matter similarly: Apollo was "the one glaring exception" to a pattern in which "in most other areas, Kennedy and Johnson continued in the same general direction that Eisenhower pointed them." Herb York also described overall continuity among administrations.[24]

That narrative, however, forgets some important points in space and national security history. *Presidents* Eisenhower, Kennedy, and Johnson ultimately opposed the weaponization of space. But *Senators* Kennedy and Johnson often sounded much friendlier to the Air Force's space outlook during the Eisenhower presidency. Air Force officers looked forward to a Kennedy presidency supporting their aerospace vision, and in March 1961, while urging greater funding for Dyna-Soar, Kennedy rebuffed NASA administrator James Webb when Webb asked for more money to accelerate the Apollo program.[25] The record also shows that the early groundwork on Apollo began in anticipation of executive support during Eisenhower's tenure, prior to Kennedy's inauguration. Finally, the idea that Kennedy followed Eisenhower's path in space except regarding Apollo is to ignore the main thrust of Kennedy's space program and to misread his approach to space policymaking.

One of the critical challenges between 1954 and 1961 was to determine the scope and contours of the Cold War. This was not a unilateral process, since the Soviets were very active players. Furthermore, the Cold War was more than a bilateral dynamic, since myriad other countries and organizations were involved in an overarching contest that swept up and sometimes fueled other unrelated tensions around the world.[26] The scope of the Cold War and the arms race was also contested at home. As elected leaders and their staffs worked to develop national policy, technological determinists in the Air Force pressed for steps they felt were crucial to prevent an otherwise inevitable gap between the country's security needs and its capabilities.

The United States at midcentury was a society reveling in prosperity and demanding security. Airpower had provided this security during much of Eisenhower's term, because it was best able to project retaliatory nuclear weapons power and dissuade Soviet aggression. It was also technologically exciting, and the Air Force took pains to ensure that this image carry forward. For Air Force officers, however, the image was useful not for its own sake but as a means to an end. Continued cultural support for airpower translated into ongoing financial and political resources for the military branch that air personnel believed was the most decisive element of the defense establishment. Some of the branch's highest ranking commanders believed in the mid-1950s that airpower research could (and therefore should) extend beyond the atmospheric realm to spaceflight. Simultaneously, policymakers placed top priority on gaining accurate intelligence information, and they wanted diplomatic tools as well as technological ones to help secure access to this information. This put Eisenhower officials and Air Force space advocates on a collision course, although the small scale of the Air Force's efforts and the secret nature of the administration's plans obscured this fact from one another and from the general public.

The Soviet launch of Sputnik I in October 1957 and Sputnik II a month later brought the issue of space activity careening into public view. The contrast between episodic Soviet accomplishments on one hand and the seemingly modest U.S. efforts (punctuated by publicly visible failures) on the other appeared in the eyes of many people to signal Eisenhower's naïveté and detachment from complicated technological issues. Because the underlying reasons for space policy could not be divulged to the public, Eisenhower appeared reactive and hesitant. Additionally, it is often difficult for the public (its attentions diverted by daily life and its access to information reduced by the demands of secrecy) to differentiate between deliberate activity and mere motion or to distinguish between individual projects and an overarching policy. Air Force initiatives were certainly deliberate, and several key officers in that branch made their doctrinal leanings public through speeches, essays, and collaboration with friends in news and entertainment media. That doctrine appeared straightforward and easy to understand—it was, at its core, an extension of airpower doctrine. The name "aerospace" was a public relations tool for expressing the idea that armed space presence was necessary and that the Air Force was the logical organization to perform the task.

Eisenhower was a big-picture man and an organization man, and he valued deliberation and coherence. He had been appalled by the haphazard approach to containment under Harry S. Truman, who had starved the military during the early escalation of the Cold War in the late 1940s and then inflated the defense budget with the outset of the Korean War. Eisenhower entered politics in 1952 because of the unacceptable danger he sensed in a return to old guard Republican isolationism on the one hand and the danger to democracy of one-party rule and the strategic and economic bankruptcy that characterized Truman's ad hoc defense postures on the other. Personally leading regular meetings of the cabinet and the National Security Council was an activity that Eisenhower believed was valuable and that neither his predecessor nor his successor practiced.[27]

Tasking Project Solarium and establishing the New Look were immediate top priorities for the new President Eisenhower because they provided a means of bringing defense spending and security policy under control in a coherent way. Alliances with dependable friends overseas (coupled with adventurist CIA gambits) were cornerstones in Eisenhower's security posture. The conspicuous centerpiece of the Eisenhower security vision lay with cost-effective nuclear retaliatory power, which required a strong Navy and especially a strong Air Force capable of wreaking devastation and havoc on the USSR in the event of war. For this, a strong domestic economy and a vigilant democratic society were vital.

In some respects, Eisenhower's own likability (his biggest potential selling point) also constrained his practical approaches to political leadership. Following five consecutive terms of Democrats in the White House, young voters could not personally remember a president who was not a Democrat. Through his own election, governing as a centrist, Eisenhower aimed to renew the Republican Party's competitiveness. He relied on his broad popularity while undertaking this delicate maneuver, and "everybody like[d] Ike." But maintaining that image of likability meant that Eisenhower needed what some confidants called "Ike's prat boy," a role that they were reluctant and resentful about filling.[28] Eisenhower's military background, culminating in the coordination of multinational forces in the

Second World War and the early stages of the Cold War, made him wary of appearing (or of being) too eager to grab political power.

Historian Fred Greenstein helped modernize the interpretation of Eisenhower as president by pointing to his "hidden-hand" style of leadership. Regarding aerospace policy, two implications were particularly significant. First, it underscored the importance of clear communication and understanding between the president and the governmental apparatus at his disposal. Many key players, such as DDR&E Herb York, PSAC chairmen James Killian and George Kistiakowsky, and budget director Maurice Stans understood and were helpful to Eisenhower in his effort to shape national space policy. Nonetheless, scholars such as Saki Dockrill have noticed the difficulty Eisenhower faced in making his New Look understood by all of those whose help he needed in exercising it. Dockrill observed that "the three services did not understand the thinking behind the New Look" and instead each pressed for greater funding.[29] The Air Force, the chief beneficiary of the New Look and the leading recipient of defense funds, fundamentally misinterpreted Eisenhower's rationales for investing in their branch. Mistaking his quest for adequacy and leveraging defense dollars for a vindication of their thesis that nuclear-armed airpower was singularly decisive, the Air Force pressed ahead with ever bolder and more expensive projects. Indeed, Chief of Staff White's confrontation with Eisenhower over the relevance of the Dyna-Soar at the NSC meeting on December 5, 1960, shows that the Air Force's top-ranking generals remained at odds with Eisenhower's perspective to the very end of his presidency. The aerospace trajectory, and the Dynamic Soarer space glider-bomber intended to pioneer spaceflight, represented the logical extension of an Air Force culture and doctrine that Eisenhower contained—but never quite controlled.

A second byproduct of Eisenhower's presidential posture was that the emphasis on a *likable* Eisenhower sacrificed attention to a *formidable* Eisenhower. Polls taken during both his election and reelection campaigns suggested that Americans admired their president. But the public admired him for his personality, approachability, and his smile (the origins of which were in fact more of a poker face than anything else). In short, it was as if Americans liked Eisenhower and voted for him because he seemed like the kind of person they could trust as a neighbor. His lengthy military resume could not be referred to frequently without drawing attention to his status as a general-president, which may well have caused Eisenhower inner discomfort and turmoil.[30] Eisenhower continued to take public pride in bringing the hot war in Korea to an armistice, but he remained a soldier at heart, even asking Congress to reactivate his commission as a general after he left the White House.

The dynamic that Eisenhower established was resilient enough to let the nation enjoy unparalleled prosperity and to foster a measure of tranquility during what was a profoundly dangerous time on the international stage and a more tumultuous one in society at home than is often remembered. The Achilles' heel of Eisenhower's New Look and of his hidden-hand leadership was that he was particularly vulnerable to criticism once a defense or security issue emerged for which he appeared unprepared. The Sputnik launch created that image of unpreparedness, and it provided a rallying point for his security policy critics. Economic slump opened the door for opponents to argue that defense spending, across the board, could propel the U.S. economy. This was political dynamite, wedding the two previously separate wings of security policy critics (airpower advocates and ground forces adherents) and grafting an economic aspect onto the critique as well. The Air Force was only too ready to answer the call for space activity, and their confidence in service doctrine led Air Force commanders to believe that if the current president did not agree with their outlook, then certainly the next president would.

Eisenhower fought a "retrograde" action in space for more than three years. He had decided on a space policy whose underpinnings had to be kept from view, and his presidential style foreclosed the narrow chance that aerospace advocates would lend him the benefit of the doubt. Concern about the possibility of surprise attack reinforced the urgent need for reliable intelligence information, and high-altitude, high-tech photography seemed the best method for attaining that. Top-secret aircraft and satellite reconnaissance programs could not be divulged, so Eisenhower's most important steps were essentially invisible. Space activity of the late 1950s was a curious mix of three kinds of projects: those that aligned with the president's policy intentions, those that he permitted in an effort to mollify the public, and those that contradicted his priorities but were politically protected by the public disquiet that was fueled by aerospace advocates in the Air Force, media, and Congress.

The Cold War demand for secrecy placed a strain on the democratic process that even the long-haul approach of the New Look could not completely address. Eisenhower insisted that "public opinion is the only motivating force there is in a republic or in a democracy." And he was equally confident that "public opinion must be an informed one if it is going to be effective in solving" national or international challenges.[31]

But the dynamics of the Cold War meant that ordinary citizens possessed only unclassified elements of the overall picture; aggressive and partisan advocacy—by elements of the press, Congress, or the military—sought to control the message and the information that was available. Further, national security concerns and familiarity with global politics tended to register with Americans only episodically. More immediate issues of everyday life were generally more top of mind. Eisenhower's hidden hand and the New Look complemented this environment, but only so long as Americans were generally satisfied that crisis was not imminent.

Eisenhower's farewell address has been interpreted in different lights. Contemporary aerospace advocates were chagrined at what they saw as the worries of an antiquated eccentric. Others at the time treated the warnings about the military-industrial complex and the scientific-technological elite with a good deal more respect. Herb York, who had clipped Dyna-Soar's wings, recalled being taken aback by Eisenhower's warning against dispro-portionate influence by scientists. Eisenhower assured York that his conster-nation had not been directed at the scientific advisers whom he trusted, but at self-serving bomb advocates like Edward Teller and space paladins like Wernher von Braun. Near the end of the Cold War, H. W. Brands suggested that the messages in the address represented an admission of defeat more than a warning for the future.[32]

Recently, scholars such as Ben Greene and Zuoyue Wang have explored the division within the scientific community and the hesita-tion that even Eisenhower's own trusted scientists had initially felt when hearing his farewell address. Evan Thomas notes that Eisenhower's address "had already been eclipsed" by the time of Kennedy's inauguration three days later. Many histories mistakenly presume that the farewell address originated with journalist Norman Cousins' suggestion, but in fact new additions to the historical record show that preparation of Eisenhower's farewell began much earlier, in the summer of 1959. The speech was the subject of more painstaking revision by the president and his writers than had been imagined. Two decades after the speech, Samuel Huntington observed that "apart from a few pro forma nods, these words provoked little immediate reaction," although they were "apposite and timely" and would be upheld a decade later when the military-industrial complex was held in derision as the war in Vietnam dragged on and faith in government and social structure eroded.[33] As with beauty, it would appear that wisdom is in the eye of the beholder.

Kennedy's active public persona and his rhetorical determination to explore space and to achieve both national security and international prestige in the process provided some early inoculation against the charges of lethargy that had dogged Eisenhower. Kennedy would not make the political mistake of ignoring the sense of a space "race," because he had himself helped frame it in these terms on his route to the White House. Executive office personnel during Eisenhower's tenure had discovered that because "the press insisted that we were in a race . . . [we were] in a race whether [we] want to be or not."[34] The race was a double-edged sword for Kennedy when he entered office, however. The American public's frustration, in the early 1950s, with the lingering war in Korea had helped sink Truman's reputation and pave the way for the Eisenhower presidency. Kennedy's vows to "pay any price, bear any burden" were soaring rhetoric, but easily forgotten in the compelling prose is the fact that Kennedy was propounding a philosophy even more stridently active than Truman's approach to the Cold War. President Johnson would in the late 1960s pay the political price of Kennedy's rhetoric in the rivers, highlands, and jungles of Vietnam. Sixty thousand Americans paid a far higher price than Johnson.

In some important ways, Kennedy was a prisoner of a context he had helped frame. At times, Kennedy's campaign-trail critiques of Eisenhower's defense policy complicated or foreclosed strategic options that Kennedy's team desired. Historian Desmond Ball acknowledged a "hasty" character to early Kennedy defense decisions, but he argued that "while the haste was unnecessary and quite avoidable, the politics were perhaps inevitable."[35] If the politics surrounding aerospace seemed to have become "inevitable," it was because figures such as Senator Johnson and Senator Kennedy had influenced attentions and capitalized on Air Force and media initiatives during the Eisenhower years. Writing in the wake of the Vietnam War, historian Richard Aliano criticized Kennedy's advocacy of both strategic and conventional buildup and of "confronting crises" as each being "part of the final climactic Communist assault upon world freedom." As a result, Aliano asserted that Kennedy "failed to define the limits of the United States's commitments and responsibilities" as his predecessor had.[36] Kennedy's insistence in 1961 that there would be "equal attention to both" civilian and military space efforts similarly indicated a belief that a compromise was a substitute for a decision.

Certainly, Eisenhower was a prisoner of the context he had helped build for himself as well. The hidden-hand approach to the presidency served him poorly regarding space issues. He had embraced it as a means of preserving his almost universally popular persona as a general, and it squared the circle on a soldier's political career and in his attempt to reconstruct the Republican Party. Space issues made Eisenhower appear out of touch and ill prepared, and the hidden hand became a manacle.

More than any other single individual in the United States, a president sets national policy. As Richard Neustadt asserted, this policymaking is not necessarily interrupted by momentary shifts in the current of the national mood. However, public opinions galvanize when they mesh with culturally valued attitudes. Policymakers ignore cultural views at the peril of their own careers, and in the long run cultural tenets will outlast the lives of individual people. Policymakers can and do, however, seek to interpret, reflect, and reshape these values in ways to complement the objectives and methods that the policymakers themselves pursue.

Policy, like military doctrine, must abide by the norms and values broadly set by a society. Air Force doctrine worked to keep in step with the beliefs of that contemporary culture in the late 1950s and early 1960s, and its efforts at advertising itself sought in turn to keep the national culture in step with the Air Force. Sputnik magnified the cleft between Eisenhower's rationale for a strong Air Force and the Air Force's own self-image. For the remainder of his presidency, Eisenhower worked to contain the Air Force and prevent it from upsetting his space policy. He delivered major messages to the nation (after Sputnik I and II and again at the close of his presidency) seeking to persuade the populace to accept his perspective of national security and responsible citizenship. Kennedy, rather than appear to contradict the vocal fervor about space activity, swam with the current and reached the White House. Once there, he had to decide what he intended to do.

In the long run, the most potent power lay not with transient policymakers or the military guarding the society, but with the people who cumulatively shaped the national public culture. This amorphous power, possessed by people who are typically consumed by more immediately personal toils and interests than national policy, was the reason that Eisenhower called for an "alert and knowledgeable citizenry." Until a crisis seemed to appear, Americans were more concerned with keeping the lawn mowed, the car washed, and the family fed than they were with what might be accomplished during the International Geophysical Year or whether a rocket-propelled glider could circle the earth. Officials and opinion leaders shape policy and seek to frame and focus public attention and attitudes on particular topics in particular ways. Eisenhower's affable smile and Kennedy's conscious exuberance were two imperfect ways that mortals sought to reach office and wield power in the midst of the Cold War and the dawn of the space age.

NOTES

Epigraph

1. Irvin Stewart, *Organizing Scientific Research for War* (Boston: Little, Brown and Company, 1948.

2. H. R. McMaster, "The Pipe Dream of Easy War," *New York Times*, July 20, 2013.

Chapter 1. "What's a Heaven For?"

Epigraph. Lee Bowen, *Threshold of Space, 1945–1959*, USAF Historical Division Liaison Office, August 1964, K168.01–7, AFHRA.

1. Attachment to memo by Gordon Gray, Jan 4, 1961, fldr 1960—Meetings with President—Volume 2 (1), box 5, Presidential Subseries, Special Asst National Security Affairs Special Asst ser, DDEL.

2. Gordon Gray, "Memorandum of Meeting with the President (Thursday, 29 December 1960 at 8:50am)," December 31, 1960, fldr 1960—Meetings with President—Volume 2 (2), box 5, Presidential Subseries, Special Asst National Security Affairs Special Asst ser, DDEL.

3. "Memorandum of Discussion of the 469th National Security Council," *FRUS*, 1958–1960, vol III, doc 129, 494–508.

4. Gray, "Memorandum of Conversation with the President (Tuesday, January 3, 1961 at 2:35pm)," January 4, 1961, fldr 1960—Meetings with President—Volume 2 (1), box 5, Presidential Subseries, Special Asst National Security Affairs Special Asst ser, DDEL.

5. "Memorandum of Conference with the President, February 17, 1959," February 24, 1959, fldr National Aeronautics and Space Administration (September 1958–January 1961) (5), box 18, Alpha subser, Staff Sec Subject ser, DDEL.

6. Franklyn W. Phillips, "Estimated Funding Requirements of Military Programs Using Space Sub-Systems," December 22, 1959, fldr Space Council (11), box 24, Staff Sec Alpha, DDEL.

7. In his landmark work, . . . *the Heavens and the Earth*, Walter McDougall noted that the history of Dyna-Soar "deserves a telling." Walter A. McDougall, . . . *the Heavens and the Earth: A Political History of the Space Age* (New York: Basic Books, 1985), 339.

8. Michael Sherry, *The Rise of American Airpower: The Creation of Armageddon* (New Haven: Yale University Press, 1987), 31; Mark Clodfelter, *Beneficial Bombing: The Progressive Foundations of American Air Power, 1917–1945* (Lincoln: University of Nebraska Press, 2011), 44.

9. Phillip S. Meilinger, *Bomber: The Formation and Early Years of Strategic Air Command* (Maxwell Air Force Base, AL: Air University Press, 2012), 19.

10. Stephen L. McFarland, *America's Pursuit of Precision Bombing, 1910–1945* (Washington, DC: Smithsonian Institution Press, 1995), 82; Meilinger, *Bomber*, 22–28.

11. Sherry, *Rise of American Airpower*, 22; Phillip S. Meilinger, "Alexander P. de Seversky and American Airpower," in *The Paths of Heaven: The Evolution of Airpower Theory*, ed. Phillip S. Meilinger (Maxwell Air Force Base, AL: Air University Press, 1997), 257.

12. I. B. Holley, *Technology and Military Doctrine: Essays on a Challenging Relationship* (Maxwell Air Force Base, AL: Air University Press, 2004), 95.

13. Written in early 1941, Air War Plans Division-1 boldly outlined the destruction of 154 targets in six months. This equates to almost one target per day being destroyed for half a year, although weather conditions in Europe proved a factor capable of limiting precision bombing attempts to one day out of three.

14. James R. Cody, "AWPD-42 to Instant Thunder: Consistent, Evolutionary Thought or Revolutionary Change?" master's thesis, Air University, 1996, 17; Meilinger, *Bomber*, 57; Colin S. Gray, *Airpower for Strategic Effect* (Maxwell Air Force Base, AL: Air University Press, 2012), 148–49.

15. Dik Daso, *Architects of American Air Superiority: General Hap Arnold and Dr. Theodore von Kármán* (Maxwell Air Force Base, AL: Air University Press, 1997); Meilinger, *Bomber*, 281.

16. Samuel Huntington, *The Soldier and the State: The Theory and Politics of Civil-Military Relations* (New York: Vintage Press, 1957), 304.

17. Presidents can and do alter the organizational models of these supporting groups, creating, abolishing, and consolidating offices and agencies. Eisenhower adopted a more regularly organized model of government than he had inherited from the long period under Franklin D.

Roosevelt and Harry S. Truman. Space politics prompted Eisenhower to transform the National Advisory Committee on Aeronautics into the National Aeronautics and Space Administration and to establish the President's Scientific Advisory Board; it coincided and contributed to his reorganization of the Defense Department and the establishment of new civilian-controlled offices supervising defense research.

18. Joan W. Sloan, "The Management and Decision-Making Style of President Eisenhower," *Presidential Studies Quarterly* 20, no. 2 (Spring 1990): 305.

19. Saki Dockrill, *Eisenhower's New-Look National Security Policy, 1953–61* (New York: St. Martin's Press, 1996), 47, 277, 279.

20. Arthur W. Radford, "The 'New Look,'" in *The Impact of Air Power*, ed. Eugene Emme (Princeton: Van Nostrand Press, 1959), 656; Andreas Wegner, *Living with Peril: Eisenhower, Kennedy, and Nuclear Weapons* (New York: Rowman and Littlefield Press, 1997), 314.

21. H. Gordon Hoxie, "Eisenhower and Presidential Leadership," *Presidential Studies Quarterly* 13, no. 4 (Fall 1983): 590; H. W. Brands, *Cold Warriors: Eisenhower's Generation and American Foreign Policy* (New York: Columbia University Press, 1988), 187–90.

22. Steven Wagner, *Eisenhower Republicanism: Pursuing the Middle Way* (DeKalb: Northern Illinois University Press, 2006).

23. Huntington, *The Soldier and the State*, 367.

24. Fred I. Greenstein, "Eisenhower as an Activist President: A Look at New Evidence," *Political Science Quarterly* 94, no. 4 (Winter 1979–80): 584.

25. Mark White, "Apparent Perfection: The Image of John Kennedy," *History* 98, no. 330 (April 2013): 228–29; David L. Snead, *John F. Kennedy: The New Frontier President* (Hauppauge, NY: Nova Science Publishers, 2012), 43–44; Theodore Sorensen, *Kennedy* (New York: Harper and Row, 1965), 528.

26. Richard Reeves, *President Kennedy: Profile of Power* (New York: Simon and Schuster, 1993), 19; Robert Dallek, *Unfinished Life: John F. Kennedy, 1917–1963* (New York: Little, Brown and Company, 2003), 226.

27. W. J. Rorabough, *The Real Making of the President: Kennedy, Nixon, and the 1960 Election* (Lawrence: University of Kansas Press, 2009), 193; Nick Bryant, *The Bystander: John F. Kennedy and the Struggle for Black Equality* (New York: Basic Books, 2006); Reeves, *President Kennedy*, 62; Snead, *John F. Kennedy*, 57, 90, 162, 225.

28. Jeremy Black, *War and Technology* (Indianapolis: Indiana University Press, 2014), 246.

29. Alex Roland, "The Technological Fix: Weapons and the Cost of War" (Carlisle, PA: Strategic Studies Institute, U.S. Army War College, 1995), xviii–xix.

30. Max Boot, *War Made New: Technology, Warfare, and the Course of History, 1500 to Today* (New York: Gotham Books, 2006), 15–16; Barry R. Posen, *The Sources of Military Doctrine: France, Britain, and Germany Between the World Wars* (Ithaca: Cornell University Press, 1984), 16.

31. Neil Sheehan, *A Fiery Peace in a Cold War: Bernard Schriever and the Ultimate Weapon* (New York: Random House, 2009), 217, 150, 217; Meilinger, *Bomber*, 288.

32. Von Hardesty and Ilya Grinberg, *Red Phoenix Rising: The Soviet Air Force in World War II* (Lawrence: University of Kansas Press, 2012), 344–57; Frank Futrell, *Ideas, Concepts, Doctrine: Basic Thinking in the United States Air Force, Volume I, 1907–1960* (Maxwell Air Force Base, AL: Air University Press, 1989), 544.

33. Michael H. Gorn, *Harnessing the Genie: Science and Technology Forecasting for the Air Force 1944–1986* (Washington, DC: Office of Air Force History, 1988), 66; Stephen B. Johnson, *The United States Air Force and the Culture of Innovation* (Washington, DC: U.S. Government Printing Office, 2002).

34. Matthew Evangelista, *Innovation and the Arms Race: How the United States and the Soviet Union Develop New Military Technologies* (Ithaca: Cornell University Press, 1988), x, 25, 61, 65.

35. Jeffrey Frank, *Ike and Dick: Portrait of a Strange Political Marriage* (New York: Simon and Schuster, 2013), 160.

36. Michael E. Heberling, "Project Termination," in *Military Project Management Handbook*, ed. David I. Cleland et al. (New York: McGraw-Hill, 1993), 35–36.

37. Geoffrey L. Herrera, *Technology and International Transformation: The Railroad, the Atom Bomb, and the Politics of Technological Change* (Albany: SUNY Press, 2006), 193.

Chapter 2. "Symbol of . . . Longing and Hope"

Epigraph. James R. Killian, *Sputnik, Scientists, and Eisenhower: A Memoir of the First Special Assistant to the President for Science and Technology* (Cambridge: MIT Press, 1977), 73.

1. Marty Jezer, *The Dark Ages: Life in the United States, 1945–1960* (Boston: South End Press, 1982), 125; Adam Rome, *The Bulldozer in the Countryside: Suburban Sprawl and the Rise of American Environmentalism* (Cambridge, UK: Cambridge University Press, 2001), 19.

2. Dwight Eisenhower, "Memorandum for Directors and Chiefs of War Department General and Special Staff Divisions and Bureaus and the Commanding Generals of the Major Commands," 30 April 1946, fldr 6 Correspondence Feb-Apr 1946, box 34, Edward L. Bowles Papers,

LoC; G. Pascal Zachary, *Endless Frontier: Vannevar Bush, Engineer of the American Century* (Cambridge: MIT Press, 1999), 315–16.

3. James Phinney Baxter, *Scientists Against Time* (Boston: Little, Brown and Company, 1946), 211.

4. Vannevar Bush, "Foreword," in Irvin Stewart, *Organizing Scientific Research for War: The Administrative History of the Office of Scientific Research and Development* (Boston: Little, Brown and Company, 1948), ix; G. C. Kenney, "Survival in the Atomic Age," in *The Impact of Air Power*, ed. Eugene Emme (Princeton: Van Nostrand Press, 1959), 426.

5. Alex Roland, "Science and War," *Osiris* 2, no. 1 (1985): 265–66; Paul Boyer, *By the Bomb's Early Light: American Thought and Culture at the Dawn of the Atomic Age* (New York: Pantheon, 1985), 183.

6. OSRD, which linked university lab researchers with Army priorities, had had a hand in the deployment of penicillin, DDT, more effective flamethrowers, advanced amphibious landing craft, and myriad other medicines, weapons, and vehicles. It was a significant step toward developing the military-industrial complex. Lawrence R. Samuel, *Future: A Recent History* (Austin: University of Texas, 2009), 38, 54, 85.

7. Boyer, *By the Bomb's Early Light*, 35.

8. K. Booth, "The Concept of Strategic Culture Affirmed," in *Strategic Power: USA/USSR*, ed. Carl G. Jacobsen (New York: St. Martin's, 1990), 121.

9. Andrew Dowdy, *The Films of the Fifties: The American State of Mind* (New York: William Morrow and Company, 1975), 26.

10. Boyer, *By the Bomb's Early Light*, 29.

11. Ibid., 109; Kenneth Osgood, *Total Cold War: Eisenhower's Secret Propaganda Battle at Home and Abroad* (Lawrence: University of Kansas Press, 2006), 179.

12. Boyer, *By the Bomb's Early Light*, 293.

13. Albert Wohlstetter, "Strategy and the National Scientists," in *Scientists and National Policy-Making*, ed. Robert Gilpin and Christopher Wright (New York: Columbia University Press, 1964), 223–25.

14. Dowdy, *The Films of the Fifties*, 160, 163.

15. Samuel, *Future: A Recent History*, 78.

16. Howard E. McCurdy, "Observations on the Robotic Versus Human Issue in Spaceflight," in *Critical Issues in the History of Spaceflight*, ed. Steven J. Dick and Roger D. Launius (Washington, DC: NASA Office of External Relations, 2006), 78.

17. Michael J. Neufeld, "The 'Von Braun Paradigm' and NASA's Long-Term Planning for Human Spaceflight," in *NASA's First 50 Years: Historical Perspectives*, ed. Steven J. Dick (Washington, DC: U.S. Government Printing Office, 2009), 328–32.

18. George M. Low, "Biological Payloads and Manned Space Flight," November 16, 1959, Document III-3, *Exploring the Unknown*, vol VI, *Space and Earth Science*, ed. John M. Logsdon (Washington, DC: NASA History Office, 2004), 307–8; Joan Vernikos, "Life Sciences in Space," in Logsdon, *Exploring the Unknown*, vol VI, *Space and Earth Science*, 269.

19. Arthur C. Clarke, "Extra-Terrestrial Relays: Can Rocket Stations Give World-Wide Radio Coverage?" October 1945, Document I-2, in *Exploring the Unknown, Selected Documents in the History of the U.S. Civilian Space Program*, vol III, *Using Space*, ed. John M. Logsdon (Washington, DC: NASA History Office, 1998), 16–22.

20. David Eisenhower and Julie Nixon Eisenhower, *Going Home to Glory: A Memoir of Life with Dwight D. Eisenhower, 1961–1969* (New York: Simon and Schuster, 2010); Ambrose, *Eisenhower: Soldier and President*, 20–21, 26–28; Evan Thomas, *Ike's Bluff: President Eisenhower's Secret Battle to Save the World* (New York: Little, Brown and Company, 2012), 30.

21. Thomas, *Ike's Bluff*, 27.

22. Ambrose, *Eisenhower: Soldier and President*, 251.

23. William Pickett, *Eisenhower Decides to Run: Presidential Politics and Cold War Strategy* (Chicago: Ivan R. Dee, 2000).

24. Jean Edward Smith, *Eisenhower in War and Peace* (New York: Random House, 2013), 470; Thomas, *Ike's Bluff*, 10; Halberstam, *The Fifties*, 184, 195.

25. The supporters portrayed in this advertisement were all Caucasian. An African-American man did appear in a separate ad from the "Eisenhower Answers America" series, which addressed the Korean War. Herbert H. Hyman and Paul B. Sheatsley, "The Political Appeal of President Eisenhower," *The Public Opinion Quarterly* 17, no. 4 (Winter 1953–54): 443–60; "Eisenhower for President (1952)," YouTube.com, accessed June 3, 2013, http://www.youtube.com/watch?v=Y9RAxAgksSE; "Eisenhower Answers America," YouTube.com, accessed June 3, 2013, http://www.youtube.com/watch?v=OJ3BngTJJS4.

26. Quoted in Tom Wicker, *Dwight D. Eisenhower: The American Presidents Series* (New York: Henry Holt, and Company, 2002), 17.

27. Ambrose, *Eisenhower: Soldier and President*, 383.

28. Robert Pape, *Bombing to Win: Air Power and Coercion in War* (Ithaca: Cornell University Press, 1996), 167; Dockrill, *Eisenhower's New-Look National Security Policy*, 50.

29. George Kennan's "long telegram" of 1947 was another formative document in shaping U.S. Cold War attitudes. It analyzed continuities in Soviet and historic Russian expansionism and predicted that the Soviet structure was beset by internal contradictions that would preclude the permanent existence of the Soviet Union.

30. Dockrill, *Eisenhower's New-Look National Security Policy*, 19.

31. Federal Civil Defense Administration, *Duck and Cover*, 1951, YouTube .com, accessed July 5, 2013, http://www.youtube.com/watch?v=IKqXu 5jw60.

32. Encyclopedia Britannica Films, *Atomic Alert*, 1951, YouTube.com, accessed July 5, 2013, http://www.youtube.com/watch?v=i4k2skbJDm8.

33. Cascade Pictures, *Self-Preservation in an Atomic Bomb Attack*, 1950, YouTube.com, accessed July 5, 2013, http://www.youtube.com/watch ?v=NCzUcwS_rPI.

34. Dwight Eisenhower, "'The Chance for Peace' Address Delivered before the American Society of Newspaper Editors, April 16, 1953," accessed July 5, 2013, http://www.eisenhower.archives.gov/all_about_ike/speech es/chance_for_peace.pdf.

35. Osgood, *Total Cold War* 57, 194.

36. Dockrill, *Eisenhower's New-Look National Security Policy*, 49–59.

37. Osgood, *Total Cold War*.

38. Dwight Eisenhower, "Text of the Address Delivered by the President of the United States Before the General Assembly of the United Nations in New York City Tuesday Afternoon, December 8, 1953," accessed July 5, 2013, http://www.eisenhower.archives.gov/research/online_documents /atoms_for_peace/Binder13.pdf.

39. Dockrill, *Eisenhower's New-Look National Security Policy*, 54.

40. Osgood, *Total Cold War*, 54; Wegner, *Living with Peril*, 14.

41. Richard Leighton, *History of the Office of the Secretary of Defense*, Vol III: *Strategy, Money, and the New Look, 1953–1956* (Washington, DC: Office of the Secretary of Defense Historical Office, 2001), 33.

42. Valerie L. Adams, *Eisenhower's Fine Group of Fellows: Crafting a National Security Policy to Uphold the Great Equation* (Lanham, MD: Lexington Books, 2006), 59, 82; Thomas, *Ike's Bluff*, 202, 270.

43. Encyclopedia Brittanica Films, *Atomic Alert*.

44. Fred I. Greenstein, *The Hidden-Hand Presidency: Eisenhower as Leader* (New York: Basic Books, 1982), 188, 212–24.

45. Dockrill, *Eisenhower's New-Look National Security Policy*, 27, 69.

46. Ibid., 71, 118.

47. In the middle of the decade, the Air Force received 47 percent of the DOD budget. The Navy and the Marine Corps combined got 29 percent, and the Army oversaw 22 percent. Gary Wills, *Bomb Power: The Modern Presidency and the National Security State* (London: Penguin Books, 2010), 44–45.

48. "Pictures Reveal Reds' New 'Sunday Punch,'" *Aviation Week* (February 15, 1954), 12–13.

49. Linda McFarland, *Cold War Strategist: Stuart Symington and the Search for National Security* (Westport, CT: Praeger, 2001), 55–60; Stuart Symington, "Our Air Force Policy," in Emme, *The Impact of Air Power*, 629; James C. Olson, *Stuart Symington: A Life* (Columbia: University of Missouri Press, 2003), 152.

50. Ambrose, *Eisenhower: Soldier and President*, 394–400; David A. Nichols, *Eisenhower 1956: The President's Year of Crisis—Suez and the Brink of War* (New York: Simon and Schuster, 2011), 55–56; Robert E. Gilbert, "Eisenhower's 1955 Heart Attack: Medical Treatment, Political Effects, and the 'Behind the Scenes' Leadership Style," *Politics and the Life Sciences* 27, no. 1 (March 2008): 2–21.

51. Eisenhower's solution to the quandary brought an element of passivity to his handling of the office. His occasional and somewhat surprising hesitance to discard problematic subordinates reflected a limitation that would again appear during the public struggle to control space policy. Wagner, *Eisenhower Republicanism*; Ambrose, *Eisenhower: Soldier and President*, 400.

52. G. K. Zhukov, "Western Policy and Soviet Armed Forces," in Emme, *The Impact of Air Power*, 592–93.

53. "U-2 Overflights of Soviet Bloc," August 18, 1960, Document #0022, declassified 2000, DDRS.

54. Thomas, *Ike's Bluff*, 215; McFarland, *Cold War Strategist*, 78.

55. "U-2 Overflights of Soviet Bloc."

56. Nichols, *Eisenhower 1956*; Benjamin P. Greene, *Eisenhower, Science Advice, and the Nuclear Test-Ban Debate, 1945–1963* (Palo Alto: Stanford University Press, 2007), 87–90.

57. Nichols, *Eisenhower 1956*; Brands, *Cold Warriors*, 19–20; Dockrill, *Eisenhower's New-Look National Security Policy*, 167; Dwight Eisenhower, "State of the Union Message," January 12, 1961; James D. Marchio, "Risking General War in Pursuit of Limited Objectives: U.S. Military Contingency Planning for Poland in the Wake of the 1956 Hungarian Uprising," *The Journal of Military History* 66, no. 3 (June 2002): 792.

58. Arthur Krock, "In the Nation," *New York Times*, November 1, 1956.

59. William L. Miller, *Piety Along the Potomac: Notes on Politics and Morals in the '50s* (Boston: Houghton Mifflin, 1964), 9.

60. Wagner, *Eisenhower Republicanism*, 4.

61. Quoted in Thomas, *Ike's Bluff*, 413.

Chapter 3. "What an Impressive Idea!"

Epigraph. Killian, *Sputnik, Scientists, and Eisenhower*, 6.

Second epigraph. "Project Research System: Hypersonic Glide Rocket Research System; Hypersonic Weapons Research & Development Supporting System," December 28, 1956, 11924 "Round Three Background Correspondence," NASA HQ.

1. Daso, *Architects of American Air Superiority*.

2. Boris Chertok, *Rockets and People,* vol II (Washington, DC: NASA History Office, 2006), 44–45.

3. Ibid., 21, 31, 52–53.

4. Boris Chertok, *Rockets and People,* vol I (Washington, DC: NASA History Office, 2005), 263, 265; Chertok, *Rockets and People,* vol II, 66.

5. Quoted in Kenneth R. Whiting, "The Past and Present of Soviet Military Doctrine," *Air University Quarterly Review* 11, no. 1 (Washington, DC: U.S. Government Printing Office, Spring 1959): 38–60.

6. Bowen, *Threshold of Space*; "Abbreviated System Development Plan," August 23, 1957.

7. Walter C. Williams, "Project Dyna-Soar: the Roots of the Shuttle—A Memoir," 43rd Congress of the International Astronautical Federation, August 28–September 5, 1992, Washington, DC, K146.3–7, AFHRA.

8. Ibid.

9. Eugene Emme, "Astronautics and the Future," in Emme, *The Impact of Air Power*, 844.

10. Adams, *Eisenhower's Fine Group of Fellows*, 1, 4.

11. Greenstein, "Eisenhower as an Activist President," 575–99; Adams, *Eisenhower's Fine Group of Fellows*, 6; R. Cargill Hall, "Sputnik, Eisenhower, and the Formation of the U.S. Space Program," *Quest* 14, no. 4 (2007): 34.

12. Adams, *Eisenhower's Fine Group of Fellows*, 6; Killian, *Sputnik, Scientists, and Eisenhower*; Francis H. Heller, "The Eisenhower White House," *Presidential Studies Quarterly* 23, no. 3 (Summer 1993): 512; Bradley H. Patterson Jr., "Teams and Staff: Dwight Eisenhower's Innovations in the Structure and Operations of the Modern White House," *Presidential Studies Quarterly* 24, no. 2 (Spring 1994): 277–98; Raymond Millen, "Cultivating Strategic Thinking: The Eisenhower Model," *Parameters* 42, no. 2 (Summer 2012): 56–70.

13. Millen, "Cultivating Strategic Thinking," 56–70.

14. Nicholas Michael Sambaluk, "Policymakers Confront Aerospace Doctrine, 1957–59," *Cold War History* 14, no. 1 (Spring 2014): 91–107.

15. Killian, *Sputnik, Scientists, and Eisenhower,* 72–74; Adams, *Eisenhower's Fine Group of Fellows,* 122–23.

16. Killian, *Sputnik, Scientists, and Eisenhower,* 74–75.

17. Ibid., 75–77.

18. Christopher Bright, *Continental Defense in the Eisenhower Era: Nuclear Antiaircraft Arms and the Cold War* (Basingstoke, UK: Palgrave Macmillan, 2010).

19. Curtis Peebles, *Shadow Flights: America's Secret Air War Against the Soviet Union* (New York: Presidio Press, 2000), 29; Michael Beschloss, *Mayday: Eisenhower, Khrushchev, and the U-2 Affair* (New York: Harper and Row, 1986), 363.

20. Zuoyue Wang, *In Sputnik's Shadow: The President's Science Advisory Committee and Cold War America* (New Brunswick, NJ: Rutgers University Press, 2008), 53.

21. R. Cargill Hall, "Clandestine Victory: Eisenhower and Overhead Reconnaissance in the Cold War," in *Forging the Shield: Eisenhower and National Security for the 21st Century,* ed. Dennis Showalter (Chicago: Imprint Books, 2005), 3, 7–10.

22. R. Cargill Hall, "Postwar Strategic Reconnaissance and the Genesis of CORONA," in *Eye in the Sky: The Story of the CORONA Spy Satellites,* ed. Dwayne A. Day, John M. Logsdon, and Brian Latell (Washington, DC: Smithsonian Institution Press, 1998), 87–92.

23. Hall, "Sputnik, Eisenhower, and the Formation of the U.S. Space Program," 34; Wang, *In Sputnik's Shadow,* 5.

24. Wang, *In Sputnik's Shadow,* 53.

25. Hall, "Sputnik, Eisenhower, and the Formation of the U.S. Space Program," 34; Wang, *In Sputnik's Shadow,* 5.

26. Greene, *Eisenhower, Science Advice, and the Nuclear Test-Ban Debate;* Adams, *Eisenhower's Fine Group of Fellows,* 132–33; Killian, *Sputnik, Scientists, and Eisenhower,* 224.

27. Rip Bulkeley, "Aspects of the Soviet International Geophysical Year," *Russian Journal of Earth Sciences* 10 (2008), accessed November 26, 2011, http://elpub.wdcb.ru/journals/rjes/v10/2007ES000249/6.shtml; National Security Council, Document 5520 (NSC 5520), "U.S. Scientific Satellite Program," accessed November 23, 2011, http://www.marshall .org/pdf/materials/805.pdf; IGY World Data Center A, *IGY Satellite Report* (Washington, DC: National Research Council, 1958).

28. Dockrill, *Eisenhower's New-Look National Security Policy,* 145; Herbert S. Dinerstein, "The Soviet Military Posture as a Reflection of Strategy," in Emme, *The Impact of Air Power,* 559; Dwight Eisenhower, "Open Skies Reconnaissance," in Emme, *The Impact of Air Power,* 665–67.

29. Wang, *In Sputnik's Shadow*, 53.

30. Ibid., 53.

31. NSC 5520, "U.S. Scientific Satellite Program."

32. Joan Lisa Bromberg, *NASA and the Space Industry* (Baltimore: Johns Hopkins University Press, 1999), 22.

33. NSC 5520, "U.S. Scientific Satellite Program."

34. Frank H. Winter, "The American Rocket Society Story, 1930–1962," *Journal of the British Interplanetary Society* 33, no. 8 (August 1980): 303–11.

35. David Beckler to William Elliot, "Earth Satellite Program," April 18, 1956, 11130 "DOD in Space," NASA HQ; David Beckler to Arthur Flemming, "The Earth Satellite Program," April 30, 1956, 11130 "DOD in Space," NASA HQ.

36. Homer J. Stewart, "Vanguard and Redstone," June 22, 1956, 11130 "DOD in Space, NASA HQ; E. V. Murphree, "U.S. of the Jupiter Reentry Test Vehicle as a Satellite," July 5, 1956, 11130 "DOD in Space," NASA HQ.

37. "Project Research System: Hypersonic Glide Rocket Research System; Hypersonic Weapons Research & Development Supporting System," December 28, 1956, 11924 "Round Three Background Correspondence," NASA HQ.

38. Ibid.

39. John V. Becker, "Hypersonic research airplane study," January 17, 1957, 11924 "Round Three Background Correspondence," NASA HQ; Memo for NACA Director, "Meeting of NACA Personnel Held at NACA Headquarters February 6, 1957, to Discuss Possible Hypersonic Research Airplane," February 6, 1957, 11924 "Round Three Background Correspondence," NASA HQ; Memo for Record, "Meeting of NACA Personnel at NACA Headquarters February 6, 1957 to Discuss Possible Hypersonic Research Airplane," February 6, 1957, 11924 "Round Three Background Correspondence," NASA HQ.

40. Memo for Record, "Meeting of NACA Personnel at NACA Headquarters February 6, 1957 to Discuss Possible Hypersonic Research Airplane"; David Beckler to Arthur Flemming, "Scientific Satellites," January 23, 1957, 11130 "DOD in Space," NASA HQ.

41. M. C. Gerson to Fred Whipple, memo, February 15, 1957, 11130 "DOD in Space," NASA HQ.

42. Clotaire Wood to Hugh Dryden, "Round Three—Areas of Agreement of New Research Airplane Steering Committee at Initial Meeting," February 26, 1957, 11924 "Round Three Background Correspondence," NASA HQ.

43. Memo for Record, "Meeting of NACA Personnel at NACA Headquarters February 6, 1957 to Discuss Possible Hypersonic Research Airplane"; Hartley Soule to J. W. Crowley, "NACA Let to Langley, Ames, and Lewis, Feb 14, 1957," March 5, 1957, 11924 "Round Three Background Correspondence," NASA HQ.

44. Walter C. Williams, "Project Dyna-Soar: the Roots of the Shuttle—a Memoir," K146.3–7, AFHRA; David N. Spires, *Beyond Horizons: A Half Century of Air Force Space Leadership* (Washington, DC: U.S. Government Printing Office, 1998), 74.

45. "Round III Presentation at ARDC Headquarters, Md," June 10, 1957, 11924 "Round Three Background Correspondence," NASA HQ.

46. Wood to Dryden, "Presentation to Air Force Headquarters on Round III," July 11, 1957, 11924 "Round Three Background Correspondence," NASA HQ.

47. Dryden to Smith J. DeFrance, memo, July 11, 1957, 11924 "Round Three Background Correspondence," NASA HQ.

48. "Department of State Paper: U.S. Position on First Phase of Disarmament," June 11, 1957, *FRUS*, 1955–1957, vol XX, doc 237, 620–625.

49. Memo by David Beckler to Gordon Gray, "Basic Research Policy," Sept 19, 1957, fldr Science Advisory Committee (2), box 23, Staff Sec Alpha ser, DDEL.

50. "Memorandum of Conference with the President, August 16, 1957," Aug 20 1957, fldr Department of Defense, Vol. II (1) August 1957, box 1, Staff Secretary Subject ser, DDEL.

51. Bowen, *Threshold of Space*, 6–7; NACA to Asst Sec USAF for R&D, ARDC Commander, "Study of the Feasibility of a Hypersonic Research Airplane," September 3, 1957, 11924 "Round Three Background Correspondence," NASA HQ; NACA to S. J. DeFrance, memo, September 13, 1957, 11924 "Round Three Background Correspondence," NASA HQ; NACA to Asst Sec USAF for R&D, "NACA Study of the Feasibility of a Hypersonic Research Airplane," September 26, 1957, 11924 "Round Three Background Correspondence," NASA HQ; NACA to ARDC Commander, "NACA Comments on ARDC Development Plan 133/LA (Dyna Soar)," September 26, 1957, 11924 "Round Three Background Correspondence," NASA HQ.

Chapter 4. "Adjacent to the Abyss"

Epigraph. Eugene Emme, "Thoughts on Future Warfare," in Emme, *The Impact of Air Power*, 420.

Second epigraph. Killian, *Sputnik, Scientists, and Eisenhower*, 6.

1. William Mitchell, "The Development of Air Power," in Emme, *The Impact of Air Power*, 175; Alexander Graham Bell, "Preparedness for Aerial Defense," in Emme, *The Impact of Air Power*, 32; Stanley Baldwin, "The Bomber Will Always Get Through," in Emme, *The Impact of Air Power*, 51–52.

2. *Bombers B-52*, directed by Gordon Douglas (Burbank, CA: Warner Brothers, 1957), DVD.

3. This dynamic was not new, and the Air Force's utilization of it was not unique. Alfred Vagts, *A History of Militarism: Civilian and Military* (New York: W. W. Norton, 1959), 13–15; Posen, *Sources of Military Doctrine*, 45.

4. Steve Call, *Selling Air Power: Military Aviation and American Popular Culture after World War II* (College Station: Texas A&M Press, 2009), 1; Gerard Clarfield, *Security with Solvency: Dwight D. Eisenhower and the Shaping of the American Military Establishment* (Westport, CT: Praeger, 1999), 1; Nancy E. Bernard, *Television News and Cold War Propaganda, 1947–1960* (Cambridge, UK: Cambridge University Press, 1999), 152.

5. Nathan Twining, quoted in Emme, *The Impact of Air Power*, 512; Curtis LeMay, "Strategic Air Command and World Peace," in Emme, *The Impact of Air Power*, 669.

6. Dockrill, *Eisenhower's New-Look National Security Policy*, 41; Robert C. Richardson III, "Balance of Power in the Atomic Age," in Emme, *The Impact of Air Power*, 461–62.

7. Arthur W. Radford, "The 'New Look,'" in Emme, *The Impact of Air Power*, 660.

8. Henry Kissinger, *Nuclear Weapons and Foreign Policy* (New York: Harper and Brothers, 1957), 130, 155, 247.

9. Ibid., 231–32.

10. Dockrill, *Eisenhower's New-Look National Security Policy*, 205; Hall, "Clandestine Victory: Eisenhower and Overhead Reconnaissance in the Cold War," in Showalter, *Forging the Shield*, 7–10; Donald Quarles, "The Uses of Modern Arms," in Emme, *The Impact of Air Power*, 706–7.

11. Smith, *Eisenhower in War and Peace*, 713–14.

12. Wagner, *Eisenhower Republicanism*, 65.

13. Smith, *Eisenhower in War and Peace*, 715–25, 727; Ambrose, *Eisenhower: Soldier and President*, 447; Thomas, *Ike's Bluff*, 248.

14. Paul Dickson, *Sputnik: The Shock of the Century* (New York: Walker and Co., 2001), 9; NASA Langley Cultural Resources Geographic Information Systems, "Flutter Test of X-15," filmed September-October 1957, YouTube.com, accessed July 19, 2013, http://www.youtube.com/watch?v=Re-YTsPifZY&list=PL5B2A1E4970DD12CA.

15. Ambrose, *Eisenhower: Soldier and President*, 447.

16. Robert T. Davis II, *The U.S. Army and the Media in the 20th Century* (Leavenworth, KS: Combat Studies Institute Press, 2009), 1, 39; Osgood, *Total Cold War*.

17. "Memorandum for the President, Subject: Earth Satellite," Oct 7, 1957, document #0745, declassified 1999 DOD, DDRS.

18. Robert Divine, *The Sputnik Challenge: Eisenhower's Response to the Soviet Satellite* (Oxford, UK: Oxford University Press, 1993), 3–8; Richard M. Bissell with Jonathan E. Lewis and Frances T. Pudlo, *Reflections of a Cold Warrior: From Yalta to the Bay of Pigs* (New Haven: Yale University Press, 1996), 134–35.

19. Special Assistant for Scientific Liaison Beckler to Acting Director ODM Victor Cooley, "Satellites and Missiles," Oct 8, 1957, fldr Space—Oct 1957–Oct 1959, box 3, Special Asst Sci & Tech ser, DDEL.

20. Memo from Scientific Advisory Committee President Rabi to ODM Director Arthur S. Flemming, Oct 10 1956, included as an attachment by David Beckler to Andrew Goodpaster, Oct 9, 1957, fldr Science Advisory Committee (1), box 23, Staff Sec Alpha ser, DDEL.

21. "Memorandum of Conference with the President," Oct 8, 1957, DDE, fldr Department of Defense, Vol. II (2) Sept-Oct 1957, box 1, Staff Secretary Subject ser, DDEL.

22. "Memorandum for the President, Subject: Earth Satellite," Oct 7, 1957.

23. "Discussion at the 339th Meeting of the National Security Council, Thursday, October 10, 1957," Oct 11, 1957, document #0850, declassified 1987 NSC, DDRS.

24. Diary, Oct 9 1957, fldr October 57—ACW Diary (2), box 9, PP (AWD), DDEL.

25. "Discussion at the 339th Meeting of the National Security Council, Thursday, October 10, 1957."

26. Goodpaster, memo for the record, Oct 15, 1957, fldr Satellites October 1957–February 1960 (1), box 23, Staff Sec Alpha ser, DDEL. An optimistic appraisal anticipated a possible "limited reconnaissance capability" as early as the second quarter of 1959, document #2541, declassified 1997 DOD, DDRS; Dickson, *Sputnik: The Shock of the Century*,113.

27. Walter Lippmann, *The Essential Lippmann: A Political Philosophy for Liberal Democracy*, ed. Clinton Rossiter and James Lare (Cambridge: Harvard University Press, 1963), 67–68.

28. ODM Science Advisory Committee meeting with President, October 15, 1957, 11130 "DOD in Space," NASA HQ.

29. Ibid.

30. "Address of the Vice President of the United States before the International Industrial Development Conference, San Francisco, California," October 15, 1957, fldr Vice President Nixon, box 2, Moos Papers, DDEL; Eisenhower to Senator Charles Potter (R–MI), October 21, 1957, fldr President's Speeches: Science and Security Background Material used for Nov 7 and Nov 13 1957 (3), box 3, Moos Papers, DDEL.

31. Memo from Larson to Eisenhower, Oct 15, 1957, document #2098, declassified 1990 USIA, DDRS.

32. Ibid.; Thomas A. Sturm, *The USAF Scientific Advisory Board: Its First Twenty Years, 1944–1964* (Washington, DC: Office of Air Force History, 1986), 81.

33. "Acceptance Remarks by Deputy Secretary of Defense Donald Quarles upon Presentation of Award of Merit, American Institute of Consulting Engineers, New York," Oct 15, 1957, fldr Department of Defense, Vol. II (2) Sept–Oct 1957, box 1, Staff Sec Subject ser, DDEL.

34. "Subjects discussed with the President by Bernard Baruch, New York, October 22 1957," fldr October '57—ACW Diary (1), box 9, PP AWD, DDEL.

35. "Shoot the Moon, Ike," October 29, 1957, *New York World-Telegram.*

36. "Facts from Secretary McElroy's Memorandum to the President ref the Symington Inquiry," October 31, 1957, fldr President's Speeches: Science and Security Background Material used for Nov 7 and Nov 13 1957 (4), box 3, Moos Papers, DDEL; Notes on Interview with Deputy Secretary of Defense Quarles Regarding the President's Speech, October 31, 1957, fldr President's Speeches: Science and Security Background Material used for Nov 7 and Nov 13 1957 (4), box 3, Moos Papers, DDEL.

37. Notes on Interview with Deputy Secretary of Defense Quarles Regarding the President's Speech, October 31, 1957, fldr President's Speeches: Science and Security Background Material used for Nov 7 and Nov 13 1957 (4), box 3, Moos Papers, DDEL. Unfortunately, the author of the handwritten remonstration has not been identified.

38. OCB draft memo, "Guide Lines for Public Information on the U.S. and Soviet Scientific Earth Satellite Program," Nov 1, 1957, fldr National Aeronautics and Space Administration–National Aeronautics and Space Council (3), box 4, PSAC ser, DDEL.

39. Draft "Science and National Security" speech, November 1, 1957, fldr President's Speeches: Sciences in National Security–Drafts Nov 7, 1957, box 2, Moos Papers, DDEL.

40. Draft "Science and National Security" speech, n.d. (between Nov 1 and Nov 5), fldr President's Speeches: Sciences in National Security–Drafts Nov 7, 1957, box 2, Moos Papers, DDEL.

41. Eisenhower's approval rating would slip further, to 52 percent, by March 1958, although Divine acknowledges that the contemporary economic recession contributed to this trend. Divine, *The Sputnik Challenge*, 44–45, 119.

42. Ibid., 43; Richard A. Aliano, *American Defense Policy from Eisenhower to Kennedy: The Politics of Changing Military Requirements, 1957–1961* (Athens: Ohio University Press, 1975), 1; Osgood, *Total Cold War*, 336; Peter J. Roman, *Eisenhower and the Missile Gap* (Ithaca: Cornell University Press, 1995), 28–29.

43. Note to Larson, November 4, 1957, fldr President's Speeches: Science and Security Background Material used for Nov 7 and Nov 13 1957 (3), box 3, Moos Papers, DDEL.

44. News clipping, "An open letter to President Eisenhower," Nov 7, 1957, fldr October 1957–February 1960 (1), box 23, Staff Sec alpha, DDEL.

45. Letter from ARS president Robert C. Truax to II Rabi, Nov 6, 1957, fldr Space November 1957 (2), Special Asst Sci & Tech, box 15, DDEL; Winter, "The American Rocket Society Story, 1930–1962," 303–11.

46. Adams, *Eisenhower's Fine Group of Fellows*, 4.

47. Dockrill, *Eisenhower's New-Look National Security Policy*, 135.

48. "Civil Defense and Survival," in Emme, *The Impact of Air Power*, 691.

49. Adams, *Eisenhower's Fine Group of Fellows*, 4; Edward Teller, "The Nature of Nuclear Warfare," in Emme, *The Impact of Air Power*, 466–68; Divine, *The Sputnik Challenge*, 35.

50. Adams, *Eisenhower's Fine Group of Fellows* 99, 162, 191.

51. Snead, *John F. Kennedy*; Adams, *Eisenhower's Fine Group of Fellows* 171–73, 177.

52. Adams, *Eisenhower's Fine Group of Fellows*, 164–65; Divine, *The Sputnik Challenge*, 36–37; David L. Snead, *The Gaither Committee, Eisenhower, and the Cold War* (Columbus: Ohio State University Press, 1999), 64–71.

53. Draft "Science and National Security" speech, n.d. (between Nov 4 and Nov 5), fldr President's Speeches: Sciences in National Security—Drafts Nov 7, 1957, box 2, Moos Papers, DDEL.

54. Draft, "Sciences in National Security," November 7, 1957, fldr President's Speeches: Sciences in National Security—Final Nov 7 1957, box 3, Moos Papers, DDEL.

55. "Southwest Research Institute Conference San Antonio, Texas, November 7–8, 1957," fldr President's Speeches: Science and Security Background Material used for Nov 7 and Nov 13 1957 (3), box 3, Moos Papers, DDEL.

56. Draft, "Our Future Security," November 9, 1957, second draft, fldr President's Speeches: Our Future Security—Oklahoma—Nov 13 1957—Drafts, box 3, Moos Papers, DDEL; Draft, Quarles's changes to "Our Future Scientists," n.d. (about Nov 10, 1957), fldr President's Speeches: Science and Security Background Material used for Nov 7 and Nov 13 1957 (2), box 2, Moos Papers, DDEL; Draft, Killian's changes to "Our Future Scientists," n.d. (about Nov 10, 1957), fldr President's Speeches: Science and Security Background Material used for Nov 7 and Nov 13 1957 (2), box 2, Moos Papers, DDEL; Draft, "Our Future Security," n.d. (about Nov 10 or 11), fldr President's Speeches: Our Future Security—Oklahoma—Nov 13, 1957—Drafts, box 3, Moos Papers, DDEL.

57. Ambrose, *Eisenhower: Soldier and President*, 451; Osgood, *Total Cold War*, 336.

58. Divine, *The Sputnik Challenge*, 58–59.

59. McFarland, *Cold War Strategist*, 86.

60. Ibid., 108.

61. Dickson, *Sputnik: The Shock of the Century*, 225–26; Adams, *Eisenhower's Fine Group of Fellows*, 118–19; Divine, *The Sputnik Challenge*, 56, 91, 164.

62. Dickson, *Sputnik: The Shock of the Century*, 23, 93.

63. Osgood, *Total Cold War*, 286; John Krige, "Atoms for Peace, Scientific Internationalism, and Scientific Intelligence," *Osiris* 2, no. 21 (2006): 178; Asif A. Siddiqi, "Competing Technologies, National(ist) Narratives, and Universal Claims: Toward a Global History of Space Exploration," *Technology and Culture* 51, no. 2 (April 2010): 428.

Chapter 5. "The First of a New Generation"

Epigraph. LeMay, article for American Ordnance Association's *Ordnance*, May 7, 1958, fldr Cleared Copies Oct 1957–Sept 1958, General LeMay, box B171, LeMay Papers, LoC.

Second epigraph. Memo from White to AF Sec Douglas, "Priority Program Augmentations to Basic 1959 Budget Estimate," White Oct 3 1958, fldr Secretary of the Air Force #2, box 18, White Papers, LoC.

1. Thomas Power, "The Air Atomic Race," in Emme, *The Impact of Air Power*, 687.

2. Thomas Power, "Strategic Air Command and the Ballistic Missile," in Emme, *The Impact of Air Power*, 488.

3. Johnson, *The United States Air Force and the Culture of Innovation*.

4. Kissinger, *Nuclear Weapons and Foreign Policy*, 123.

5. Power, "Strategic Air Command and the Ballistic Missile," in Emme, *The Impact of Air Power*, 493.

6. Bernard A. Schriever, speech, "ICBM—A Step Toward Space Conquest, San Diego, California, 19 Feb 57," Schriever, B.A., Public Statements on Important Military Issues, Roll 35253, IRIS 1040169, AFHRA.

7. This speed, in Earth atmosphere, would be the equivalent of about Mach 22. As an orbital velocity, however, Dyna-Soar III would travel through space, where sound (and therefore mach numbers) loses effective meaning.

8. Bowen, *Threshold of Space*; "Abbreviated System Development Plan," August 23, 1957, in *Dyna-Soar: Hypersonic Strategic Weapons System*, ed. Robert Godwin (Burlington, Ontario: Apogee Books, 2003).

9. "Abbreviated System Development Plan," in Godwin, *Dyna-Soar*.

10. Bowen, *Threshold of Space*.

11. Quoted in McFarland, *America's Pursuit of Precision Bombing*, 88.

12. Nathan F. Twining, letter to the Advisory Group for Aeronautical Research & Development (NATO), Nov 25, 1957, box 129, Twining Papers, LoC.

13. Curtis LeMay, "Remarks by Curtis E. LeMay to the USAF Scientific Advisory Board at Patrick Air Force Base, Florida, 21 May 1957," boxes B206, B207, LeMay Papers, LoC.

14. Nathan F. Twining, speech to the American Ordnance Association in Cleveland, OH, November 12, 1957, box 129, Twining Papers, LoC.

15. "Report of the Scientific Advisory Board Ad Hoc Committee on Space Technology," Dec 6, 1957, fldr Scientific Advisory Board, box 7, White Papers, LoC.

16. "National Advisory Committee for Aeronautics, minutes of meeting, October 10, 1957," fldr NACA, box 7, White Papers, LoC.

17. Thomas D. White, "Perspective at the Dawn of the Space Age," in Emme, *The Impact of Air Power*, 498; Eugene Emme, "Astronautics and the Future," in Emme, *The Impact of Air Power*, 844.

18. White, "Perspective at the Dawn of the Space Age," in Emme, *The Impact of Air Power*, 500; "Memorandum for Secretary of Defense, Attention: Special Assistant Armed Forces Policy Council," undated, sometime after November 21, 1957, fldr Secretary of Defense, box 7, White Papers, LoC.

19. Bowen, *Threshold of Space*; Schriever quotes, Senate Armed Services, Dec 57–Jan 58.

20. Bowen, *Threshold of Space*.

21. Ibid.

22. Ibid.

23. Power, "Strategic Air Command and the Ballistic Missile," in Emme, *The Impact of Air Power*, 494.

24. "Development Requirement: Hypersonic Conceptual Test Vehicle System 464L Project Dyna-Soar," January 24, 1958, fldr 11924 "Round Three Background Correspondence," NASA HQ; Clotaire Wood, memo for files, "Visit of Boeing Personnel to NACA Headquarters to Discuss Advanced Research Airplanes (Dyna Soar I)," January 21, 1958, fldr 11324 "X-20 Correspondence Martin 1962–64," NASA HQ.

25. Hartley Soule, memo, "Suggested Memorandum of Understanding Between Air Force and NACA for Conduct of Dyna Soar I Project," February 13, 1958, 11924 "Round Three Background Correspondence," NASA HQ; Untitled notes, February 13, 1958, 11924 "Round Three Background Correspondence," NASA HQ; Clotaire Wood to NASA Director Glennan, "Need for Clarification of Role of NACA in Development of Dyna Soar I," February 25, 1958, 11924 "Round Three Background Correspondence," NASA HQ; Draft memo of understanding, March 19, 1958, 11924 "Round Three Background Correspondence," NASA HQ.

26. Memo to Dryden and J. W. Crowley, "Recommendations for changes in DynaSoar I memorandum of understanding," March 24, 1958, 11327 "Dyna Soar Proposals & Eval," NASA HQ.

27. "Memorandum of Understanding: Principles for Participation of NACA in Development and Testing of the 'Air Force System 464L Hypersonic Boost Glide Vehicle (Dyna Soar I),'" May 20, 1958, 11327 "Dyna Soar Proposals & Eval," NASA HQ.

28. Ibid.; "NACA memo for clarification of NACA role," in Godwin, *Dyna-Soar*, 89; "Memorandum of Understanding," fldr NASA, box 18, White Papers, LoC.

29. Dockrill, *Eisenhower's New-Look National Security Policy*, 243; McFarland, *America's Pursuit of Precision Bombing*, 89.

30. "Part of Chron[ology] for Bruce Byers, re: Red Socks," 1957, 11130 "DOD in Space," NASA HQ; Lt Gen Putt, memo, January 31, 1958, 11130 "DOD in Space," NASA HQ.

31. Dennis Jenkins, *X-15: Extending the Frontiers of Flight* (Seattle: CreateSpace Independent Publishing Platform, 2010).

32. Hartley Soule, "Meeting of the Steering Committee for a New Research Vehicle System at Ames Laboratory on March 11, 1958," 11924 "Round Three Background Correspondence," NASA HQ; Bromberg, *NASA and the Space Industry*, 42.

33. Twining, "Speech to the Industrial College of the Armed Forces and the National War College, Washington, DC," March 18, 1958, box 129, Twining Papers, LoC.

34. LeMay, article for *Ordnance*.

35. Ibid.

36. White, "Perspective at the Dawn of the Space Age," in Emme, *The Impact of Air Power*, 499.

37. Col. Claude E. Putnam, USAF, "Missiles in Perspective," *Air University Quarterly Review* 10, no. 1 (Spring 1958): 3–10; Col. Allen W. Stephens, USAF, "Missilemen—Present and Future," *Air University Quarterly Review* 10, no. 1 (Spring 1958): 11–19.

38. Schriever, speech to the National Geographic Society, Washington, DC, 21 Nov 1958; Schriever, B.A., Public Statements on Important Military Issues, Roll 35253, IRIS 1040169, AFHRA.

39. Col. Martin B. Schofield, USAF, "In My Opinion . . . Control of Outer Space," *Air University Quarterly Review* 10, no. 1 (Spring 1958): 93–112.

40. Maj. Gen. Lloyd P. Hopwood, USAF, "The Military Impact of Space Operations," *Air University Quarterly Review* 10, no. 2 (Summer 1958): 142–46; Maj. Gen. Dan C. Ogle, USAF, "The Threshold of Space," *Air University Quarterly Review* 10, no. 2 (Summer 1958): 2–6.

41. Brig. Gen. Homer A. Boushey, USAF, "Lunar Base Vital," *Army-Navy-Air Force Register* 79 (February 8, 1958): 11.

42. Bowen, *Threshold of Space*.

43. Ibid.

44. "Proposal for Memo of Understanding between NACA & USAF," in Godwin, *Dyna-Soar*, 87.

45. "NACA memo for clarification of NACA role," in Godwin, *Dyna-Soar*, 88.

46. "Memorandum of understanding between NACA & USAF," in Godwin, *Dyna-Soar*, 89; "Principles for Participation of NASA in Development and Testing the 'Air Force System 464L Hypersonic Boost Glide Conceptual Test Vehicle (Dyna Soar I),'" fldr NASA, box 18, White papers, LoC.

47. *An Air Force History of Space Activities, 1945–1959*, USAF Historical Division Liaison Office, online, accessed October 4, 2011, http://www.google.com/url?sa=t&rct=j&q=&esrc=s&source=web&cd=1&cad=rja&uact=8&ved=0CB4QFjAA&url=http%3A%2F%2Fwww.nro.gov%2Ffoia%2Fdeclass%2FWS117L_Records%2F277.PDF&ei=54uPVKKhEojasAS4oILYDw&usg=AFQjCNFHTdnSN36FU08_KrxLJZVNUZtcEg&sig2=DWC64YdiMxIj6O8nX6tqSw.

48. White, "Memorandum for DCS/Development," Sept 8, 1958, fldr Chief of Staff Signed Memos Jan 58-Dec 58, box 15, White Papers, LoC.

49. Memo from White to AF Sec Douglas, "Priority Program Augmentations to Basic 1959 Budget Estimate," Oct 3 1958, fldr Secretary of the Air Force #2, box 18, White Papers, LoC.

50. Memo from AF Sec Douglas to Chief of Staff White, Oct 13, 1958, fldr Secretary of the Air Force #2, box 18, White Papers, LoC.

51. Ibid.

52. LeMay, speech at the Public Information Seminar Banquet at the Manhattan Hotel, New York City, Nov 7, 1958, fldr Cleared Copies Oct 1958–1960, box B171, LeMay Papers, LoC.

53. LeMay, approved draft remarks on NBC Monitor Program in Washington, DC, Feb 17, 1959, fldr Cleared Copies Oct 1958–1960, box B171, LeMay Papers, LoC.

54. Ira Abbott, "Statement of Ira H. Abbott, Assistant Director of Research, National Aeronautics and Space Administration before Subcommittee on Manpower Utilization of the Committee on Post Office and Civil Service, U.S. House of Representatives," December 5, 1958, 7; "Abbott: Congressional Testimony (Commtes)," NASA HQ.

55. Staff Study, "The Spiral Toward Space," *Air University Quarterly Review* 10, no. 3 (Fall 1958): 10–21.

56. NASA was a changing organization, as noted by Robert S. Arrighi in *Revolutionary Atmosphere: The Story of the Altitude Wind Tunnel and the Space Power Chambers* (Washington, DC: NASA History Division, 2010).

57. Staff Study, "The Spiral Toward Space," 10–21.

58. Ibid.

59. Bowen, *Threshold of Space.*

60. Memo from White to Deputy Chief of Staff for Development, Mar 30, 1959, fldr Chief of Staff "Signed" Memos, box 26, White Papers, LoC.

61. Ibid.

62. Issues Relevant to 1960 Budget Items, May 28, 1959, fldr Chief of Staff Memos, box 26, White Papers, LoC.

63. LeMay, "approved draft speech at SUAF Symposium at the Air Force Association Convention in Miami Beach, FL, 5 Sept 1959," Sept 2, 1959, fldr Cleared Copies Oct 1958–1960, box B171, LeMay Papers, LoC; LeMay, radio interview on "Capitol Assignment" at the Pentagon on Sept 8, 1959, Sept 4, 1959, fldr Cleared Copies Oct 1958–1960, box B171, LeMay Papers, LoC.

64. White, handwritten notation on a memo from Asst Exec to Chief of Staff Lt. Col. Robert N. Ginsburgh to White, Sept 2, 1959, fldr Chief of Staff Memos, box 26, White Papers, LoC; Lt. Col. Benjamin H. Ferer, "Actions on Dyna-Soar—5 Sept 59 thru 6 Nov 59," Nov 10, 1959, K140.8636–4, AFHRA; Lt. Col. Robert N. Ginsburgh, asst exec to C/S USAF, "Memorandum for Deputy Chief of Staff, Development," fldr Chief of Staff Memos, box 26, White Papers, LoC.

65. Letter from Schriever to White, Sept 15 1959, fldr Command ARDC, box 26, White Papers, LoC.

66. Ferer, "Actions on Dyna-Soar—5 Sept 59 thru 6 Nov 59."

67. "Cancellation of Dyna Soar Selection," Oct 9, 1959, K140.8636–4, AFHRA.

68. "Dyna Soar Source Selection" and "Required Action on Dyna Soar," Oct 29, 1959, K140.8636–4, AFHRA.

69. "Status of Action on Dyna Soar," Nov 19, 1959, K140.8636–4, AFHRA.

Chapter 6. "The Air Force Must *Not* Lose Dynasoar"

Epigraph. United States Air Force, "Basic Doctrine AFM 1-2," December 1, 1959.

Second epigraph. Booklet on Dyna-Soar, about April 1960, K140.8636–4, AFHRA.

1. Ibid.

2. Memo from White to Charyk, Oct 27, 1959, fldr Chief of Staff "Signed" Memos, box 26, White Papers, LoC.

3. Ibid.

4. Memo from White to Deputy to the Chief of Staff for Development, Oct 28, 1959, fldr Chief of Staff "Signed" Memos, box 26, White Papers, LoC. The "internal hassles" to which White referred may have been the authority squabble between the Space Systems Division and the Ballistic Missile Division.

5. Memo from White to Wilson, Oct 30, 1959, fldr Chief of Staff "Signed" Memos, box 26, White Papers, LoC.

6. Memo from ARDC Arnold Air Force Station in Tennessee to Chief of Staff White, Oct 30, 1959, K140–8636.4, AFHRA; Ferer, "Actions on the Dyna Soar—5 Sep 59 thru 6 Nov 59," K140.8636–4, AFHRA.

7. "Statement of Critical Problem Concerning SATURN, DYNA SOAR, and Air Force Space Responsibilities," about Nov 2, 1959, fldr Chief of Staff "Signed" Memos, box 26, White Papers, LoC.

8. Ibid.

9. Glennan and Gates, "Memo for the President: Responsibility and Organization for Certain Space Activities," Oct 21, 1959, fldr McElroy, Neil H. Secy. of Defense, 1959 (1), box 23, PP AWD, DDEL; Notes from Oct 1, 1959 meeting, Oct 2 ,1959, fldr Space January–June 1959 (6), box 15, Special Asst Sci & Tech ser, DDEL.

10. "Statement of Critical Problem Concerning SATURN, DYNA SOAR, and Air Force Space Responsibilities."

11. Bowen, *Threshold of Space,* 160.

12. Ibid.

13. Brig. Gen. Homer A. Boushey, "Blueprints for Space," *Air University Quarterly Review* 11, no. 1 (Spring 1959): 30–37.

14. Ibid.

15. Lt. Col. S. E. Singer, "The Military Potential of the Moon," *Air University Quarterly Review* 11, no. 2 (Summer 1959): 31–53.

16. Boushey, "Blueprints for Space"; Singer, "The Military Potential of the Moon."

17. Boushey, "Blueprints for Space."

18. Gen. Frederic H. Smith, "Nuclear Weapons and Limited War," *Air University Quarterly Review* 12, no. 1 (Spring 1960): 3–27.

19. Singer, "The Military Potential of the Moon."

20. Ibid.

21. Max Rosenberg, *The Air Force in Space, 1959–1960*, USAF Historical Division Liaison Office, June 1962, K168.01–7, AFHRA.

22. Ibid.

23. Ibid.

24. Ibid.

25. Memo from White to AFCCS, "Relations with NASA," Dec 15, 1959, fldr NASA, box 28, White Papers, LoC.

26. T. A. Heppenheimer, *The Space Shuttle Decision, 1965–1972* (Washington, DC: Smithsonian Institution Press, 2002), 51.

27. Roy Houchin, *U.S. Hypersonic Research and Development: The Rise and Fall of Dyna-Soar, 1944–1963* (New York: Routledge, 2006), 122–24.

28. Ibid.

29. "Status of Dyna Soar," Dec 15, 1959, fldr 4–5 Missiles/Space/Nuclear, box 36, White Papers, LoC.

30. "Basic Doctrine AFM" 1–2, 6–8, 13.

31. Maj. Robert D. Newberry, "Space Doctrine for the 21st Century," research paper, Air Command and Staff College, accessed November 26, 2011, http://www.au.af.mil/au/awc/awcgate/acsc/97–0427.pdf; "Basic Doctrine AFM 1–2," 8.

32. Memo to White, "Importance Category List," Feb 5, 1960, fldr 5–1 Air Force Council (Folder #1) Jan–June 1960 correspondence, box 36, White Papers, LoC.

33. Letter from Ennis C. Whitehead to White, March 6, 1960, fldr 4–5 Missiles/Space/Nuclear, box 36, White Papers, LoC.

34. Memo for White, "SAINT Program," Mar 25, 1960, fldr 5–1 Air Force Council (Folder #1) Jan–June 1960 correspondence, box 36, White

Papers, LoC; "Importance Category Lists," August 1, 1960, Air Force Council (Folder #2) Jul–Dec 1960 Correspondence, box 36, White Papers, LoC.

35. Memo for White, "The Threat," Mar 27, 1960, fldr 5–1 Air Force Council (Folder #1) Jan–June 1960 correspondence, box 36, White Papers, LoC.

36. Dr. Kenneth R. Whiting, "The Past and Present of Soviet Military Doctrine," *Air University Quarterly Review* 11, no. 1 (Spring 1959): 38–60.

37. "Current Status Report: System 464L—Dyna Soar," Dec 18, 1959, K140.8636–4, AFHRA; "Current Status Report: 620A Dyna Soar," Feb 1960, K140.8636–4, AFHRA; "Current Status Report: 620A Dyna Soar," April 1960, K140.8636–4, AFHRA; "Dyna Soar Study Phase Completed: Funds Released for Development," April 27, 1960, K140.8636–4, AFHRA.

38. "Information for Dr. Charyk," May 5, 1960, in Godwin, *Dyna-Soar*, 117–19.

39. Memo to White, "Dyna Soar—Phase Alpha Study," fldr 5–1 Air Force Council (Folder #1) Jan–June 1960 correspondence, box 36, White Papers, LoC.

40. Memo to White, "Dyna Soar—Phase Alpha Study."

41. Houchin, *U.S. Hypersonic Research and Development*, 127.

42. Rosenberg, *The Air Force in Space, 1959–1960*, 42.

43. Memo for White, "Space Defense, FY 1961–1970," fldr 5–1 Air Force Council (Folder #1) Jan-June 1960 correspondence, box 36, White Papers, LoC.

44. Houchin, *U.S. Hypersonic Research and Development*, 128.

45. Eugene M. Emme, "Brief on NASA," Spring 1960, K140.8636–5, AFHRA.

46. Ibid.

47. Booklet on Dyna-Soar.

48. Ibid.

49. Memo to White, "Program Review Exercise 62-E," n.d., fldr 5–1 Air Force Council (Folder #2) correspondence, box 36, White Papers, LoC.

50. Ibid.

51. Peebles, *Shadow Flights*.

52. Dwayne A. Day, "The Development and Improvement of the Corona Satellite," in *Eye in the Sky: The Story of the Corona Spy Satellites*, ed. Dwayne A. Day, John M. Logsdon, and Brian Latell (Washington, DC: Smithsonian Institution Press, 1998), 52–59; Rosenberg, *The Air Force in Space, 1959–1960*, 33.

53. Memo for White, "Air Force Exploration of Space Capsule," Aug 12, 1960, fldr 4–6 Information Services Correspondence, box 36, White Papers, LoC.

54. "DOD Office of Public Affairs Fact Sheet: Discoverer Program," Aug 10, 1960, fldr 4–6 Information Services Correspondence, box 36, White Papers, LoC.

55. "Remarks by General Thomas D. White, Chief of Staff, USAF," about Aug 12 1960, fldr 4–6 Information Services Correspondence, box 36, White Papers, LoC.

56. Memo for White, "Air Force Exploration of Space Capsule."

57. Ibid.

58. National Reconnaissance Office (NRO), "Organization and Functions of the Office of Missile and Satellite Systems," accessed May 2, 2011, http://www.nro.gov/foia/declass/NROStaffRecords/736.PDF. The Library of Congress holds a similar, but more complete, organizational chart. "Organization and Functions of the Office of Missile and Satellite Systems," attachment, 4–5 Missiles/Space/Nuclear, box 36, White Papers, LoC. Strangely, sources cannot agree on who first headed the SAFMS. NRO, "NRO Directors," accessed May 2, 2011, http://www.nro.gov/history/directors/dir1.html. This author concludes that SAFMS was headed by Brigadier General Curtin until Charyk and Bissell were handed the organization when it was redubbed NRO.

59. Dudley Sharp, "Memorandum for the Chief of Staff, USAF," Sept 13, 1960, fldr 4–5 Missiles/Space/Nuclear, box 36, White Papers, LoC.

60. Houchin pointed to a specific exchange in October 1963 between Dyna-Soar program engineer Lamar and Defense Secretary McNamara that underscored this exact point: when kept out of the loop, personnel could not effectively respond to the needs or questions of policymakers. Houchin, U.S. Hypersonic Research and Development, 176.

61. "Major General Richard D. Curtin," U.S. Air Force Biographies, accessed June 16, 2011, http://www.af.mil/information/bios/bio.asp?bioID=5139.

62. Houchin, U.S. Hypersonic Research and Development, 132; Sheehan, A Fiery Peace in a Cold War, 455.

63. Memo to White, "Important Category Lists," Aug 1, 1960, fldr 5–1 Air Force Council (Folder #2) Jul-Dec 1960 correspondence, box 36, White Papers, LoC; "Development Directive for 620A DYNA SOAR (Step I) Hypersonic Glider System," Oct 12, 1960, K140.8636–4, AFHRA.

64. Boeing Company, "This is Dyna-Soar," 168.7127–38, AFHRA.

65. Booklet on Dyna-Soar.

Chapter 7. "A Capacity . . . Adequate for Our Own U.S. Purposes"

Epigraph. "Draft No. 2. 'The Scientific Exploration of Outer Space,'" Mar 12, 1958, fldr National Aeronautics and Space Administration–National Aeronautics and Space Council (3), box 4, PSAC, DDEL.

Second epigraph. Notes, "Responsibility for Activities Directed Toward the Development of a Manned Satellite," July 23, 1958, fldr Space Notebook Killian 1958–59 (3), box 15, Special Asst Sci & Tech ser, DDEL.

1. Power, "Perspective at the Dawn of the Space Age," in Emme, *The Impact of Air Power*, 486.

2. "Briefing on Army Satellite Program," Nov 1957, fldr Space November 1957 (2), box 15, Special Asst Sci & Tech ser, DDEL.

3. Notes taken at NSC Meeting, May 1, 1958, document #0433 declassified 2000 NSC, DDRS.

4. "Space Flight Program: Report by the Space Flight Technical Committee of the American Rocket Society," revised Oct 10, 1957, fldr Space October 1957 (1), box 15, Special Asst Sci & Tech ser, DDEL.

5. Letter from Alan Waterman to James Killian, Dec 6, 1957, fldr Space December 1957 (3), box 15, Special Asst Sci & Tech ser, DDEL.

6. Confidential NSF position paper, Dec 6, 1957, fldr Space December 1957 (3), box 15, Special Asst Sci & Tech ser, DDEL.

7. Letter and attachments from Bursk to Cutler, Dec 2, 1957, fldr Department of Defense, Vol. II (3) November-December 1957, box 1, Department of Defense subseries, Staff Secretary Subject ser, DDEL.

8. Ibid.

9. "Memorandum on Organizational Alternatives for Space Research and Development," Dec 30, 1957, fldr National Aeronautics and Space Administration–National Aeronautics and Space Council, box 4, PSAC, DDEL.

10. Telegram, Dec 12, 1957, fldr Department of Defense, Vol. II (3) November-December 1957, box 1, Department of Defense subseries, Staff Secretary Subject ser, DDEL.

11. Ibid.

12. H. Julian Allen and A. J. Eggers, "Research Memorandum: A Study of the Motion and Aerodynamic Heating of Missiles Entering the Earth's Atmosphere at High Supersonic Speeds," 25 August 1953, document I-2, in *Exploring the Unknown*, vol VII, *Human Spaceflight: Projects Mercury, Gemini, and Apollo*, ed. John M. Logsdon (Washington, DC: NASA History Office, 2008), 50–51.

13. Message and attachment from James Douglas to Sherman Adams, Dec 15, 1957, fldr Department of Defense, Vol. II (3) November-December 1957, box 1, Department of Defense subseries, Staff Secretary ser, DDEL.

14. Thomas D. White, "The Inevitable Climb to Space," *Air University Quarterly Review* (Winter 1958–59): 3–4.

15. "Space Flight Program: Report by the Space Flight Technical Committee of the American Rocket Society."

16. Ibid.

17. Emme, *The Impact of Air Power*, 846.

18. Adams, *Eisenhower's Fine Group of Fellows*, 164–65, 185.

19. "Memorandum of Conversation, Washington," Jan 3, 1958, *FRUS*, 1958–1960, vol III, doc 1, 1–3.

20. "Memorandum of Discussion at the 350th Meeting of the National Security Council," Jan 6, 1958, *FRUS*, 1958–1960, vol III, doc 2, 4–7.

21. Ibid.

22. "Statement of Secretary of Defense Neil McElroy before the Committee on Armed Services House of Representatives, 13 January 1958," fldr Preparedness Investigation December 1957, box 22, Staff Sec Alpha ser, DDEL.

23. Quoted in "American Air Policy," in Emme, *The Impact of Air Power*, 604.

24. "Scientific and Reconnaissance Satellites," memo from Robert Cutler to James Killian, Jan 20, 1958, fldr Space—Satellites July 1956–February 1960, box 3, Special Asst Sci & Tech ser, DDEL. n.d. document #2547, declassified 1997, DDRS.

25. Gen. T. D. White and Maj. Gen. B. A. Schriever, "Immediate Prospects in Astronautics," quoted in Emme, *The Impact of Air Power*, 857–61.

26. "Meeting of the Atomic Energy Working Group of the Disarmament Panel," Feb 5, 1958, fldr President's Science Advisory Committee October 1957–March 1958 (2), box 3, Special Asst Sci & Tech ser, DDEL.

27. "Memorandum of Conference with the President, Feb 4, 1958," Feb 6, 1958, document #Ret 914C, DDRS.

28. Ibid.

29. Notes of conversation, Feb 7 1958, fldr Meeting Notes February 1958, box 1, PSAC, DDEL.

30. "Preliminary Observations on the Organization for the Exploitation of Outer Space," memo from S. P. Johnson to Killian, Feb 21, 1958, fldr National Aeronautics and Space Administration–National Aeronautics and Space Council (3), box 4, PSAC, DDEL.

31. Ibid.

32. "Draft No. 2. 'The Scientific Exploration of Outer Space.'"

33. Ibid.

34. Ibid.

35. Draft of "The Scientific Exploration of Outer Space," Mar 7, 1958, fldr Space January-May 1958 (4), box 15, Special Asst Sci & Tech ser, DDEL.

36. Letter from Killian to Eisenhower, Mar 8, 1958, fldr Killian James R. 1957 (1), box 23, PP (AWD), DDEL.

37. Roman, *Eisenhower and the Missile Gap,* 186.

38. When Eisenhower made the $15 million missile comment, Deputy Secretary of Defense Quarles answered that the Titan missile itself cost $1 million–$2 million. Basing cost a further $5 million per missile. The total allocation of $454 million also took into account "years of active development still lying ahead of the Titan missiles." "Memorandum of Discussion at the 363d Meeting of the National Security Council," April 28, 1958, *FRUS,* 1958–1960, vol III, doc 21, 70–78.

39. "Memorandum of Conference with the President," Mar 6, 1958, document #Ret904B, DDRS.

40. "Diary, Mar 26, 1958," fldr ACW Diary March 1958 (1), box 9, PP (AWD), DDEL.

41. "Memorandum of Conference with the President, June 18, 1958, 11:45," document #Ret914A, DDRS.

42. Futrell, *Ideas, Concepts, Doctrine,* 553–54.

43. Edmund Halley had, in the early 18th century, estimated the atmosphere as existing up to forty-five miles in altitude. By the 1950s, scientists measured an "air envelope extend[ing] much more than the odd 200 miles" cited in earlier works. Some contemporary estimates extended to seven hundred or even one thousand miles. Furthermore, "If the atmosphere rotated as a body with the Earth it would be easy to obtain the maximum height. It would be that at which centrifugal force would balance gravitational force. In the case of the Earth this comes to a height of about 21,000 miles (34,000 km.) above sea level at the equator." Eric Burgess, *Frontier to Space* (New York: Macmillan, 1956), 47.

44. "LS No. 10662," April 21, 1958, document #0300, declassified 2001, DDRS.

45. Bissell, *Reflections of a Cold Warrior,* 113–14.

46. The first U-2 flight over the Soviet bloc occurred on June 20, 1956, when a plane overflew Poland, East Germany, and Czechoslovakia in order to test the Soviet detection ability and air defense reaction. From the start, the Soviets were able to detect the flights, though until Gary Francis Powers' May 1 mission, the Soviets were unable to bring one of these planes down. A CIA document following the shootdown lists all of the U-2 missions over the Soviet bloc. A total of fifty missions were completed during this period. Fourteen were directed solely toward the USSR, and eleven overflew the USSR as well as at least one other communist state. The other twenty-five overflew communist China,

other portions of Southeast Asia, or communist states in Eastern Europe without passing over the USSR itself. "U-2 Overflights of Soviet Bloc," Aug 18, 1960, document #0022, declassified 2000, DDRS.

47. "Conference in the President's Office, 8:30 a.m., Oct 8, 1957," document #3444, declassified 1993, DDRS; DOD memo for the President, "Earth Satellite," Oct 7, 1957, #0745, declassified 1999 DOD, DDRS.

48. Killian believed that even an unlimited amount of aerial surveillance by itself would be insufficient to guarantee national security, asserting that such flights would not be able to extend U.S. warning of an incoming attack to more than the time taken by attacking missiles to make their journey. Letter from Killian to J. F. Dulles, July 10, 1958, document #2944, declassified 1993 WH, DDRS.

49. A prohibition on space weapons and territorial claims on heavenly bodies would be agreed in 1967, and a ban on atmospheric and in-space nuclear tests would occur in 1963, but in the late 1950s, neither could be taken for granted.

50. Robert J. Watson, *History of the Office of the Secretary of Defense,* vol IV, *Into the Missile Age, 1956–1960* (Washington, DC: Office of the Secretary of Defense Historical Office, 1997), 391.

51. "National Security Council: U.S. Policy on Outer Space, NSC 5814," June 20, 1958, document #345, declassified 1992 NSC, DDRS.

52. Historian Robert Divine deemed these objectives to be "contradictory." Divine, *The Sputnik Challenge,* 154.

53. "Discussion at the 371st NSC Meeting on Thursday, July 3, 1958," July 5, 1958, document #2801, declassified 1996 NSC, DDRS.

54. Ibid.

55. Ibid.

56. "National Security Council: U.S. Policy on Outer Space, NSC 5814."

57. Adelbert O. Tischler to NACA Associate Director, "Minimum Man-In-Space Proposals Presented at WADC, 29 January 1958 to 1 February 1958," 10 April 1958, document I-3, in Logsdon, *Exploring the Unknown,* vol VII, *Human Spaceflight,* 51–67.

58. Memo for Killian, notes of the Space Panel Meeting of July 2, 1958, fldr Space Notebook Killian, 1958–59 (2), box 15, Special Asst Sci & Tech ser, DDEL.

59. Maurice H. Stans to Dwight Eisenhower, "Responsibility for 'Space' Programs," 10 May 1958, Document I-5, in Logsdon, *Exploring the Unknown,* vol VII, *Human Spaceflight,* 73; "Outer Space Notes (Talks with Schriever, ARPA, [illegible], Purcell, Dryden)," June 10, 1958, fldr June 1958 (1), box 5, Special Asst National Security Affairs Special Asst ser, DDEL.

60. "Outer Space Notes (Talks with Schriever, ARPA, [illegible], Purcell, Dryden)."

61. Memo from Piland to Killian, July 7, 1958, fldr Space Notebook Killian, 1958059 (1), box 15, Special Asst Sci & Tech ser, DDEL.

62. York explained that "if manned exploration to Mars, for instance, was a national objective, then it would be reasonable to put 'man-in-space' in the civil category." "NASA-ARPA Combined Budget for Space Activities Exclusive of Military Reconnaissance (117L): Fiscal Year 1959," July 10, 1958, fldr Space Notebook Killian, 1958–59 (3), box 15, Special Asst Sci & Tech ser, DDEL.

63. Notes, "Responsibility for Activities Directed Toward the Development of a Manned Satellite"; Memo from Piland to Killian, "Notes of Space Budget and Responsibility Meeting of July 10, 1958," July 11, 1958, fldr Space Notebook Killian 1958–59 (3), box 15, Special Asst Sci & Tech ser, DDEL; Space Notebook, approximately July 28, 1958, fldr Space Notebook Killian, 1958–59 (1), box 15, Special Asst Sci & Tech ser, DDEL.

64. Memo from Piland to Killian, notes of "Possible DOD-NASA Responsibility Relationships," July 16, 1958, fldr Space Notebook Killian, 1958–59 (3), box 15, Special Asst Sci & Tech ser, DDEL.

65. Notes, "Responsibility for Activities Directed Toward the Development of a Manned Satellite."

66. Memo from Leghorn to PSAC Beckler, "Political Action and Unauthorized Overflight of the USSR Field: Space-Satellite," July 26, 1958, fldr Space—Satellites July 1956–February 1960, box 3, Special Asst Sci & Tech ser, DDEL.

67. Sylvia Kraemer, *Science and Technology Policy in the United States: Open Systems in Action* (New Brunswick: Rutgers University Press, 2006), 152.

68. "Some Aspects of the Problem of Surprise Attack," Aug 12, 1958, fldr Surprise Attack August-September 1958 (2), box 24, Staff Sec Alpha ser, DDEL.

69. Memo from Leghorn to PSAC Beckler, "Political Action and Unauthorized Overflight of the USSR Field: Space-Satellite."

70. "Discussion at the 374th Meeting of the National Security Council, Thursday, July 31, 1958," Aug 1, 1958, document #1026, declassified 197 NSC, DDRS.

71. "Diary, August 25, 1958," fldr August ACW Diary–1958 (1), box 10, PP (AWD), DDEL; letter attachment from Symington to Eisenhower, Aug 29, 1958, fldr Symington Letter, box 24, Staff Sec Alpha ser, DDEL.

72. "Effects of Nuclear Weapon Explosions in the Outer Atmosphere: A Brief Status Report," July 24, 1958, document #0285, declassified 2000, DDRS.

73. "Discussion at the 376th Meeting of the National Security Council, Thursday, August 14, 1958," Aug 15, 1958, document #2771, declassified 1997, DDRS.

Chapter 8. "Satellites Are Our Last Chance"

Epigraph. Gordon Gray, "Handwritten notes of meeting at Augusta, Georgia, Nov. 28, 1958 re defense expenditures," Nov 28, 1958, fldr Meetings with the President 1958 (4), box 3, Special Assistant for National Security Affairs, Special Assistant ser, DDEL.

1. "Diary: October 13, November 2 period [1958]," fldr ACW Diary—Nov 1958, box 10, PP AWD, DDEL; Divine, *The Sputnik Challenge*, 185, 196–97.

2. Wagner, *Eisenhower Republicanism*, 123.

3. Memo from Piland to Killian, "Dyna-Soar," Nov 3, 1958, fldr Missiles April-December 1958 (3), box 12, Special Asst Sci & Tech ser, DDEL.

4. Ibid.

5. Ibid.

6. Killian, speech script, Sept 26, 1958, fldr Space—Speeches September 1958, box 3, Special Asst Sci & Tech ser, DDEL.

7. "National Space Activities: A Brief Summary Prepared for the National Aeronautics and Space Council," Sept 12, 1958, document #0974, declassified 1998 NASA, DDRS; "Report of the Working Group on Space Research Objectives, Special Committee on Space Technology," 14 November 1958, Document I-2, in Logsdon, *Exploring the Unknown*, vol VI, *Space and Earth Science*, 52–53.

8. Memo by Edwin H. Land, Courtland D. Perkins, Edward M. Purcell, Allen F. Donovan, and H. Guyford Stever to James R. Killian, Nov 15, 1958, document #2366, declassified 2000 WH, DDRS.

9. The D-21 drone and sole remaining M-21 modified SR-71 are on display at the Museum of Flight in Seattle, Washington. Chertok, *Rockets and People,* vol II, 546.

10. White and Schriever, "Immediate Prospects in Astronautics," in Emme, *The Impact of Air Power*, 857; Lyndon Johnson, "International Control of Outer Space," November 17, 1958, in Emme, *The Impact of Air Power,* 719.

11. Gray, "Handwritten notes of meeting at Augusta, Georgia, Nov. 28, 1958 re defense expenditures."

12. Ibid.

13. Letter from Quarles to NASA Administrator Glennan, Nov 28, 1958, fldr Space Council (10), box 24, Staff Sec Alpha ser, DDEL; Letter from NASA Administrator Glennan to Sec Def McElroy, Dec 1, 1958,

fldr Space Council (10), box 24, Staff Sec Alpha ser, DDEL; Memo by Goodpaster to Hagerty, Nov 26, 1958, fldr National Aeronautics and Space Administration September 1958–January 1961 (3), box 18, Staff Sec Alpha ser, DDEL; "Memorandum of Discussion at the 375th Meeting of the National Security Council," Aug 7, 1958, *FRUS, 1958–1960*, vol III, doc 32, 132–35.

14. Memo from Colonel Seedlock to Murray Snyder, "Jupiter Shot on 12 December," Dec 9, 1958, fldr National Aeronautics and Space Administration September 1958–January 1961 (5), box 18, Staff Sec Alpha ser, DDEL; undated letter by Asst Sec of Near Eastern, South Asian and African Affairs William M. Roundtree to Asst Sec of Defense for International Security Affairs Mansfield D. Sprague; Reply letter by J. N. Irwin to Roundtree, May 19, 1958, fldr National Aeronautics and Space Administration September 1958–January 1961 (5), box 18, Staff Sec Alpha ser, DDEL; "Public Information Handling of Biomedical Experiments for Jupiter IRBM Test Vehicle No. 18," May 27, 1959, fldr National Aeronautics and Space Administration September 1958–January 1961 (5), box 18, Staff Sec Alpha ser, DDEL; "Notes on pre-press briefing, June 3, 1959," fldr ACW Diary—June 1959 (2), box 10, PP AWD, DDEL.

15. Letter from Quarles to Eisenhower, Aug 7, 1958, fldr McElroy, Neil 1957–58 (1), box 23, PP AWD, DDEL.

16. Memo from Piland to Killian, "Atlas 'Carcass' Satellite," Nov 6, 1958, fldr Missiles April-December 1958 (3), box 12, Special Asst Sci & Tech ser, DDEL.

17. "Intelligence Notes," Jan 9, 1959, document #2095, declassified 1985 WH, DDRS.

18. "DOD minutes of press conference held by Mr Roy Johnson, Director, ARPA," Dec 3, 1958, fldr National Aeronautics and Space Administration September 1958–January 1961 (3), box 18, Staff Sec Alpha ser, DDEL.

19. Letter from C. G. Villard to V. L. Berkner, Jan 22, 1959, fldr Space January-June 1959 (6), box 15, Special Asst Sci & Tech ser, DDEL.

20. Dockrill, *Eisenhower's New-Look National Security Policy*, 212.

21. "Memorandum of Conversation: Subject: DCI Briefing of Senator Stuart Symington on the Soviet Ballistic Missile Programs and Capabilities," Dec 16, 1958, fldr Symington Letter, box 24, Staff Sec Alpha ser, DDEL.

22. Roman, *Eisenhower and the Missile Gap*, 131.

23. John Keegan, *Intelligence in War: Knowledge of the Enemy from Napoleon to Al-Qaeda* (New York: Alfred A. Knopf, 2003).

24. Herbert York, *Making Weapons, Talking Peace: A Physicist's Odyssey from Hiroshima to Geneva* (New York: Basic Books, 1987), 123.

25. "Editorial Note" for Jan 12, 1959, *FRUS,* 1958–1960, vol III, 172–73.

26. "Memorandum of Conference with the President, February 17, 1959," Feb 24, 1959, fldr National Aeronautics and Space Administration (September 1958–January 1961) (5), box 18, Alpha subser, Staff Sec Subject ser, DDEL.

27. "Operations Coordinating Board: Operating Plan for Outer Space," Mar 18, 1959, document #3540, declassified 2000 WH, DDRS.

28. Ibid.

29. Ibid.

30. George Low, Memo to House Committee on Science and Astronautics, "Urgency of Project Mercury," 27 April 1959, Document I-23, in Logsdon, *Exploring the Unknown,* vol VII, *Human Spaceflight,* 137; NASA and DOD, "A National Program to Meet Satellite and Space Vehicle Tracking and Surveillance Requirements for FY 1959 and 1960," unclassified April 1959, document #2759, declassified 2001 DOS, DDRS; "Minutes of Meeting of Research Steering Committee on Manned Space Flight," 25–26 May 1959, Document II-2, in Logsdon, *Exploring the Unknown,* vol VII, *Human Spaceflight,* 441.

31. "Political Action and Satellite Reconnaissance," April 24, 1959, fldr Space January-June 1959 (6), box 15, Special Asst Sci & Tech ser, DDEL.

32. Memo from Killian to Eisenhower, "Memorandum on Some Technical Factors Involved in Policy Decisions on Arms Limitations and Specifically on the Limitation of Nuclear Testing," Mar 11, 1959, fldr Killian, James R. 1957 (1), box 23, PP AWD, DDEL.

33. National Aeronautics and Space Council, "Meeting of 27 April 1959," document #0390, declassified 2000 NASA, DDRS.

34. Ibid.

35. Ibid.

36. "Memorandum of Conference with the President, June 30, 1959," July 2 1959, fldr National Aeronautics and Space Administration September 1958–January 1961 (6), box 18, Staff Sec Alpha ser, DDEL.

37. "Meeting of 30 June 1959: the sixth meeting of the National Aeronautics and Space Council," July 8, 1959, fldr Space Council (9), box 24, Staff Sec Alpha ser, DDEL.

38. "Memorandum of Conference with President Eisenhower," July 14, 1959, *FRUS,* 1958–1960, vol III, doc 65, 253–355.

39. For their own part, NASA leaders abided by the president's wishes. Glennan recalled that he had decided against sending a probe to Venus the previous June because the technological difficulties made the project "an unwise gamble." "Memorandum of Conference with the President,

July 15, 1959, 12:15 PM," July 15, 1959, fldr National Aeronautics and Space Administration September 1958–January 1961 (6), box 18, Staff Sec Alpha ser, DDEL.

40. "Memorandum of Discussion at the 417th Meeting of the National Security Council," Aug 18, 1959, *FRUS, 1958–1960*, vol III, doc 72, 319–23.

41. Memo from Thomas Gates to Andrew Goodpaster, "Military Department Contracts for Space Activities," July 27, 1959, fldr Department of Defense, Vol. III (7) July-August 1959, box 1, Secretary Subject Series Department of Defense, DDEL; "Notes of Matters Discussed with the President at Meeting on July 14, 1959," July 15, 1959, fldr Science Advisory Committee (6), box 23, Staff Sec Alpha ser, DDEL; "Consolidated List of Military Department Contracts for Space Activities Under ARPA Cognizance within $500,000," July 2, 1959, fldr Department of Defense, Vol. III (7) July-August 1959, box 1, Secretary Subject Series Department of Defense, DDEL.

42. "Memorandum of Conference with the President, July 8, 1959," document #1774, declassified 2001 WH, DDRS.

43. "Memorandum of Conference with the President, July 20, 1959," document #2941, declassified 2000 WH, DDRS.

44. Killian, *Sputnik, Scientists, and Eisenhower*, 142, 205, 207, 212.

45. W. H. Shapley to Reid, "Comments on revision of Outer Space policy," July 16, 1959, 12456 BOB, NASA HQ; "Discussion at the 415th Meeting of the National Security Council, Thursday, July 30, 1959," document #2275, declassified 1999 NSC, DDRS.

46. "Discussion at the 415th Meeting of the National Security Council, Thursday, July 30, 1959," document #2275, declassified 1999 NSC, DDRS.

47. "Operations Coordinating Board, Operating Plan for Outer Space," March 18, 1959, document #3540, declassified 2000 WH, DDRS.

48. "Discussion at the 415th Meeting of the National Security Council, Thursday, July 30, 1959." Outer space would be addressed in paragraph 63 of the resulting document, NSC 5906/1. No comparative terms, "comparable," "equal," or any other, appeared in the final version of the space policy paragraph, which stated an interest in "broad-based" programs "consistent with U.S. security" in ways conforming with factors including international cooperation and NASA authority. "National Security Council Report: NSC 5906/1," August 5, 1959, *FRUS, 1958–1960*, vol III, doc 70, 314–15.

49. Ibid.

50. "To Race or Not to Race? (A Discussion Paper)," Oct 16, 1959, fldr Space January-June 1959 (6), box 15, Special Asst Sci & Tech ser, DDEL.

51. "Memorandum for the Chairman, Joint Chiefs of Staff: Subject: Coordination of Satellite and Study Vehicle Operations," Sept 15, 1959, fldr Kistiakowsky, Dr G.B. (2), box 23, PP AWD, DDEL.

52. Gordon Gray, "Memorandum for the President," Sept 21, 1959, fldr 1960—Meetings with President—Volume 2 (6), box 5, Presidential subser, Special Asst National Security Affairs Special Asst ser, DDEL.

53. "Memo for the President: Responsibility and Organization for Certain Space Activities," Oct 21, 1959, fldr McElroy, Neil H. Secy. of Defense, 1959 (1), box 23, PP AWD, DDEL; Notes from Oct 1, 1959 meeting by Kistiakowsky, Glennan, Horner, Silverstein, Purcell, and Hornig, Oct 2, 1959, fldr Space January-June 1959 (6), box 15, Special Asst Sci & Tech ser, DDEL.

54. Ibid.

55. A. J. Goodpaster, Memo of Conference with the President, 29 September 1959, Document I-28, in Logsdon, *Exploring the Unknown,* vol VII, *Human Spaceflight,* 146; letter from Glennan to Eisenhower, Nov 16, 1959, fldr Space Council (11), box 24, Staff Sec Alpha ser, DDEL.

56. Letter from Glennan to Eisenhower, Dec 3, 1959, fldr Space Council (11), box 24, Staff Sec Alpha ser, DDEL; Goodpaster, "Memorandum of Conference with the President, November 17, 1959," Dec 1, 1959, fldr Space Council (11), box 24, Staff Sec Alpha ser, DDEL.

57. Goodpaster, "Memorandum of Conference with the President, November 17, 1959"; "Memorandum of Discussion at the 425th Meeting of the National Security Council," Nov 25, 1959, *FRUS,* 1958–1960, vol III, doc 79, 339–52.

58. Low, "Biological Payloads and Manned Space Flight," in Logsdon, *Exploring the Unknown,* vol VI, *Space and Earth Science,* 309; "Final Draft, U.S. Policy on Outer Space," Nov 12, 1959, document #1141, declassified 1997 WH, DDRS.

59. Memo from James Lay to Gordon Gray, "Dynasoar," Nov 25, 1959, document #2953, declassified 1998 WH, DDRS.

60. Ibid.

61. Ibid.

Chapter 9. "Slipping Out of Control"

Epigraph. Edward H. Kolcum, "Services Challenge NASA's Dominant Role," November 28, 1960, *Aviation Week.*

Second epigraph. Philip Coombs, "Public Opinion Pressures on Education Today," *Public Opinion Quarterly* 24, no. 3 (1960): 470–73.

1. Evert Clark, "U.S. Space Officials Rap Lack of Urgency," November 23, 1959, *Aviation Week.*

2. "Information for Dr. Charyk," May 5, 1960, in Godwin, *Dyna-Soar: Hypersonic Strategic Weapons System*, 117–19.

3. Clark, "U.S. Space Officials Rap Lack of Urgency."

4. Ibid.

5. "Memorandum of Conference with President Eisenhower," November 16, 1959, *FRUS, 1958–1960*, vol III, doc 78, 333–39.

6. John F. Kennedy, *The Strategy of Peace*, ed. Allan Nevins (New York: Harper, 1960), 37–38, 169–71,183–86.

7. "Russians Credited with Space Bomber," *New York Times*, January 9, 1960, 3; Edward Gamarekian, "Reds Raise Specter of Space Arm," *Washington Post*, January 15, 1960, 8.

8. Robert Hotz, "Editorial: Military Space Problems," *Aviation Week*, November 30, 1959, and "Editorial: A Dangerous Decision," *Aviation Week*, December 7, 1959.

9. Franklyn W. Phillips, "Estimated Funding Requirements of Military Programs Using Space Sub-Systems," December 22, 1959, fldr Space Council (11), box 24, Staff Sec Alpha, DDEL.

10. "U.S. Policy on Outer Space," December 16, 1959, fldr Space Council (11), box 24, Staff Sec Alpha, DDEL; National Security Council, "U.S. Policy on Outer Space," December 17, 1959, doc #0347 declass. 1992 NSC, DDRS.

11. "President's Advanced Planning Views," *Aviation Week*, December 7, 1959; J. S. Butz, "Budget Cuts Force Stretchout of B-70," *Aviation Week*, December 7, 1959.

12. Ibid. Transport plane, an all-weather interceptor, cruise missile carrier, parasite bomber carrier, and bomber were among the other identified tasks that a B-70 could fulfill.

13. Robert Hotz, "Editorial: Beginning a New Decade," *Aviation Week*, January 4, 1960; Butz, "Budget Cuts Force Stretchout of B-70."

14. Evert Clark, "Costs Crimp Nation's Top Space Project," *Aviation Week*, December 28, 1959; Butz, "Budget Cuts Force Stretchout of B-70;" Dwight B. Schears, "Cutbacks Darken Scene in Seattle," *New York Times*, January 11, 1960, 150.

15. "Briefing Note," January 14, 1960, *FRUS, 1958–1960*, vol III, doc 83, 360–61; "Memorandum of Conference with the President," January 14, 1960, fldr National Aeronautics and Space Administration September 1958–January 1961 (7), box 18, Staff Sec Alpha, DDEL.

16. Robert Hotz, "Editorial: Small Steps Taken Slowly," *Aviation Week*, January 25, 1960; "B-70 Rescue Operations," *Aviation Week*, January 18, 1960; Kistiakowsky to Eisenhower, "Problems of the B-70 Project," *Aviation Week*, February 12, 1960.

17. Special Asst to Sec of State Philip J Farley to Kistiakowsky, "Proposed Prohibition of Weapons in Space Vehicles," Feb 2, 1960, doc #1934 declassified 1983 DOS, DDRS; Robert Hotz, "Editorial: Changing the Mixture," *Aviation Week*, April 4, 1960.

18. Robert Hotz, "Editorial: The Space Debate," *Aviation Week*, February 1, 1960; "Soviet Bomber Threat," *Aviation Week*, February 1, 1960; Goodpaster, memo, February 8, 1960, fldr Missiles January 1956–January 1960 (8), box 18, Staff Sec Alpha, DDEL; Ford Eastman, "Attacks Increase on U.S. Defense Policy," *Aviation Week*, February 15, 1960; "Washington Roundup," *Aviation Week*, February 22, 1960; Ford Eastman, "Space Lag Blamed on Conservative Effort," *Aviation Week*, February 29, 1960.

19. The others were General Dynamics vice president Thomas Lanphier and former Gaither Committee member James P. Baxter III. Ford, "Space Lag Blamed on Conservative Effort"; "Washington Roundup," *Aviation Week*, March 21, 1960.

20. Kistiakowsky to General Persons, memo, March, 12, 1960, doc #1155 declassified 2000 WH, DDRS.

21. Wang, *In Sputnik's Shadow*; Greene, *Eisenhower, Science Advice, and the Nuclear Test-Ban Debate*; Killian, *Sputnik, Scientists, and Eisenhower*.

22. Rod Serling, *The Twilight Zone* episodes, on DVD (Hollywood: Cayuga Productions, 2010): "Where Is Everybody," October 2, 1959; "The Lonely," November 13, 1959; "And When the Sky Was Opened," December 11, 1959; "Third from the Sun," January 8, 1960; "I Shot an Arrow into the Air," January 15, 1960; "The Monsters are Due on Maple Street," March 4, 1960; "People are Alike All Over," March 25, 1960.

23. Lewis J. Rachmil, creator, *Men into Space*, "Burnout," December 9, 1959, YouTube.com, https://www.youtube.com/watch?v=MfIIzE5CFDk.

24. *Men into Space* episodes, "Moon Landing," October 7, 1959; "Moonquake," November 11, 1959; "Earthbound," January 27, 1960.

25. Greene, *Eisenhower, Science Advice, and the Nuclear Test-Ban Debate*.

26. Ibid., 3–7, 87–111.

27. Burton Crane, "Stocks End Week in Retreat Again," *New York Times*, April 23, 1960, 27, and "Market Set Back as Rally Expires," *New York Times*, April 28, 1960, 49.

28. *Men into Space* episodes, "Verdict in Orbit," March 16, 1960; "Is There Another Civilization?" March 23, 1960.

29. "Washington Roundup," *Aviation Week*, April 25, 1960; "Dyna-Soar Flight Test Schedule Detailed," *Aviation Week*, May 9, 1960.

30. Geiger, "History of the X-20 Dyna-Soar," October 1963, in Godwin, *Dyna-Soar: Hypersonic Strategic Weapons System*, 353; "Washington

Roundup," *Aviation Week*, April 11, 1960; "Washington Roundup," *Aviation Week*, May 16, 1960; "Industry Observer," *Aviation Week*, May 2, 1960.

31. Memo from James Lay to Gordon Gray, "Dynasoar," Nov 25, 1959, doc #2953, declassified 1998 WH, DDRS.

32. Notes, May 11, 1960, fldr Space January-June 1959 (6), box 15, Special Asst Sci & Tech, DDEL.

33. Editorial note, "The President's Science Advisory Committee, 'Report of the Ad Hoc Panel on Man-in-Space,'" December 16, 1960, NASA Historical Reference Collection, History Office, NASA Headquarters, Washington, DC, accessed October 3, 2011, http://www.hq.nasa.gov/office/pao/History/report60.html. Hornig's associates on the panel were nuclear scientist and PSAC member Dr. Malcolm Hebb, PSAC consultant Lawrence Hyland, Dr. Donald Ling, Bell Telephone director of military research Dr. Brockway McMillan, and astronomer Dr. Martin Schwarzschild. Douglas Lord served as technical assistant.

34. "Lockheed U-2 Research Aircraft Based in Japan," *Aviation Week*, May 2, 1960; "U-2 Missing; Soviets Say U.S. Plane Down," *Aviation Week*, May 9, 1960.

35. Robert Hotz, "Editorial: Lockheed U-2 Over Sverdlovsk: A Study in Fabrication," *Aviation Week*, May 16, 1960; "U-2 Missions Approved by Eisenhower," *Aviation Week*, May 16, 1960.

36. Robert Hotz, "Editorial: Military vs. Civil Space Requirements," *Aviation Week*, June 6, 1960; Robert Hotz, "Editorial: Top Priority for Military Space Systems," *Aviation Week*, May 30, 1960; "Soviets Boast of Impregnable Defenses," *Aviation Week*, June 6, 1960; Katherine Johnsen, "President Blamed for Loss of U-2 Program," *Aviation Week*, July 4, 1960; Evert Clark, "U-2 Data Aided U.S. Military Planning," *Aviation Week*, June 6, 1960; Gray, "Memorandum of Conversation with the President, Tuesday May 24, 1960 at 8:40am," *Aviation Week*, May 25, 1960.

37. Hotz, "Editorial: Top Priority for Military Space Systems."

38. Goodpaster, "Memorandum of Conference with the President, May 26, 1960," doc #2973 declassified 1999 WH, DDRS.

39. Eisenhower to Gates, memo, June 10, 1960, doc #1269 declassified 1997 DOD, DDRS; "The Samos Program," July 29, 1960, doc #0308 declassified 2000 DOS, DDRS.

40. Series of stories on the Strategic Air Command, *Aviation Week*, June 20, 1960; "USAF Space Pilots," *Aviation Week*, June 13, 1960; "Dyna-Soar Research," *Aviation Week*, July 11, 1960; J. S. Butz, "USAF Builds Capability for Hypersonic Development," *Aviation Week*, July 25, 1960;

Russell Hawkes, "Scout Variant to Have Dyna-Soar Role," *Aviation Week*, July 1, 1960.

41. Philip J. Klass, "Value of Bombardment Satellites Debated," *Aviation Week*, July 11, 1960.

42. Norstad to Herter, memo, August 9, 1960, doc #1239 declassified 2002 DOD, DDRS.

43. Gates to Eisenhower, memo, August 18, 1960, doc #0195 declass 1983 DOD, DDRS; "Summary: June, July, August 1960: Discoverer Project (Research and Development Satellite)," October 24, 1960, fldr Gates Thomas S Jr 1959–61 (2), box 15, AWD, DDEL.

44. Richard Witkin, "Russian Expected as First Orbit; U.S. Scientists Call Retrieval of Animals Major Stride Toward Human Trip," *New York Times*, August 21, 1960, 32; Larry Booda, "First Capsule Recovered from Satellite," *Aviation Week*, August 22, 1960; "U.S. Plans to Put Monkey in Space," *Aviation Week*, August 27, 1960.

45. Robert Hotz, "Editorial: Defense is Now a Key Issue," *Aviation Week*, August 1, 1960; "Washington Roundup," *Aviation Week*, August 8, 1960; "Washington Roundup," *Aviation Week*, August 29, 1960; Ford Eastman, "Democrats Study Defense Reorganization," *Aviation Week*, August 15, 1960. Former ambassador to France David K. E. Bruce and Gaither Committee member James Perkins would also help advise Kennedy. Gilpatric was to do double duty as a defense reviewer, accompanied by former Air Force Secretary Thomas Finletter, Truman's special council Clark Clifford, and former counsel to the Secretary of Defense Marx Leva. Ford Eastman, "Kennedy Creates Defense Planning Group," *Aviation Week*, September 5, 1960; "Kennedy Names Defense Reviewers," *Aviation Week*, September 19, 1960; McFarland, *Cold War Strategist: Stuart Symington*, 110.

46. Text box, *Aviation Week*, August 15, 1960.

47. Robert Hotz, "Editorial: Space Technology Comes of Age," *Aviation Week*, September 5, 1960.

48. "USAF Space Required Operational Capabilities," September 16, 1960, fldr 5–1 Air Force Council (Folder #2) Jul-Dec 1960 correspondence, box 36, White Papers, LoC.

49. Maj. Abbott C. Greenleaf to White, memo, n.d. (late 1960), fldr 7–3 Congressional, box 39, White Papers, LoC; Olson, *Stuart Symington*, 352.

50. Ibid.

51. Lawrence E. Davies, "Air Chief Terms Space Lead Vital," *Aviation Week*, September 23, 1960.

52. "Comments on Dynasoar," September 17, 1960, fldr Missiles July-September 1960 (6), box 12, Special Asst Sci & Tech, DDEL.

53. Rathjens to Kistiakowsky, "Dinosaur," September 19, 1960, fldr Missiles July-September 1960 (6), box 12, Special Asst Sci & Tech, DDEL.

54. Rathjens to Strategic Systems Panel, memo, September 23, 1960, doc #1143 declassified 1988 WH, DDRS.

55. "Memorandum of Conference with President Eisenhower," September 19, 1960, *FRUS*, 1958–1960, vol III, doc 263, 909–10; Larry Booda, "New Capsule to Be Developed for Samos," *Aviation Week*, September 19, 1960; Gray, "Memorandum of Meeting with the President (Thursday, 6 October 1960 at 8:45am)," October 6, 1960, fldr 1960—Meetings with President—Volume 2 (4), box 5, Presidential Subseries, Special Assistant for National Security Affairs, Special Asst ser, DDEL.

56. Larry Booda, "Hardware, Training, Operation Plan Takes Shape in Dyna-Soar Project," *Aviation Week*, September 26, 1960.

57. Coombs, "Public Opinion Pressures on Education Today."

58. Leo Egan, "Upstate Crowds Cheer Kennedy," *New York Times*, September 29, 1960; "Washington Roundup," *Aviation Week*, October 3, 1960.

59. "256: The President's News Conference, August 10, 1960," *The American Presidency Project*, accessed July 29, 2013, http://www.presidency.ucsb.edu/ws/index.php?pid=11902.

60. "268: The President's News Conference, August 24, 1960," *The American Presidency Project*, accessed July 29, 2013, http://www.presidency.ucsb.edu/ws/index.php?pid=11915.

61. Robert Hotz, "Editorial: Where Do They Stand?" *Aviation Week*, October 10, 1960.

62. Robert Hotz, "Editorial: Nixon Takes a Stand," *Aviation Week*, October 17, 1960; Robert Hotz, "Kennedy on National Defense," *Aviation Week*, October 24, 1960; "Washington Roundup," *Aviation Week*, October 31, 1960.

63. "Nixon Claims U.S. Dominates Space Race," *Aviation Week*, October 31, 1960.

64. Gabriel A. Almond, "Public Opinion and the Development of Space Technology," *Public Opinion Quarterly* 24, no. 4 (1960): 553–72.

65. Geiger, "History of the X-20 Dyna-Soar," in Godwin, *Dyna-Soar: Hypersonic Strategic Weapons System*, 354.

66. To date, the United States had launched twenty-five satellites; thirteen remained in orbit, and seven still functioned. The Soviet Union was known to have launched eight; two were still in orbit, and neither was working. "Excerpts of Remarks by Secretary of Defense Thomas S. Gates

to the Annual Convention of the U.S. Postmasters of America in Miami, Florida, at about 10:30am, EST, 25 October 1960," fldr Gates Thomas S Jr 1959–61 (2), box 15, AWD, DDEL.

67. Ibid.

68. Ibid.

Chapter 10. "A Thousand Drawing Boards"

Epigraph. "Memorandum of Discussion of the 469th National Security Council," *FRUS*, 1958–1960, vol III, doc 129, 494–508.

Second epigraph. Killian, *Sputnik, Scientists, and Eisenhower*, 238.

1. Parthian archers of ancient Iran were known for having shot arrows at their enemies and escaping on horseback. Robert Hotz, "Editorial: The Parthian Shots," *Aviation Week*, January 23, 1961.

2. Although two B-70 prototypes were built and one crashed after a smaller plane was sucked into its slipstream and collided with it, no other B-70s were ever built. Aside from a half-built prototype, no Dyna-Soar ever appeared, either.

3. Larry Booda, "USAF Plans Radical New Space Plane," *Aviation Week*, October 31, 1960.

4. Ibid.

5. Ultimately, *Aviation Week* would publish five articles about the aerospace plane: on October 31 and December 26, 1960, on June 19 and November 6, 1961, and on April 23, 1962. Heppenheimer, *The Space Shuttle Decision*, 79.

6. D. R. Lord to Kistiakowsky, "NSC Planning Board briefing on DOD Space Activities," Nov 3, 1960, fldr Space July-December 1960 (9), box 15, Special Asst Sci & Tech ser, DDEL; D. Z. Beckler to Kistiakowsky, "Comments on Lord's memorandum re NSC Planning Board Briefing on DOD Space Activities," Nov 4, 1960, fldr Space July-December 1960 (9), box 15, Special Asst Sci & Tech ser, DDEL.

7. Robert Hotz, "Editorial: Gathering Storm Over Space," *Aviation Week*, November 7, 1960; Booda, "USAF Plans Radical New Space Plane."

8. "Memorandum of Meeting with the President (Thursday, November 8, 1960 at 9:15am)," Nov 8, 1960, fldr 1960—Meetings with President—Volume 2 (4), box 5, Presidential Subseries, Special Asst National Security Affairs Special Asst ser, DDEL; "Discussion at the 466th Meeting of the National Security Council, Monday, November 7, 1960," November 8, 1960, doc #1031 declassified 1999 NSC, DDRS.

9. "Sharp Defense, Space Challenges Expected," *Aviation Week*, November 14, 1960.

10. General Electric research focused on a direct cycle nuclear engine, which was comparatively lightweight and straightforward from an engineering perspective. However, it risked contaminating the airspace in which it flew. Westinghouse research aimed toward an indirect cycle nuclear engine that vastly reduced contamination risks but increased the complexity and weight of the overall system. "Possible Further Adjustments—FY 1962 Budget," Dec 5, 1960, fldr Gates, Thomas S Jr 1959–61 (2), box 15, AWD, DDEL.

11. The meeting notes did not clarify how the totals had been calculated, but this was the figure requested by the Air Force for fiscal year 1962.

12. Kolcum, "Services Challenge NASA's Dominant Role."

13. Ibid.; memo, technical asst to director of Launch Vehicle Programs Col. USAF Daniel D. McKee to Mr. Hyatt, "Pursuing the Space Program," Nov 25, 1960, fldr 11130: DOD in Space, NASA HQ Archives.

14. John W. Finney, "Air Force Seeks Top Role in Space; Drafts Publicity Offensive Keyed to the Change in Administrations," *New York Times*, December 11, 1960, 68; "Washington Roundup," *Aviation Week*, December 5, 1960.

15. "Memorandum of Discussion of the 469th National Security Council," *FRUS*, 1958–1960, vol III, doc 129, 494–508.

16. Ibid.

17. "Washington Roundup," *Aviation Week*, December 19, 1960; "Killian Cautions Against Excessive Stress on Man-in-Space Program," *Aviation Week*, December 19, 1960; "Washington Roundup," *Aviation Week*, December 26, 1960.

18. Memo to C[hief of] S[taff] General White, "Concept for the B-70 Weapon System," Dec 19, 1960, fldr 5–1 Air Force Council (Folder #2) Jul-Dec 1960 correspondence, box 36, White Papers, LoC; Memo to CS General White, "Commanders' Reaction to Exercise 62-E," 1960 (n.d.), fldr 5 Conferences/Boards/Exercises, box 36, White Papers, LoC; Letters between Senator Stuart Symington and CS General White: White to Symington, April 19, 1960; Symington to White, May 6, 1960; White to Symington, Dec 20, 1960, fldr 7–3 Congressional, box 39, White Papers, LoC; Memo to CS General White, "Commanders' Reaction to Exercise 62-E," 1960 (n.d.), fldr 5 Conferences/Boards/Exercises, box 36, White Papers, LoC.

19. "Draft Record of Actions, 470th NSC Meeting," Dec 20, 1960, fldr 1960—Meetings with President—Volume 2 (1), box 5, Presidential Subseries, Special Asst National Security Affairs Special Asst ser, DDEL; "Draft Record of Actions, 471st NSC Meeting," Dec 22, 1960, fldr 1960—Meetings with President—Volume 2 (1), box 5, Presidential Subseries, Special Asst National Security Affairs Special Asst ser, DDEL.

20. Memo for record, May 20, 1959, fldr Farewell Address (1) box 16, Moos Papers, DDEL; Fox, Frederic, to Malcolm Moos, April 5, 1960, fldr Farewell Address (1) box 16, Moos Papers, DDEL.

21. "Commencement," n.d., fldr Farewell Address (2), box 16, Moos Papers, DDEL.

22. Memo to "Mac," unsigned, undated, attached to "Commencement" draft, fldr Farewell Address (2), box 16, Moos Papers, DDEL; Farewell Address draft, Dec 16, 1960, fldr Farewell Address (3), box 16, Moos Papers, DDEL.

23. Farewell Address draft, Dec 16, 1960.

24. Ibid.

25. Farewell Address draft, Dec 19, 1960, third draft, fldr Farewell Address (6), box 16, Moos Papers, DDEL.

26. Farewell Address draft, Dec 29, 1960, second draft, fldr Farewell Address (8), box 16, Moos Papers, DDEL.

27. Farewell Address draft, Dec 30, 1960, first draft, fldr Farewell Address (9), box 16, Moos Papers, DDEL.

28. Gordon Gray, "Memorandum of Meeting with the President (Thursday, 29 December 1960 at 8:50am)," Dec 31, 1960, fldr 1960—Meetings with President—Volume 2 (2), box 5, Presidential Subseries, Special Asst National Security Affairs Special Asst ser, DDEL.

29. Ibid.

30. Kolcum, "Services Challenge NASA's Dominant Role"; Keith Glennan, "Accomplishments of the National Aeronautics and Space Administration, 1958–1960," Dec 28, 1960, fldr Glennan Dr Keith—NASA, box 15, AWD, DDEL.

31. Glennan, "Accomplishments of the National Aeronautics and Space Administration, 1958–1960."

32. The sections were the Introduction; Foreign Policy; National Defense; Commerce (later renamed the Economy); Government Finance and Administration; Agriculture; National Resources; Education, Science, and Technology; Civil Rights; Health and Welfare; Housing and Urban Development; Immigration; Veterans; and the Conclusion. Drafting envisioned a statement of "spiritual goals" in the introduction, a "prayer for the future" in the conclusion, and a "statement of principles" closing each of the sections, although the outline does not suggest this planned for the Health and Welfare section.

33. State of the Union draft #1, Dec 31, 1960, fldr 1961 State of the Union Message Draft #1, box 18, Moos Papers, DDEL; State of the Union draft #1A, Dec 31, 1960, fldr 1961 State of the Union Message Draft #1A, box 18, Moos Papers, DDEL.

34. State of the Union draft #1A, Dec 31, 1960.

35. Attachment to memo by Gordon Gray, Jan 4, 1961, fldr 1960—Meetings with President—Volume 2 (1), box 5, Presidential Subseries, Special Asst National Security Affairs Special Asst ser, DDEL.

36. Gray, "Memorandum of Conversation with the President (Tuesday, January 3, 1961 at 2:35pm)," Jan 4, 1961, fldr 1960—Meetings with President—Volume 2 (1), box 5, Presidential Subseries, Special Asst National Security Affairs Special Asst ser, DDEL.

37. Ibid.

38. State of the Union draft #1B (1), Jan 5, 1961, fldr 1961 State of the Union Message Draft #1B (1), box 18, Moos Papers, DDEL; State of the Union draft #1B (2), Jan 5, 1961, fldr 1961 State of the Union Message Draft #1B (2), box 18, Moos Papers, DDEL.

39. Draft "Farewell Address," n.d. (about January 1, 1961), first and second drafts, fldr Farewell Address (12), box 16, Moos Papers, DDEL.

40. Draft "Farewell Address," January 9, 1961, second draft, fldr Farewell Address (16), box 16, Moos Papers, DDEL.

41. Ibid.

42. Dwight D. Eisenhower, "Radio and Television Address to the American People on Science in National Security," *The American Presidency Project*, November 7, 1957, http://www.presidency.ucsb.edu/ws/?pid=10946; Donald Quarles, "The Uses of Modern Arms," in Emme, *The Impact of Air Power*, 706–7.

43. Draft "State of the Union," January 8, 1961, draft 3, fldr 1961 State of the Union Message Draft #3, box 18, Moos Papers, DDEL; Draft "State of the Union," January 9–10, 1961, draft #4/5, fldr 1961 State of the Union Message Draft #4/5, box 18, Moos Papers, DDEL.

44. Draft "State of the Union," January 9–10, 1961, draft #4/5.

45. Dwight D. Eisenhower, "Annual Budget Message to the Congress: Fiscal Year 1962. January 16, 1961," *Public Papers of the Presidents*, accessed November 2, 2011, http://www.presidency.ucsb.edu/ws/?pid=12078.

46. Ibid.

47. Dwight D. Eisenhower, "Farewell Address," January 17, 1961, accessed August 1, 2011, http://www.americanrhetoric.com/speeches/dwightdeisenhowerfarewell.html.

48. TV listing, *New York Times*, January 15, 1961, X14; TV advertisement, *New York Times*, January 17, 1961, 75; "Farewell Speech on Air Tonight," *Washington Post*, January 17, 1961, A1; Cecil Smith, "The TV Scene: Video Farewell Seems Fitting," *Los Angeles Times*, January 17, 1961.

49. "Ike Offers Budget of Record Size," *Washington Post*, January 17, 1961; Robert Bedingfield, "1960, a Year of Great Promise, Developed 'Hidden Recession,'" *New York Times*, January 1, 1961, F1.

50. Robert T. Hartmann, "The President: First in War, Peace, Hearts," *Los Angeles Times*, January 13, 1961, B4.

51. "Parting Words," *Wall Street Journal*, January 19, 1961, 16.

52. Stuart Leslie, *The Cold War and American Science: The Military-Industrial Academic Complex at MIT and Stanford* (New York: Columbia University Press, 1993), 33; Hotz, "Editorial: The Parthian Shots."

53. Alex Roland, *The Military-Industrial Complex* (Washington, DC: American Historical Association, 2001), 3, 9.

54. Alexander P. De Seversky, *America: Too Young to Die!* (New York: McGraw-Hill, 1961), 63.

55. "422: The President's News Conference, January 18, 1961," *The American Presidency Project*, accessed July 25, 2013, http://www.presidency.ucsb.edu/ws/index.php?pid=12087.

56. Ibid.

57. Robert Hotz, "Editorial: The Arms Control Problem," *Aviation Week*, January 30, 1961.

58. Chalmers M. Roberts, "Kennedy and Eisenhower Find Big Contrast in State of Union," *Washington Post*, January 31, 1961, A1; "How Kennedy, Eisenhower Differ," *Los Angeles Times*, January 31, 1961, 14.

59. Arthur Krock, "Inaugural Contrast: Kennedy Dramatizes the Change but the Basic Aspirations Remain," *New York Times*, January 22, 1961, E11.

60. Walter Lippmann, "Today and Tomorrow," *Washington Post,* January 19, 1961.

61. Ibid.

62. Ibid.; Howell John Harris, *The Right to Manage: Industrial Relations Policies of American Business in the 1940s* (Madison: University of Wisconsin, 1982), 193.

63. "McNamara, Named Defense Secretary, Rose from Ranks," *St. Joseph Gazette*, December 14, 1960, accessed November 2, 2011, http://news.google.com/newspapers?id=ZhRdAAAAIBAJ&sjid=HFoNAAAAIBAJ&pg=836,4387904&dq=mcnamara+named+defense+secretary&hl=en.

64. Lippmann, "Today and Tomorrow."

Chapter 11. "Equal Attention to Both"

Epigraph. John W. Finney, "Call for Urgency is Expected in U.S.; Demand to be Renewed for Military Man-in-Space," *New York Times*, August 7, 1961, 7.

Second epigraph. Frederick Pilcher, letter to the editor, *Aviation Week*, November 27, 1961.

1. Robert Hotz, "Editorial: New Vigor for Space Program," *Aviation Week*, January 16, 1961; Evert Clark, "Stronger Space, Missile Efforts Urged," *Aviation Week*, January 16, 1961.

2. "Washington Roundup," *Aviation Week*, January 9, 1961; "President Affirms NASA Support of Industry Satellite Operations," *Aviation Week*, January 9, 1961; "Washington Roundup," *Aviation Week*, January 2, 1961; Spires, *Beyond Horizons*, 88.

3. "Washington Roundup," *Aviation Week*, January 16, 1961; "Industry Observer," *Aviation Week*, January 9, 1961; Larry Booda, "NASA, Defense Seek More Space Funds," *Aviation Week*, January 2, 1961; Dwight B. Schear, "Seattle Pins Hope on Boeing Orders," *New York Times*, January 9, 1961.

4. John F. Kennedy, 1961 inaugural speech, OurDocuments.gov, accessed November 4, 2011, http://ourdocuments.gov/doc.php?doc=91&page=transcript.

5. Ibid.

6. "Editorial Comment Across the Nation on President Kennedy's Inauguration," *New York Times*, January 21, 1961.

7. "Kennedy Group Criticizes Space Effort," *Aviation Week*, January 23, 1961.

8. John F. Kennedy, State of the Union address, January 30, 1961, JFKLibrary.org, accessed November 5, 2011, http://www.jfklibrary.org/Asset-Viewer/Archives/JFKWHA-006.aspx; "Democrats Hail Speech; Republicans Critical," *Washington Post*, January 31, 1961, A6; "Speech Triggers Wall St. Surge," *Washington Post*, January 31, 1961, A1.

9. Landis S. Gephart to Seamans, "General Management Instructions Pertaining to the NASA Reliability Program," January 26, 1961, 3059 "Glennan Select Correspondence," NASA HQ; Saturn Reliability Steering Committee to Seamans, "ARINC Study of S-1 Reliability," January 25, 1961, 3059 "Glennan Select Correspondence," NASA HQ.

10. The agenda items were joint research on dynamic loads' effects on launch vehicles; Centaur project review; housing near Ballistic Missile Division; NASA and/or Dyna-Soar use of the Saturn; "Scout relationships"; life sciences and Discoverer satellites; manned lunar landing; NASA orbital docking and Air Force rendezvous projects; Atlas pad modifications; flight safety on Atlantic and Pacific Missile Ranges. Seamans to Ritland, February 13, 1961, 11324 "X-20 Correspondence Martin 1962–64," NASA HQ.

11. "Aeronautics and Astronautics Coordinating Board Activities," February 6, 1961, 11130 "DOD in Space Thru 1972," NASA HQ.

12. John M. Logsdon, *John F. Kennedy and the Race to the Moon* (Basingstoke, UK: Palgrave Macmillan, 2010).

13. Jack Raymond, "Kennedy Defense Study Finds No Evidence of a 'Missile Gap,'" *New York Times*, February 7, 1961, 1.

14. "Industry Observer," *Aviation Week*, February 13, 1961; Robert Hotz, "Editorial: The Stern Chase in Space," *Aviation Week*, February 20, 1961; "Washington Roundup," *Aviation Week*, February 20, 1961; Robert Hotz, "Editorial: Good Luck, Astronauts," *Aviation Week*, February 27, 1961.

15. Hotz, "Editorial: Good Luck, Astronauts"; Ira H. Abbott, "Advanced Research Programs," March 1961, 6 "Abbott, Ira H. (Speeches)," NASA HQ; "Global Surveillance System Study Planned," *Aviation Week*, February 27, 1961.

16. Krafft Ehricke, statement, March 1, 1961, 559 "Ehricke, Krafft A. (General Bio)," NASA HQ; "USAF May Gain Stronger Role in Space," *Aviation Week*, March 6, 1961; Hotz, "Editorial: Good Luck, Astronauts"; "Washington Roundup," *Aviation Week*, February 27, 1961.

17. "Washington Roundup," *Aviation Week*, March 6, 1961; John G. Norris, "News Leak Crackdown Ordered by McNamara," *Washington Post*, March 3, 1961; "Washington Roundup," *Aviation Week*, April 3, 1960.

18. "Statement of the Administrator, National Aeronautics and Space Administration on Selection of a Contractor for the Apollo Spacecraft," March 16, 1961, 03799 "Chronological Reading File June 1962," NASA HQ; David Bell, to Kennedy, March 22, 1961, 12456 "BOB," NASA HQ.

19. John F. Kennedy, "Special Message to Congress on Defense Spending," March 28, 1961, *The American Presidency Project*, accessed November 5, 2011, http://www.presidency.ucsb.edu/ws/index.php?pid=8554#axzz 1crwqH0MG.

20. Ibid.; "Soviets Predict First Manned Flight on Basis of Sputnik IX Recovery," *Aviation Week*, March 20, 1961.

21. "Fund Notes Innovation; Aerospace Category is Added by Keystone S-4 Company," *New York Times*, March 9, 1961; "Kennedy Budget Shifts May Accelerate Slight Rise in Aerospace Sales," *Aviation Week*, March 12, 1961; Bill Beckner, "New Unit Studies Rocket of Future; Aerospace Corporation Has No Stockholders—Works on Air Force Projects," *New York Times*, March 26, 1961, 42.

22. Bergen Evans, "New World, New Words," *New York Times*, April 9, 1961.

23. Saturn would become a lunar booster, Centaur pioneered high-energy thrust in a rocket upper stage, the F-1 booster would be a Saturn component, and Rover was a nuclear rocket research project. "Washington Roundup," *Aviation Week*, April 3, 1961; Edward H. Kolcum, "Propulsion Key to NASA Budget Increase," *Aviation Week*, April 3, 1961; Martin Caiden, *Rendezvous in Space: The Story of America's Man-in-Space Programs—Project Mercury, Gemini, Dyna-Soar, and Apollo* (New York: E. P. Dutton and Co., 1962), 268.

24. Richard Witkin, "U.S. Project Mercury Hopes to Send an Astronaut Aloft by the End of this Year; Efforts Dogged by Test Failures, Space Agency to Need Some Luck to Achieve Its First Manned Operation," *New York Times*, April 12, 1961.

25. United Press International, "187-Mile Heigh [*sic*]: Yuri Gagarin, A Major Makes the Flight in 5-Ton Vehicle," *New York Times*, April 12, 1961, 1; Evert Clark, "Space Strategy Hinges on Time Factors," *Aviation Week*, March 12, 1961.

26. Boris Chertok, *Rockets and People*, vol III (Washington, DC: NASA History Office, 2009), 44–46, 63, 67.

27. Robert Hotz, "Editorial: First Man in Space," *Aviation Week*, April 17, 1961; "Washington Roundup," *Aviation Week*, April 17, 1961; "Soviet Space Rise Keyed to Management," *Aviation Week*, April 17, 1961; "Washington Roundup," *Aviation Week*, May 1, 1961.

28. Ford Eastman, "Kennedy Seeks U.S. Space Race Gains," *Aviation Week*, May 1, 1961.

29. Edward H. Kolcum, "Flickinger Warns of Soviet Space Threat," *Aviation Week*, May 22, 1961.

30. Logsdon, *John F. Kennedy and the Race to the Moon*, 101.

31. Wernher von Braun to Lyndon Johnson, letter, 29 April 1961, doc II-10, in Logsdon, *Exploring the Unknown*, vol VII, *Human Spaceflight*, 486; Michael Beschloss, "Kennedy and the Decision to Go to the Moon," in *Spaceflight and the Myth of Presidential Leadership*, ed. Roger D. Launius and Howard E. McCurdy (Champaign: University of Illinois Press, 1997), 63; Logsdon, *John F. Kennedy and the Race to the Moon*, 80; Hotz, "Editorial: First Man in Space"; "Washington Roundup," *Aviation Week*, April 17, 1961; "Soviet Space Rise Keyed to Management," *Aviation Week*, April 17, 1961; "Washington Roundup," *Aviation Week*, May 1, 1961.

32. "Dyna-Soar: Project Streamline," May 4, 1961, 11325 "X-20 Dyna-Soar Documentation," NASA HQ.

33. Ibid.

34. The lattice tower and emergency escape rockets, to be discarded during the mission, accounted for about two-thirds of the difference between

the launch weight and the splashdown weight. "Mercury Capsule No. 2 Configuration Specification (Mercury-Redstone No 1)," accessed November 7, 2011, http://ntrs.nasa.gov/archive/nasa/casi.ntrs.nasa.gov/19740075935_1974075935.pdf.

35. John W. Finney, "Air Force Sets up Space-Man Course; Pilot Training to be Given at California Base," *Aviation Week*, May 13, 1961, 3; "Dyna-Soar: Project Streamline"; John W. Finney, "U.S. Pushes Study of Lunar Landing," *Aviation Week*, May 20, 1961; George C. Wilson, "Kennedy to Launch Major Space Effort: New budget request, expected to total $600 million, seen as beginning of attempt to outdistance Soviets," *Aviation Week*, May 15, 1961; "Randt Cites Manned Space Flight Hazards," *Aviation Week*, May 15, 1961.

36. A day before Kennedy's speech, the *New York Times* cited estimates ranging from $20 billion to $40 billion.

37. John F. Kennedy, State of the Union address, May 25, 1961, JFKLibrary.org, accessed November 6, 2011, http://www.jfklibrary.org/Research/Ready-Reference/JFK-Speeches/Special-Message-to-the-Congress-on-Urgent-National-Needs-May-25–1961.aspx.

38. Logsdon, *John F. Kennedy and the Race to the Moon*, 113–15.

39. W. H. Lawrence, "President to Ask an Urgent Effort to Land on Moon," *Aviation Week*, May 24, 1961; Alvin Shuster, "Congress Wary on Cost, But Likes Kennedy Goals," *New York Times*, May 26, 1961, 1; George C. Wilson, "Congress is Expected to Support Kennedy's Space Race Program," *Aviation Week*, June 5, 1961.

40. The *New York Times* noted that "until recently, Mr. Kennedy seemed to have been largely influenced by these scientists." John W. Finney, "A 3-Man Trip to Moon by 1967 Projected by White House Aides," *New York Times*, May 26, 1961, 13.

41. Logsdon, *John F. Kennedy and the Race to the Moon*, 87; "Contract Awards," *New York Times*, May 31, 1961.

42. Katherine Johnsen, "Missiles Dominate Research Budget Plan," *Aviation Week*, June 12, 1961.

43. Craig Lewis, "Defense in Evolving New Weapon Plans," *Aviation Week*, June 19, 1961.

44. Francis J. Muller, letter to the editor, *Aviation Week*, June 12, 1961; General Thomas White, letter to the editor, *Aviation Week*, July 3, 1961; William K. Callam, letter to the editor, *Aviation Week*, July 3, 1961; Katherine Johnsen, "House Votes Bomber, Dyna-Soar Increase," *Aviation Week*, July 3, 1961.

45. C. B. Palmer, "Search for Spacemen," *New York Times*, June 25, 1961, SM28; "50 Airmen Picked for Space Flights; Would Fly Military Missions—Projects Still Debated," *New York Times*, July 23, 1961, 33.

46. "Space Award Set Up; Medal of Geographic Society Is in Honor of Gen White," *New York Times*, June 30, 1961, 3.

47. "NASA Attempts to Sell Program as Broader than Lunar Goals," *Aviation Week*, June 26, 1961; Trevor Gardner, "Gardner Analyzes U.S. Lunar Race Plan," *Aviation Week*, July 3, 1961. A retrospective estimate made for the House Subcommittee for Science and Astronautics in 1973 identified Apollo's cost as $25.4 billion. "America Starts for the Moon, 1957–1963: Project Apollo: The Debate," NASA.gov, accessed November 11, 2011, http://history.nasa.gov/SP-4214/ch1–4.html.

48. Honeywell Aero advertisement, *Aviation Week*, July 17, 1961; Boeing advertisement, *Aviation Week*, July 17, 1961; "Washington Roundup," *Aviation Week*, July 31, 1961.

49. "Soviet 'Space Skipper' Use Is Suspected," *Washington Post*, June 17, 1961, D14; "Scientific Goals Spur Expenditure of Millions in Space Drive," *New York Times*, July 22, 1961.

50. Advertisement, *Aviation Week*, July 24, 1961.

51. John W. Finney, "Call for Urgency Is Expected in U.S.; Demand to be Renewed for Military Man-in-Space," *New York Times*, August 7, 1961, 7.

52. Eugene Zuckert to Robert McNamara, August 1, 1961, 11271 "USAF-Docs/Correspondence 1957–1961," NASA HQ; Roswell Gilpatric to McNamara, August 15, 1961, 11271 "USAF-Docs/Correspondence 1957–1961," NASA HQ; Jack Raymond, "Sharp Reduction in Troops Likely if Crisis Recedes; Prospect Disturbs the Army—Pentagon Said to Plan 53-Billion Budget for '63 Troop Cut Likely if Crisis Recedes," *New York Times*, October 7, 1961, 1.

53. Robert Hotz, "Editorial: Defense Problems," *Aviation Week*, August 7, 1961; Robert Hotz, "Editorial: The Nuclear Shoe Drops," *Aviation Week*, September 4, 1961.

54. "Industry Observer," *Aviation Week*, August 7, 1961; "Industry Observer," *Aviation Week*, August 14, 1961.

55. Garrett Corporation advertisement, *Aviation Week*, September 11, 1961.

56. "Dyna-Soar Speeded; Air Force Awards Contracts Totaling 36.5 Million," *New York Times*, September 30, 1961; Harold M. Schmeck, "What's in a Name? How the Many Rockets Receive Designations," *New York Times*, October 8, 1961, M22; Richard Witkin, "U.S. Pushes Rocket Tests; Firing Schedule Puts Nation in the Space Race in Earnest," *New York Times*, October 8, 1961, M3.

57. Robert Hotz, "Editorial: The Military Space Role—I," *Aviation Week*, September 11, 1961.

58. Robert Hotz, "Editorial: The Military Space Role—II," *Aviation Week*, September 18, 1961; Robert Hotz, "Editorial: The Military Space Role—III," *Aviation Week*, October 2, 1961; Hotz, "Editorial: The Military Space Role—I"; Gordon Gray, "handwritten notes of meeting at Augusta, Georgia, Nov. 28, 1958 re defense expenditures," Nov 28, 1958, fldr Meetings with the President 1958 (4), box 3, Special Assistant for National Security Affairs, Special Assistant ser, DDEL; "Washington Roundup," *Aviation Week*, September 11, 1961; "Industry Observer," *Aviation Week*, September 18, 1961; "Dornberger Sees Space as Military Area," *Aviation Week*, September 18, 1961.

59. "Variety of Programs Typifies ASD Effort," *Aviation Week*, September 25, 1961; "USAF Charting Future Spacecraft Needs," *Aviation Week*, September 25, 1961.

60. "Bioastronautics May Set Pace for Space," *Aviation Week*, September 25, 1961; Martin advertisement, "Booster Ability," *Aviation Week*, September 25, 1961; "Houston Picked for Space Flight Center," *Aviation Week*, September 25, 1961; "Brooks' Death to Alter Space Group's Course," *Aviation Week*, September 25, 1961; George C. Wilson, "Miller Opposes Stress on Race to Moon," *Aviation Week*, October 2, 1961.

61. Richard Witkin, "U.S. Pushes Rocket Tests; Firing Schedule Puts Nation in the Space Race in Earnest," *New York Times*, October 8, 1961, M3; Gladwin Hill, "Funds are Urged for Atom Rocket; AEC Chairman Warns of Program 'Bottleneck,'" *New York Times*, October 25, 1961, 20; William L. Laurence, "Space Propulsion; Nuclear and Electrical Thrust Held the Best Means for Flights," *New York Times*, October 29, 1961; John W. Finney, "U.S. Plans to Fire Saturn this Week; Giant Is Slated for 225-Mile Test Flight," *New York Times*, October 15, 1961, 46.

62. Edward H. Kolcum, "Vice President Begins Nationwide Tour of Major Space Installations," *Aviation Week*, October 9, 1961; Robert Hotz, "Editorial: Education's Role in Space," *Aviation Week*, October 16, 1961.

63. "ARS Underscores U.S. Space Goals, Gaps," *Aviation Week*, October 16, 1961; "Industry Observer," *Aviation Week*, October 16, 1961.

64. Richard Witkin, "Air Force to Push Dyna-Soar Drive; Schriever to Urge Earlier Test and Wider Capacity," *New York Times*, October 13, 1961, 15.

65. Richard Witkin, "Agreement Near on Hybrid Rocket; Solid-Fuel 'Building Blocks' Would Help Loft Titans," *New York Times*, November 1, 1961, 17.

66. "Solid Booster to be Developed for USAF," *Aviation Week*, October 30, 1961; Edward H. Kolcum, "Rendezvous Is Urged for Moon Flight," *Aviation Week*, November 6, 1961; John C. Houbolt to Robert C. Seamans, November 15, 1961, doc II-15, in Logsdon, *Exploring the Unknown*, vol VII, *Human Spaceflight*, 528.

67. Logsdon, *John F. Kennedy and the Race to the Moon*, 229; Mark Erickson, *Into the Unknown Together: The DOD, NASA, and Early Spaceflight* (Maxwell Air Force Base, AL: Air University Press, 2005), 228.

68. "Kennedy Leads Worldwide Protest on Soviet Plan to Test Huge Bomb," *Aviation Week*, October 23, 1961; "50-Megaton Blast Tops Soviet Nuclear Test Series," *Aviation Week*, November 6, 1961; "Washington Roundup," *Aviation Week*, October 16, 1961; George C. Wilson, "U.S. Plans to Control Space Dollar Impact," *Aviation Week*, November 13, 1961; Elbert V. Bowden, *Economics Through the Looking Glass* (San Francisco: Canfield Press, 1974), 108–21; "Oral History Interview with Theodore Sorensen," March 26, 1964, doc II-43, in Logsdon, *Exploring the Unknown*, vol VII, *Human Spaceflight*, 627.

69. Edward C. Welsh, "Welsh Analyzes U.S. Space Expenditures," *Aviation Week*, October 30, 1961; Warren H. Flarity, letter to the editor, *Aviation Week*, November 6, 1961.

70. "Industry Observer," *Aviation Week*, November 20, 1961.

71. Ibid.; "Complete List of All U.S. Nuclear Weapons," www.nuclear weaponarchive.com, accessed July 2, 2011, http://nuclearweaponarchive .org/Usa/Weapons/Allbombs.html.

72. Larry Booda, "Kennedy to Stand Firm on Bomber Funds," *Aviation Week*, November 6, 1961.

73. Pilcher, letter to the editor, *Aviation Week*.

74. Edward H. Kolcum, "Manned Space Flight Program Expanded: McDonnell to build 12 two-man capsules and modify four Mercury spacecraft for flights of 18 orbits," *Aviation Week*, December 18, 1961; "Washington Roundup," *Aviation Week*, January 8, 1962.

75. Barry Miller, "USAF Sees Need for 3-Man Lunar Rover," *Aviation Week*, December 18, 1961; Irving Stone, "USAF Plans Lunar Shelter Design Study," *Aviation Week*, December 25, 1961; "U.S., USSR Agree of Space, Disarmament," *Aviation Week*, December 18, 1961.

76. John W. Finney, "Clearance Given for Giant Rocket; Advanced Missile Slated for Big Load-Lifting Job," *New York Times*, December 24, 1961, 22.

77. John W. Finney, "Air Force Authorized to Orbit Its Projected Dyna Soar Glider; Air Force Plans to Orbit a Pilot," *New York Times*, December 27, 1961, 1.

78. Ibid.; John G. Norris, "AF Revises Dynasoar Program to Speed Up Orbiting of Glider," *Washington Post*, December 27, 1961, A4.

79. Jack Raymond, "Pentagon Troubled by Rise in Industry Groups Asking for Help," *New York Times*, November 14, 1961, 7.

80. Burton Crane, "Market Steady; Turnover Rises," *New York Times*, December 29, 1961.

81. "Washington Roundup," *Aviation Week*, January 1, 1962; Larry Booda, "Dyna-Soar Decision Ends Long Impasse," *Aviation Week*, January 1, 1962.

Epilogue and Conclusion

Epigraph. Richard E. Neustadt, *Presidential Power: The Politics of Leadership* (New York: Mentor, 1964), 91.

1. Dwight D. Eisenhower, "Farewell Address," January 17, 1961, accessed August 1, 2011, http://www.americanrhetoric.com/speeches/dwightdeisenhowerfarewell.html; Eisenhower and Eisenhower, *Going Home to Glory*, 40.

2. Mark D. Mandeles, *The Development of the B-52 and Jet Propulsion: A Case Study in Organizational Innovation* (Maxwell Air Force Base, AL: Air University Press, 1998), 21.

3. "Space Triumph! Glenn Flight Thrills World," Universal International News, February 22, 1962, YouTube.com, accessed July 30, 2013, http://www.youtube.com/watch?v=SA4zGhJvrv8&list=PL_hX5wLdhf_JhjIwLJjL8MHIhHVJyMm0b.

4. Logsdon, *John F. Kennedy and the Race to the Moon*, 244.

5. William Duggan, *Strategic Intuition: The Creative Spark in Human Achievement* (New York: Columbia Business School, 2007), 176–77.

6. Transcript of presidential meeting in the Cabinet Room of the White House, November 21, 1962, doc II-33, in Logsdon, *Exploring the Unknown,* vol VII, *Human Spaceflight,* 596.

7. Logsdon, *John F. Kennedy and the Race to the Moon*, 155–56; Yanek Mieczkowski, *Eisenhower's Sputnik Moment: The Race for Space and World Prestige* (Ithaca: Cornell University Press, 2013), 258; transcript of presidential meeting, in Logsdon, *Exploring the Unknown*, vol VII, *Human Spaceflight,* 598–99.

8. Transcript of presidential meeting in Logsdon, *Exploring the Unknown,* vol VII, *Human Spaceflight,* 599.

9. Logsdon, *John F. Kennedy and the Race to the Moon*, 197, 215.

10. "Project Development Plan for Rendezvous Development Utilizing the Mark II Two Man Spacecraft," 8 December 1961, doc I-49, in Logsdon, *Exploring the Unknown,* vol VII, *Human Spaceflight,* 259; Caiden, *Rendezvous in Space*, 232–33; Logsdon, *John F. Kennedy and the Race to the Moon*, 219–21.

11. Logsdon, *John F. Kennedy and the Race to the Moon*, 237.

12. "DOD Press Release: X-20 Designation," June 1962, in Godwin, *Dyna-Soar: Hypersonic Strategic Weapons System,* 248; "DOD Press Release: Unveiling X-20," September 1962, in Godwin, *Dyna-Soar: Hypersonic Strategic Weapons System,* 254.

13. McDougall, . . . the Heavens and the Earth, 340; Houchin, U.S. Hypersonic Research and Development, 200, 210–12.

14. Julian Scheer to Marvin Watson, 24 May 1965, doc I-62, in Logsdon, Exploring the Unknown, vol VII, Human Spaceflight, 289–90; NASA Manned Spacecraft Center, "Project Development Plan for Rendezvous Development Utilizing the Mark II Two Man Spacecraft," 8 December 1961, doc I-49, in Logsdon, Exploring the Unknown, vol VII, Human Spaceflight, 262, "Gemini Summary Conference," 1–2 February 1967, doc I-73, in Logsdon, Exploring the Unknown, vol VII, Human Spaceflight, 379.

15. Houchin, U.S. Hypersonic Research and Development, 218.

16. "Oral History Interview with Theodore Sorensen," 26 March 1964, doc II-43, in Logsdon, Exploring the Unknown, vol VII, Human Spaceflight, 628.

17. James V. Zimmerman to Arnold W. Frutkin et al., "Foreign Policy Issues Regarding Earth Resource Surveying by Satellite: A Report of the Secretary's Advisory Committee on Science and Foreign Affairs," 24 July 1974, doc II-24, in Logsdon, Exploring the Unknown, vol III, Using Space, 265; Willis H. Shapley to George Mueller, "Symbolic Items for the First Lunar Landing," 19 April 1969, doc II-70, in Logsdon, Exploring the Unknown, vol VII, Human Spaceflight, 732; NASA Mission Evaluation Team, "Apollo 11: Mission Report," 1971, doc II-81, in Logsdon, Exploring the Unknown, vol VII, Human Spaceflight, 767; Logsdon, John F. Kennedy and the Race to the Moon, 237.

18. For example, during a visit to Soviet space project workers in 1960, Central Committee secretary Leonid Brezhnev said, "it would be good if you launched one of these 'bugs' to cause a bit more of a stir" in world affairs. Boris Chertok, a leading scientist and administrator, noted in his memoirs that "this attitude toward space technology grated on everyone." Chertok, Rockets and People, vol II, 555; Chertok, Rockets and People, vol III, 45–46, 217, 266–67; Caiden, Rendezvous in Space, 167–69.

19. Houchin, U.S. Hypersonic Research and Development, 217–18; Maxime Faget, oral history with Dick Hallion, March 22, 1979, NASA HQ.

20. McDougall, . . . the Heavens and the Earth, 341.

21. Thomas D. White, "The Inevitable Climb to Space," Air University Quarterly Review (Winter 1958–59): 3–4; Benjamin S. Lambeth, Mastering the Ultimate High Ground: Next Steps in the Military Uses of Space (Santa Monica, CA: RAND, 2003), 27, 58, 90–91.

22. Brian Berger, "NASA Transfers X-37 Project to DARPA," Space. com, September 15, 2004, accessed May 5, 2104, http://www.space. com/337-nasa-transfers-37-project-darpa.html; Tom Burghardt, "The

Militarization of Outer Space: The Pentagon's Space Warriors," *Global Research*, May 11, 2010, accessed May 5, 2014, http://www.spacedaily .com/reports/The_Militarization_of_Outer_Space_The_Pentagon_ Space_Warriors_999.html; Kyle Mizokami, "Secret Spaceplane, Mystery Mission," *Daily Beast*, April 7, 2014, accessed April 7, 2014, http://news .yahoo.com/secret-spaceplane-mystery-mission-094500663—politics .html.

23. To date, the longest X-37B mission has extended twenty-two months, from December 11, 2012, until October 17, 2014. Stephen Clark, "Air Force X-37B Spaceplane Arrives in Florida for Launch," *Spaceflight Now*, February 25, 2010, accessed May 5, 2014, http://spaceflightnow .com/atlas/av012/100225x37arrival/; "X-37B Lands This Morning at Vandenberg AFB," *Santa Maria Times*, June 16, 2012, accessed May 1, 2014, http://santamariatimes.com/news/local/miltiary/vandenger/x— b-lands-this-morning-at-vandenberg-afb/article_31f5827c-b7bc-11e1 –80b2–0019bb2963f4.html; Leonard David, "Secretive US X-37B Space Plane Could Evolve to Carry Astronauts," *Space.com*, October 7, 2011, accessed May 1, 2014, http://www.space.com/13230-secretive-37b -space-plane-future-astronauts.html; Michael Martinez, "Unmanned X-37B space plane lands, its exact mission a mystery," *CNN*, October 19, 2014, accessed October 20, 2014, http://www.cnn.com/2014/10/18/us/ air-force-space-plane-x37b/.

24. Sean Kalic, *U.S. Presidents and the Militarization of Space, 1946–1967* (College Station: Texas A&M Press, 2012), 24, 64; Erickson, *Into the Unknown Together*, 184; Herbert York, *Race to Oblivion: A Participant's View of the Arms Race* (New York: Simon and Schuster, 1970), 149; Killian, *Sputnik, Scientists, and Eisenhower*, 129.

25. Mieczkowski, *Eisenhower's Sputnik Moment*, 259.

26. Odd Arne Westad's Bancroft Award–winning *The Global Cold War: Third World Interventions and the Making of Our Times* (Cambridge University Press, 2005) is a leading example of scholarship that draws new attention to the complex multilateral dynamics of the Cold War.

27. During his first eighty days in office, Eisenhower held ten NSC meetings; in contrast, his successor was far less enamored with such organization and held only three in the same timeframe. Greenstein, *The Hidden-Hand Presidency*, 104–5, 113.

28. Frank, *Ike and Dick*, 188.

29. Dockrill, *Eisenhower's New-Look National Security Policy*, 189.

30. Brands, *Cold Warriors*, 187–88.

31. Mieczkowski, *Eisenhower's Sputnik Moment*, 75.

32. H. W. Brands, "The Age of Vulnerability: Eisenhower and the National Insecurity State," *American Historical Review* 94 (October 1989): 988–89.

33. Thomas, *Ike's Bluff*, 4; Smith, *Eisenhower in War and Peace*, 759; Greene, *Eisenhower, Science Advice, and the Nuclear Test-Ban Debate*; Wang, *In Sputnik's Shadow*; Samuel P. Huntington, *American Politics: The Promise of Disharmony* (Cambridge, MA: Belknap Press, 1981), 77.

34. Mieczkowski, *Eisenhower's Sputnik Moment*, 22.

35. Wegner, *Living with Peril*, 189; Desmond Ball, *Politics and Force Levels: The Strategic Missile Program of the Kennedy Administration* (Oakland: University of California Press, 1981), 277.

36. Aliano, *American Defense Policy from Eisenhower to Kennedy*, 277.

BIBLIOGRAPHY

Archival Resources

AFHRA Air Force Historical Research Agency, Maxwell Air Force Base, AL

DDEL Dwight David Eisenhower Library, Abilene, KS

DDRS Declassified Document Retrieval System

FRUS *Foreign Relations of the United States* series, U.S. Department of State

LoC Library of Congress

NASA HQ National Air and Space Administration Headquarters Archive, Washington, DC

Newspapers and Periodicals

Air University Quarterly Review

Army-Navy- Air Force Register

Aviation Week

Los Angeles Times

Missiles and Rockets

New York Times

New York World-Telegram

Public Opinion Quarterly

St. Joseph Gazette (Missouri)

Wall Street Journal

Washington Post

Published Primary Resources

An Air Force History of Space Activities, 1945–1959. USAF Historical Division Liaison Office. Accessed October 4, 2011. http://www.google .com/url?sa=t&rct=j&q=&esrc=s&source=web&cd=1&cad=rja &uact=8&ved=0CB4QFjAA&url=http%3A%2F%2Fwww.nro. gov%2Ffoia%2Fdeclass%2FWS117L_Records%2F277.PDF&ei=54uPV KKhEojasAS4oILYDw&usg=AFQjCNFHTdnSN36FU08_KrxLJZVNUZtc Eg&sig2=DWC64YdiMxIj6O8nX6tqSw.

Baxter, James Phinney. *Scientists Against Time.* Boston: Little, Brown and Company, 1946.

Berger, Brian. "NASA Transfers X-37 Project to DARPA." Space.com, September 15, 2004. Accessed May 5, 2014. http://www.space.com/337-nasa-transfers-37-project-darpa.html.

Bissell, Richard M., with Jonathan E. Lewis and Frances T. Pudlo. *Reflections of a Cold Warrior: From Yalta to the Bay of Pigs.* New Haven: Yale University Press, 1996.

Burgess, Eric. *Frontier to Space.* New York: Macmillan, 1956.

Burghardt, Tom. "The Militarization of Outer Space: The Pentagon's Space Warriors." *Global Research,* May 11, 2010. Accessed May 5, 2014. http:// www.spacedaily.com/reports/The_Militarization_of_Outer_Space_ The_Pentagon_Space_Warriors_999.html.

Caiden, Martin. *Rendezvous in Space: The Story of America's Man-in-Space Programs—Projects Mercury, Gemini, Dyna-Soar, and Apollo.* New York: E. P. Dutton and Co., 1962.

Cascade Pictures. *Self-Preservation in an Atomic Bomb Attack* (film). 1950. Accessed July 5, 2013. http://www.youtube.com/watch?v=NCzUcwS_ rPI.

Chertok, Boris. *Rockets and People,* Volume I. Washington, DC: NASA History Office, 2005.

———. *Rockets and People,* Volume II. Washington, DC: NASA History Office, 2006.

———. *Rockets and People,* Volume III. Washington, DC: NASA History Office, 2009.

Clark, Stephen. "Air Force X-37B Spaceplane Arrives in Florida for Launch." *Spaceflight Now,* February 25, 2010. Accessed May 5, 2014. http:// spaceflightnow.com/atlas/av012/100225x37arrival/.

David, Leonard. "Secretive US X-37B Space Plane Could Evolve to Carry Astronauts." *Space.com,* October 7, 2011. Accessed May 1, 2014. http:// www.space.com/13230-secretive-37b-space-plane-future-astronauts .html.

De Seversky, Alexander P. *America: Too Young to Die!* New York: McGraw-Hill, 1961.

Douglas, Gordon (director). *Bombers B-52.* Burbank, CA: Warner Brothers, 1957. On DVD.

"Eisenhower Answers America." Accessed June 3, 2013. http://www.you tube.com/watch?v=OJ3BngTJJS4.

Eisenhower, David and Julie Nixon Eisenhower. *Going Home to Glory: A Memoir of Life with Dwight D. Eisenhower, 1961–1969.* New York: Simon and Schuster, 2010.

Eisenhower, Dwight. " 'The Chance for Peace' Address Delivered Before the American Society of Newspaper Editors, April 16th, 1953." Accessed July 5, 2013. http://www.eisenhower.archives.gov/all_about_ike/spee ches/chance_for_peace.pdf.

———. "Text of the Address Delivered by the President of the United States Before the General Assembly of the United Nations in New York City Tuesday Afternoon, December 8, 1953." Accessed July 5, 2013. http:// www.eisenhower.archives.gov/research/online_documents/atoms_for_ peace/Binder13.pdf.

"Eisenhower for President (1952)." Accessed June 3, 2013. http://www.you tube.com/watch?v=Y9RAxAgksSE.

Emme, Eugene, ed. *The Impact of Air Power.* Princeton: Van Nostrand, 1959.

Encyclopedia Britannica Films. *Atomic Alert* (film). 1951. Accessed July 5, 2013. http://www.youtube.com/watch?v=i4k2skbJDm8.

Federal Civil Defense Administration. *Duck and Cover* (film). 1951. Accessed July 5, 2013. http://www.youtube.com/watch?v=IKqXu-5jw60.

Gilpin, Robert, and Christopher Wright, eds. *Scientists and National Policy-Making.* New York: Columbia University Press, 1964.

Godwin, Robert, ed. *Dyna-Soar: Hypersonic Strategic Weapons System.* Burlington, Ontario, Canada: Apogee Books, 2003.

Huntington, Samuel. *The Soldier and the State: The Theory and Politics of Civil-Military Relations.* New York: Vintage, 1957.

International Geophysical Year World Data Center A. *IGY Satellite Report.* Washington, DC: National Research Council, 1958.

Kennedy, John F. Inaugural Speech. January 20, 1961. Accessed November 4, 2011. http://ourdocuments.gov/doc.php?doc=91&page=transcript.

———. Special Message to Congress on Defense Spending. March 28, 1961. Accessed November 5, 2011. http://www.presidency.ucsb.edu/ws/ index.php?pid=8554#axzz1crwqH0MG.

———. State of the Union Address. January 30, 1961. Accessed November 5, 2011. http://www.jfklibrary.org/Asset-Viewer/Archives/JFKWHA-006 .aspx.

————. State of the Union Address. May 25, 1961. Accessed November 6, 2011. http://www.jfklibrary.org/Research/Ready-Reference/JFK-Speech es/Special-Message-to-the-Congress-on-Urgent-National-Needs -May-25–1961.aspx.

————. *The Strategy of Peace.* Allan Nevins, ed. New York: Harper, 1960.

Kettering, C. F. "The Future of Science." *Science* 104, no. 2713 (December 27, 1946): 609–14.

Killian, James R. *Sputnik, Scientists, and Eisenhower: A Memoir of the First Special Assistant to the President for Science and Technology.* Cambridge: MIT Press, 1977.

Kissinger, Henry. *Nuclear Weapons and Foreign Policy.* New York: Harper and Brothers, 1957.

Lippmann, Walter. *The Essential Lippmann: A Political Philosophy for Liberal Democracy.* Clinton Rossiter and James Lare, eds. Cambridge: Harvard University Press, 1963.

Logsdon, John M., ed. *Exploring the Unknown, Selected Documents in the History of the U.S. Civilian Space Program,* Volume III: *Using Space.* Washington, DC: NASA History Office, 1998.

————. *Exploring the Unknown, Selected Documents in the History of the U.S. Civilian Space Program,* Volume VI: *Space and Earth Science.* Washington, DC: NASA History Office, 2004.

————. *Exploring the Unknown, Selected Documents in the History of the U.S. Civilian Space Program,* Volume VII: *Human Spaceflight: Projects Mercury, Gemini, and Apollo.* Washington, DC: NASA History Office, 2008.

"Mercury Capsule No. 2 Configuration Specification (Mercury-Redstone No 1)." Accessed November 7, 2011. http://ntrs.nasa.gov/archive/nasa/casi .ntrs.nasa.gov/19740075935_1974075935.pdf.

Miller, William L. *Piety Along the Potomac: Notes on Politics and Morals in the '50s.* Boston: Houghton Mifflin, 1964.

Mizokami, Kyle. "Secret Spaceplane, Mystery Mission." *Daily Beast,* April 7, 2014. Accessed April 7, 2014. http://news.yahoo.com/secret-spaceplane -mystery-mission-094500663--politics.html.

National Reconnaissance Office. "NRO Directors." Accessed May 2, 2011. http://www.nro.gov/history/directors/dir1.html.

————. "Organization and Functions of the Office of Missile and Satellite Systems." Accessed May 2, 2011. http://www.nro.gov/foia/declass/ NROStaffRecords/736.PDF.

National Security Council. "U.S. Scientific Satellite Program." Document 5520. Accessed November 23, 2011. http://www.marshall.org/pdf/ materials/805.pdf.

Neustadt, Richard E. *Presidential Power: The Politics of Leadership*. New York: Mentor, 1964.

President's Science Advisory Committee. "Report of the Ad Hoc Panel on Man-in-Space." December 16, 1960. Accessed October 3, 2011. http://www.hq.nasa.gov/office/pao/History/report60.html.

Rachmil, Lewis J., creator. *Men into Space* (television series). Columbia Broadcasting Service, 1959–60.

Serling, Rod, creator. *The Twilight Zone* (television series). Columbia Broadcasting Service, 1959–60. Hollywood, CA: Cayuga Productions, 2010. DVD.

Sorensen, Theodore. *Kennedy*. New York: Harper and Row, 1965.

"Space Triumph! Glenn Flight Thrills World." *Universal International News*. February 22, 1962. Accessed July 30, 2013. http://www.youtube.com/watch?v=SA4zGhJvrv8&list=PL_hX5wLdhf_JhjIwLJjL8MHIhHVJyMm0b.

Stewart, Irvin. *Organizing Scientific Research for War: The Administrative History of the Office of Scientific Research and Development*. Boston: Little, Brown and Company, 1948.

United States Air Force. Basic Doctrine AFM 1-2. December 1, 1959.

University of California Santa Barbara Library. *The American Presidency Project*. http://www.presidency.ucsb.edu/

"X-37B Lands This Morning at Vandenberg AFB." *Santa Maria Times*, June 16, 2012. Accessed May 1, 2014. http://santamariatimes.com/news/local/military/vandenger/x--b-lands-this-morning-at-vandenberg-afb/article_31f5827c-b7bc-11e1-80b2-0019bb2963f4.html.

York, Herbert. *Making Weapons, Talking Peace: A Physicist's Odyssey from Hiroshima to Geneva*. New York: Basic Books, 1987.

———. *Race to Oblivion: A Participant's View of the Arms Race*. New York: Simon and Schuster, 1970.

Secondary Bibliography

Adams, Valerie L. *Eisenhower's Fine Group of Fellows: Crafting a National Security Policy to Uphold the Great Equation*. Lanham, MD: Lexington Books, 2006.

Aliano, Richard A. *American Defense Policy from Eisenhower to Kennedy: The Politics of Changing Military Requirements, 1957–1961*. Athens: Ohio University Press, 1975.

Ambrose, Stephen E. *Eisenhower: Soldier and President*. New York: Simon and Schuster, 1990.

Arrighi, Robert S. *Revolutionary Atmosphere: The Story of the Altitude Wind Tunnel and the Space Power Chambers*. Washington, DC: NASA History Office, 2010.

Ball, Desmond. *Politics and Force Levels: The Strategic Missile Program of the Kennedy Administration*. Oakland: University of California Press, 1981.

Bernard, Nancy E. *Television News and Cold War Propaganda, 1947–1960*. Cambridge, UK: Cambridge University Press, 1999.

Beschloss, Michael. *Mayday: Eisenhower, Khrushchev, and the U-2 Affair*. New York: Harper and Row, 1986.

Black, Jeremy. *War and Technology*. Indianapolis: Indiana University Press, 2014.

Boorstin, Daniel. *The Democratic Experience*. New York: Random House, 1973.

Boot, Max. *War Made New: Technology, Warfare, and the Course of History, 1500 to Today*. New York: Gotham Books, 2006.

Boushey, Brig. Gen. Homer A. "Lunar Base Vital." *Army-Navy-Air Force Register* 79 (February 8, 1958): 11.

Bowden, Elbert V. *Economics Through the Looking Glass*. San Francisco: Canfield Press, 1974.

Boyer, Paul. *By the Bomb's Early Light: American Thought and Culture at the Dawn of the Atomic Age*. New York: Pantheon, 1985.

Brands, H. W. "The Age of Vulnerability: Eisenhower and the National Insecurity State." *American Historical Review* 94 (October 1989): 963–89.

———. *Cold Warriors: Eisenhower's Generation and American Foreign Policy*. New York: Columbia University Press, 1988.

Bright, Christopher. *Continental Defense in the Eisenhower Era: Nuclear Antiaircraft Arms and the Cold War*. Basingstoke, UK: Palgrave Macmillan, 2010.

Bromberg, Joan Lisa. *NASA and the Space Industry*. Baltimore: Johns Hopkins University Press, 1999.

Bryant, Nick. *The Bystander: John F. Kennedy and the Struggle for Black Equality*. New York: Basic Books, 2006.

Bulkeley, Rip. "Aspects of the Soviet International Geophysical Year." *Russian Journal of Earth Sciences* 10 (2008).

Call, Steve. *Selling Air Power: Military Aviation and American Popular Culture after World War II*. College Station: Texas A&M Press, 2009.

Clarfield, Gerard. *Security with Solvency: Dwight D. Eisenhower and the Shaping of the American Military Establishment*. Westport, CT: Praeger, 1999.

Cleland, David I., et al., eds. *Military Project Management Handbook*. New York: McGraw Hill, 1993.

Clodfelter, Mark. *Beneficial Bombing: The Progressive Foundations of American Air Power, 1917–1945.* Lincoln: University of Nebraska Press, 2011.

Cohen, Lizabeth. *A Consumer's Republic: The Politics of Mass Consumption in Postwar America.* New York: Vintage Books, 2003.

Dallek, Robert. *Unfinished Life: John F. Kennedy, 1917–1963.* New York: Little, Brown and Company, 2003.

Daso, Dik. *Architects of American Air Superiority: General Hap Arnold and Dr. Theodore von Kármán.* Maxwell Air Force Base, AL: Air University Press, 1997.

Davis II, Robert T. *The U.S. Army and the Media in the 20th Century.* Leavenworth, KS: Combat Studies Institute Press, 2009.

Day, Dwayne A., John M. Logsdon, and Brian Latell, eds. *Eye in the Sky: The Story of the Corona Spy Satellites.* Washington, DC: Smithsonian Institution Press, 1998.

Dick, Steven J., ed. *NASA's First 50 Years: Historical Perspectives.* Washington, DC: U.S. Government Printing Office, 2009.

———— and Roger D. Launius, eds. *Critical Issues in the History of Spaceflight.* Washington, DC: NASA Office of External Relations, 2006.

Dickson, Paul. *Sputnik: The Shock of the Century.* New York: Walker and Co., 2001.

Divine, Robert. *The Sputnik Challenge: Eisenhower's Response to the Soviet Satellite.* Oxford, UK: Oxford University Press, 1993.

Dockrill, Saki. *Eisenhower's New-Look National Security Policy, 1953–61.* London, UK: St. Martin's Press, 1996.

Dowdy, Andrew. *The Films of the Fifties: The American State of Mind.* New York: William Morrow and Company, 1975.

Duggan, William. *Strategic Intuition: The Creative Spark in Human Achievement.* New York: Columbia Business School Press, 2007.

Erickson, Mark. *Into the Unknown Together: The DOD, NASA, and Early Spaceflight.* Maxwell Air Force Base, AL: Air University Press, 2005.

Evangelista, Matthew. *Innovation and the Arms Race: How the United States and the Soviet Union Develop New Military Technologies.* Ithaca: Cornell University Press, 1988.

Eisenhower, David, and Julie Nixon Eisenhower. *Going Home to Glory: A Memoir of Life with Dwight D. Eisenhower, 1961–1969.* New York: Simon and Schuster, 2010.

Frank, Jeffrey. *Ike and Dick: Portrait of a Strange Political Marriage.* New York: Simon and Schuster, 2013.

Futrell, Frank. *Ideas, Concepts, Doctrine: Basic Thinking in the United States Air Force,* Volume I, *1907–1960.* Maxwell Air Force Base, AL: Air University Press, 1989.

Gilbert, Robert E. "Eisenhower's 1955 Heart Attack: Medical Treatment, Political Effects, and the 'Behind the Scenes' Leadership Style." *Politics and the Life Sciences* 27, no. 1 (March 2008): 2–21.

Gorn, Michael H. *Harnessing the Genie: Science and Technology Forecasting for the Air Force, 1944–1986.* Washington, DC: Office of Air Force History, 1988.

Gray, Colin S. *Airpower for Strategic Effect.* Maxwell Air Force Base, AL: Air University Press, 2012.

Greene, Benjamin P. *Eisenhower, Science Advice, and the Nuclear Test-Ban Debate, 1945–1963.* Palo Alto: Stanford University Press, 2007.

Greenstein, Fred I. "Eisenhower as an Activist President: A Look at New Evidence." *Political Science Quarterly* 94, no. 4 (Winter 1979–80): 575–99.

———. *The Hidden-Hand Presidency: Eisenhower as Leader.* New York: Basic Books, 1982.

Halberstam, David. *The Fifties.* New York: Villard Books, 1993.

Hall, R. Cargill. "Sputnik, Eisenhower, and the Formation of the U.S. Space Program." *Quest* 14, no. 4 (2007): 32–39.

Hardesty, Von, and Ilya Grinberg. *Red Phoenix Rising: The Soviet Air Force in World War II.* Lawrence: University of Kansas Press, 2012.

Harris, Howell John. *The Right to Manage: Industrial Relations Policies of American Business in the 1940s.* Madison: University of Wisconsin, 1982.

Heller, Francis H. "The Eisenhower White House." *Presidential Studies Quarterly* 23, no. 3 (Summer 1993): 509–17.

Heppenheimer, T. A. *The Space Shuttle Decision, 1965–1972.* Washington, DC: Smithsonian Institution Press, 2002.

Herrera, Geoffrey L. *Technology and International Transformation: The Railroad, the Atom Bomb, and the Politics of Technological Change.* Albany: SUNY Press, 2006.

Holley, I. B. *Technology and Military Doctrine: Essays on a Challenging Relationship.* Maxwell Air Force Base, AL: Air University Press, 2004.

Houchin, Roy. *U.S. Hypersonic Research and Development: The Rise and Fall of Dyna-Soar, 1944–1963.* New York: Routledge, 2006.

Hoxie, H. Gordon. "Eisenhower and Presidential Leadership." *Presidential Studies Quarterly* 13, no. 4 (Fall 1983): 589–612.

Huntington, Samuel P. *American Politics: The Promise of Disharmony.* Cambridge, MA: Belknap Press, 1981.

———. *The Soldier and the State: The Theory and Politics of Civil-Military Relations.* New York: Vintage Books, 1964.

Hyman, Herbert H., and Paul B. Sheatsley. "The Political Appeal of President Eisenhower." *The Public Opinion Quarterly* 17, no. 4 (Winter 1953–54): 443–60.

Jacobsen, Carl G., ed. *Strategic Power: USA/USSR*. New York: St. Martin's, 1990.

Jenkins, Dennis. *X-15: Extending the Frontiers of Flight*. Seattle: CreateSpace Independent Publishing Platform, 2010.

Jezer, Marty. *The Dark Ages: Life in the United States, 1945–1960*. Boston: South End Press, 1982.

Johnson, Stephen B. *The United States Air Force and the Culture of Innovation*. Washington, DC: U.S. Government Printing Office, 2002.

Kalic, Sean. *U.S. Presidents and the Militarization of Space, 1946–1967*. College Station: Texas A&M Press, 2012.

Keegan, John. *Intelligence in War: Knowledge of the Enemy from Napoleon to Al-Qaeda*. New York: Alfred A. Knopf, 2003.

Kraemer, Sylvia. *Science and Technology Policy in the United States: Open Systems in Action*. New Brunswick: Rutgers University Press, 2006.

Krige, John. "Atoms for Peace, Scientific Internationalism, and Scientific Intelligence." *Osiris* 2, no. 21 (2006): 161–81.

Lambeth, Benjamin S. *Mastering the Ultimate High Ground: Next Steps in the Military Uses of Space*. Santa Monica, CA: RAND, 2003.

Launius, Roger D., and Howard E. McCurdy, eds. *Spaceflight and the Myth of Presidential Leadership*. Champaign: University of Illinois Press, 1997.

Lears, Thomas J. Jackson. *No Place of Grace: Antimodernism and the Transformation of American Culture, 1880–1920*. Chicago: University of Chicago Press, 1994.

Leighton, Richard. *History of the Office of the Secretary of Defense, Volume III: Strategy, Money, and the New Look, 1953–1956*. Washington, DC: Office of the Secretary of Defense Historical Office, 2001.

Leslie, Stuart. *The Cold War and American Science: The Military-Industrial Academic Complex at MIT and Stanford*. New York: Columbia University Press, 1993.

Lhamon, W. T. *Deliberate Speed: The Origins of a Cultural Style in the American 1950s*. Washington, DC: Smithsonian Institution Press, 1990.

Logsdon, John M. *John F. Kennedy and the Race to the Moon*. Basingstoke, UK: Palgrave Macmillan, 2010.

"Major General Richard D. Curtin." U.S. Air Force Biographies. Accessed June 16, 2011. http://www.af.mil/information/bios/bio.asp?bioID=5139.

Mandeles, Mark D. *The Development of the B-52 and Jet Propulsion: A Case Study in Organizational Innovation*. Maxwell Air Force Base, AL: Air University Press, 1998.

Marchio, James D. "Risking General War in Pursuit of Limited Objectives: U.S. Military Contingency Planning for Poland in the Wake of the 1956

Hungarian Uprising." *The Journal of Military History* 66, no. 3 (June 2002): 783–812.

Martinez, Michael. "Unmanned X-37B space plane lands, its exact mission a mystery." *CNN*, October 19, 2014. Accessed October 20, 2014. http://www.cnn.com/2014/10/18/us/air-force-space-plane-x37b/.

McDougall, Walter A. . . . *the Heavens and the Earth: A Political History of the Space Age*. New York: Basic Books, 1985.

McFarland, Linda. *Cold War Strategist: Stuart Symington and the Search for National Security*. Westport, CT: Praeger, 2001.

McFarland, Stephen L. *America's Pursuit of Precision Bombing, 1910–1945*. Washington, DC: Smithsonian Institution Press, 1995.

Meilinger, Phillip S. *Bomber: The Formation and Early Years of Strategic Air Command*. Maxwell Air Force Base, AL: Air University Press, 2012.

———, ed. *The Paths of Heaven: The Evolution of Airpower Theory*. Maxwell Air Force Base, AL: Air University Press, 1997.

Mieczkowski, Yanek. *Eisenhower's Sputnik Moment: The Race for Space and World Prestige*. Ithaca: Cornell University Press, 2013.

Millen, Raymond. "Cultivating Strategic Thinking: The Eisenhower Model." *Parameters* 42, no. 2 (Summer 2012): 56–70.

Newberry, Robert D. "Space Doctrine for the 21st Century." Accessed November 26, 2011. http://www.au.af.mil/au/awc/awcgate/acsc/97-0427.pdf.

Nichols, David A. *Eisenhower 1956: The President's Year of Crisis—Suez and the Brink of War*. New York: Simon and Schuster, 2011.

Olson, James C. *Stuart Symington: A Life*. Columbia: University of Missouri Press, 2003.

Osgood, Kenneth. *Total Cold War: Eisenhower's Secret Propaganda Battle at Home and Abroad*. Lawrence: University of Kansas Press, 2006.

Pape, Robert. *Bombing to Win: Air Power and Coercion in War*. Ithaca: Cornell University Press, 1996.

Patterson Jr., Bradley H. "Teams and Staff: Dwight Eisenhower's Innovations in the Structure and Operations of the Modern White House." *Presidential Studies Quarterly* 24, no. 2 (Spring 1994): 277–98.

Peebles, Curtis. *Shadow Flights: America's Secret Air War Against the Soviet Union*. New York: Presidio Press, 2000.

Pickett, William. *Eisenhower Decides to Run: Presidential Politics and Cold War Strategy*. Chicago: Ivan R. Dee, 2000.

Posen, Barry R. *The Sources of Military Doctrine: France, Britain, and Germany Between the World Wars*. Ithaca: Cornell University Press, 1984.

Reeves, Richard. *President Kennedy: Profile of Power*. New York: Simon and Schuster, 1993.

Roland, Alex. *The Military-Industrial Complex*. Washington, DC: American Historical Association, 2001.

———. "Science and War." *Osiris* 2, no. 1 (1985): 247–72.

———. "The Technological Fix: Weapons and the Cost of War." Carlisle, PA: Strategic Studies Institute, U.S. Army War College, 1995.

Roman, Peter J. *Eisenhower and the Missile Gap*. Ithaca: Cornell University Press, 1995.

Rome, Adam. *The Bulldozer in the Countryside: Suburban Sprawl and the Rise of American Environmentalism*. Cambridge, UK: Cambridge University Press, 2001.

Rorabough, W. J. *The Real Making of the President: Kennedy, Nixon, and the 1960 Election*. Lawrence: University of Kansas Press, 2009.

Sambaluk, Nicholas M. "Policymakers Confront Aerospace Doctrine, 1957–59." *Cold War History* 14, no. 1 (Spring 2014): 91–107.

Samuel, Lawrence R. *Future: A Recent History*. Austin: University of Texas, 2009.

Sheehan, Neil. *A Fiery Peace in a Cold War: Bernard Schriever and the Ultimate Weapon*. New York: Random House, 2009.

Sherry, Michael. *The Rise of American Airpower: The Creation of Armageddon*. New Haven: Yale University Press, 1987.

Showalter, Dennis, ed. *Forging the Shield: Eisenhower and National Security for the 21st Century*. Chicago: Imprint, 2005.

Siddiqi, Asif A. "Competing Technologies, National(ist) Narratives, and Universal Claims: Toward a Global History of Space Exploration." *Technology and Culture* 51, no. 2 (April 2010): 425–43.

Sloan, Joan W. "The Management and Decision-Making Style of President Eisenhower." *Presidential Studies Quarterly* 20, no. 2 (Spring 1990): 295–313.

Smith, Jean Edward. *Eisenhower in War and Peace*. New York: Random House, 2013.

Snead, David L. *The Gaither Committee, Eisenhower, and the Cold War*. Columbus: Ohio State University Press, 1999.

———. *John F. Kennedy: The New Frontier President*. Hauppauge, NY: Nova Science Publishers, 2012.

Spires, David N. *Beyond Horizons: A Half Century of Air Force Space Leadership*. Washington, DC: U.S. Government Printing Office, 1998.

Sturm, Thomas A. *The USAF Scientific Advisory Board: Its First Twenty Years, 1944–1964*. Washington, DC: Office of Air Force History, 1986.

Thomas, Evan. *Ike's Bluff: President Eisenhower's Secret Battle to Save the World*. New York: Little, Brown and Company, 2012.

Vagts, Alfred. *A History of Militarism: Civilian and Military.* New York: W.W. Norton, 1959.

Wagner, Steven. *Eisenhower Republicanism: Pursuing the Middle Way.* DeKalb: Northern Illinois University Press, 2006.

Wang, Zuoyue. *In Sputnik's Shadow: The President's Science Advisory Committee and Cold War America.* New Brunswick, NJ: Rutgers University Press, 2008.

Watson, Robert J. *History of the Office of the Secretary of Defense,* Volume IV: *Into the Missile Age, 1956–1960.* Washington, DC: Office of the Secretary of Defense Historical Office, 1997.

Wegner, Andreas. *Living with Peril: Eisenhower, Kennedy, and Nuclear Weapons.* New York: Rowman and Littlefield, 1997.

White, Mark. "Apparent Perfection: The Image of John Kennedy." *History* 98, no. 330 (April 2013): 228–46.

Wicker, Tom. *Dwight D. Eisenhower: The American Presidents Series.* New York: Henry Holt and Company, 2002.

Wills, Gary. *Bomb Power: The Modern Presidency and the National Security State.* London: Penguin Books, 2010.

Winter, Frank H. "The American Rocket Society Story, 1930–1962." *Journal of the British Interplanetary Society* 33, no. 8 (August 1980): 303–11.

Zachary, G. Pascal. *Endless Frontier: Vannevar Bush, Engineer of the American Century.* Cambridge: MIT Press, 1999.

INDEX

Abbott, Ira, 88, 197

Adams, Sherman, 131

Advanced Research Projects Agency. *See* ARPA

Advanced Systems Importance Category List, 102

AEC. *See* Atomic Energy Commission

Aerobee rocket, 24

Aeronautics and Astronautics Coordinating Board, 196

aerospace, definition of, 71, 101, 121, 169, 199

aerospace security advocates, 16, 37, 60, 74, 80–81, 87, 92, 101, 110, 113, 118, 130, 135, 150, 156, 165, 177, 192, 196, 209

Air Corps Tactical School, 6

Air Force: aerospace doctrine, 101, 170; astronaut training, 202, 205, 214; astronautics initiatives, 74–76, 77, 79, 81, 83–84, 88–90, 92, 95, 101, 105, 115–16, 146, 165, 173, 175, 178, 197–98, 202, 204–5, 208, 218, 223–24, 226–28, 231; budget, 29, 44, 52, 95, 99, 105–6, 228, 230; Directorate of Astronautics, 76, 86–87, 115–16;

lunar programs, 202, 205, 208, 216; public perceptions, 52–53, 59, 66, 158–59, 209, 228–29; relationship with NACA, 43–47, 49, 70, 74, 77, 79–81, 84–86, 88–89, 176; relationship with NASA, 95–96, 99–100, 104, 161, 173–75, 181–82, 195–96; research and development, 7, 12–15, 35–37, 43–47, 49, 70–72, 75–76, 79, 90, 109, 115, 127, 130, 133, 156, 161, 167, 173, 175, 216

Air Force Directorate of Advanced Technology, 76

Air Force Scientific Advisory Board, 73–74, 173; Aero and Vehicles Panel, 100

Air Force Space Command, 210

Air Materiel Command, 70

Air Research and Development Command. *See* ARDC

aircraft, nuclear powered, 106–7, 134–35, 138, 199

airpower security advocates, 5, 9, 13, 30, 49, 51–52, 69–70, 74. *See also* aerospace security advocates

ABOUT THE AUTHOR

Nicholas Michael Sambaluk is an assistant professor of Practice in Military Science and Technology for Purdue University and a research fellow for the Army Cyber Institute at West Point. He has taught courses including military history at West Point, war technology and innovation at Purdue University, humanities courses at the University of Kansas, and European and northwestern Asian regional and strategic trends to U.S. students overseas. Beginning in January 2013, Dr. Sambaluk served two years as historian project member in studies conducted by the Office of the Assistant Secretary of the Army for Acquisitions, Logistics, and Technology.